Advanced Sensing and Safety Control for Connected and Automated Vehicles

Advanced Sensing and Safety Control for Connected and Automated Vehicles

Editors

Chao Huang
Yafei Wang
Peng Hang
Zhiqiang Zuo
Bo Leng

MDPI • Basel • Beijing • Wuhan • Barcelona • Belgrade • Manchester • Tokyo • Cluj • Tianjin

Editors

Chao Huang
The Hong Kong Polytechnical University
Hong Kong
China

Yafei Wang
Shanghai Jiao Tong University
Shanghai
China

Peng Hang
Tongji University
Shanghai
China

Zhiqiang Zuo
Tianjin University
Tianjin
China

Bo Leng
Tongji University
Shanghai
China

Editorial Office
MDPI
St. Alban-Anlage 66
4052 Basel, Switzerland

This is a reprint of articles from the Special Issue published online in the open access journal *Sensors* (ISSN 1424-8220) (available at: https://www.mdpi.com/journal/sensors/special_issues/_CAV).

For citation purposes, cite each article independently as indicated on the article page online and as indicated below:

LastName, A.A.; LastName, B.B.; LastName, C.C. Article Title. *Journal Name* **Year**, *Volume Number*, Page Range.

ISBN 978-3-0365-7330-4 (Hbk)
ISBN 978-3-0365-7331-1 (PDF)

Contents

About the Editors

Chao Huang

Dr. Huang Chao is currently a research assistant professor in the Department of Industrial and Systems Engineering at The Hong Kong Polytechnic University. She received a bachelor's degree in Control Engineering from the China University of Petroleum, Beijing, and a Ph.D. degree from the University of Wollongong, Australia. She has published 45 journal papers, 21 conference papers, and 2 books. Her number of citations is 1397 (h-index: 24, Google Scholar).

Yafei Wang

Yafei Wang (Member, IEEE) received his B.S. degree from Jilin University, Changchun; his M.S. degree from Shanghai Jiao Tong University, Shanghai, China; and his Ph.D. degree from The University of Tokyo, Tokyo, Japan. He is currently an associate professor at the School of Mechanical Engineering, Shanghai Jiao Tong University, Shanghai. His research interests include state estimation and control for connected and automated vehicles.

Peng Hang

Dr. Peng Hang received his Ph.D. degree from the School of Automotive Studies, Tongji University, Shanghai, China, in 2019. From 2020 to 2022, he served as a research fellow with the School of Mechanical and Aerospace Engineering, Nanyang Technological University, Singapore. He is currently a research professor at the Department of Traffic Engineering, Tongji University. His research interests include vehicle dynamics and control, decision making, and motion control of autonomous vehicles. He has been a visiting researcher with the Department of Electrical and Computer Engineering, National University of Singapore, and a software engineer in the Research & Advanced Technology Dept., SAIC Motor, China. He received many awards and honors, including Excellent Doctoral Thesis of Tongji University, APAC Excellent PaperAward, etc. He serves as the Topic Editor or Guest Editor for Games, Vehicles, Autonomous Intelligent Systems, and Actuators.

Zhiqiang Zuo

Prof. Zhiqiang Zuo received his M.S. degree in control theory and control engineering in 2001 from Yanshan University and his Ph.D. degree in control theory in 2004 from Peking University, China. In 2004, he joined the School of Electrical and Information Engineering, Tianjin University, where he is a full professor. From 2008 to 2010, he was a research fellow in the Department of Mathematics, City University of Hong Kong. From 2013 to 2014, he was a visiting scholar at the University of California, Riverside. His research interests include nonlinear control, robust control, and multi-agent systems. He serves as the Associate Editor of the Journal of the Franklin Institute (Elsevier).

Bo Leng

Dr. Leng received his B.S. and Ph.D. degrees in vehicle engineering from Tongji University, Shanghai, China, in 2014 and 2019, respectively. He is currently working as a postdoctoral researcher in both the School of Automotive Studies and the Postdoctoral Station of Mechanical Engineering at Tongji University. His research interests include vehicle dynamics and control, parameter estimation, and intelligent vehicle trajectory motion control. He won the first prize of the China Automobile Industry Technology Invention Award in 2020 and was selected for the 7th Young Talent Scholar Program of China Association for Science and Technology.

Preface to "Advanced Sensing and Safety Control for Connected and Automated Vehicles"

This Special Issue on Sensors aims to report on some of the recent research efforts on this increasingly important topic. The 11 accepted papers in this Issue cover vehicle position estimation [1], 3D Object Detection [2], pedestrian state sense [3], trajectory prediction [4], criticality assessment [5], active fault-tolerant control for actuator failure [6], decision-making in the scenario of highway driving out of the ramp [7] and uncertain interactive traffic scenarios [8], RRT-based path planning algorithm [9], C/GMRES motion planning algorithm [10], and lane-keeping controller design [11]. In this introduction, a brief description of the content of each contribution forming the Special Issue is provided. Localization, or a vehicle's capacity to ascertain its position and orientation, is one of the crucial capacities needed to accomplish fully autonomous driving [1]. The first article describes a general, modular image processing pipeline that improves the robustness and accuracy of feature-based VO techniques, allowing for better pose estimation. Contrastlimited adaptive histogram equalization (CLAHE), square covering (SSC), and angle-based outliers rejection (AOR) are the three stages of the proposed pipeline. Each level deals with a different problem related to posing estimate errors. In order to improve the performance of any feature-based algorithm, the proposed pipeline is designed to be generic and modular so that it can be incorporated into any existing method. The proposed pipeline offers a large improvement in VO accuracy and resilience, with just a slight increase in processing time, according to the quantitative and qualitative results. Ref. [2] proposes an easy-to-use technique for up-sampling low-resolution point clouds, a process which improves the accuracy of 3D object detection by reconstructing objects rendered in sparse point cloud data into more dense data. First, 4-Chs is used to transform the 3D point cloud dataset into a 2D range image. In order to preserve the forms of the original object throughout the reconstruction, the interpolation on the empty space is calculated based on both the pixel distance and the range values of six neighbor points. In [3], a novel pedestrian status-detecting approach, based on multi-sensor fusion, is developed for automated patrol vehicles. First, the 2D and 3D pedestrian data are acquired through YOLO V4 and Euclidean clustering algorithms, respectively. A novel multi-sensor fusion method, based on the R-Tree algorithm, is then designed to detect the pedestrian and the body temperature is detected based on thermal imaging. The experimental results on an automated patrol vehicle indicated that the single-sensor algorithm's pedestrian detection results could be improved by more than 17% and that the multi-layer fusion method provided more understandable density estimation results.

A multi-head attention-based LSTM sequence-to-sequence model is designed in [4] to predict the future trajectories of the ego vehicle's surrounding vehicles. The social and temporal interaction are successfully extracted by the authors using a multi-head transformer method, and each one is then encoded into the input as hidden information using an LSTM encoder in the encoder module. The hidden information is decoded using the decoder module, which uses an additional multi-head attention layer to increase the accuracy of the subsequent decoding. The proposed model, according to ablative studies, resolves the cumulative error brought on by the transformer's autoregressive decoding behavior. The visualization outcomes demonstrate the model's robustness in challenging congested conditions. In [5], a novel risk assessment method, based on the time-to-react measurement, is proposed to determine overall criticality. The novelty of this method lies in the introduction of variable threshold values which are converted into the time domain from minimal safety distance metrics and acceleration values for criticality level determination. The designed variable criticality threshold closes a research gap by accounting for the kinematic relationships between vehicles and, as a result, the variable nature of criticality. The proposed method is proven to have the advantages of simplifying the consideration of the surrounding ego vehicles and evaluating possible evasive actions.

In [6], an active fault-tolerant control (AFTC) system is designed for the longitudinal motion control of autonomous mobile robots (AMRs) in the condition of braking failure of actuators. Specifically, a velocity-tracking controller is designed, with integrated feedforward and feedback control for normal longitudinal driving control. Once a key actuator failure is detected, the driving/braking torques of the remaining normal actuators are then reallocated based on the weighted least-squares (WLS) method for failure compensation. The simulation results show that the proposed AFTC algorithm can deal with the braking failure of IWMs and realize braking energy recovery at the same time. Ref. [7] proposes a generalized single-vehicle-based graph neural network reinforcement learning (SGRL) algorithm that incorporates interactive information between agents in the environment into the decision-making process of autonomous driving. The proposed SGRL algorithm uses the training approach for a single agent, constructs a clearer incentive reward function, and greatly increases the dimension of the action space over the conventional deep neural network (DQN) algorithm. The simulation results show that the proposed SGRL algorithm excels in terms of network convergence, decision-making impact, and training effectiveness.

Similar to [7], a reward function matrix for training different decision-making modes is provided in [8], with emphasis placed both on decision-making styles and, additionally, on rewards and penalties. A decision-weighted coefficient matrix, an incentive punishment weighted coefficient matrix, and a reward–penalty function matrix are all included in the proposed reward function matrix. The simulation results demonstrate that the proposed reward function can significantly increase the algorithm's stability and speed of convergence. Ref. [9] proposes using an improved heuristic Bi-RRT algorithm to obtain a smooth and asymptotically optimal path, with continuous curvature possessing high efficiency and accuracy in an uncertain dynamic environment. The consideration of the driver's driving habit and the obstacle-free direct connection mode of two trees, as well as the introduction of the greedy step size and the design of the path reorganization, can expand the node more effectively, make the path smooth, and ensure the ride comfort of the vehicle. The simulation results show that the proposed algorithm can generate the smoothest path and take the shortest time compared with the current studies.

In [10], a computationally efficient motion planning method that both considers traffic interactions and accelerates calculation is proposed. A nonlinear predictive controller is designed to dynamically generate a path by considering the predicted trajectories of other traffic participants. In addition, the C/GMRES algorithm is used to accelerate the calculation of trajectory generation. The simulation experiments show that the proposed path planning algorithm can enable greater rationality of movement planning. Ref. [11] proposes a cooperative control strategy for lane-keeping by integrating driving monitoring, a variable level of assistance allocation, and human-in-the-loop control. The relationship between lateral acceleration, road curvature, and the observed maximum driver torque is used in the first stage of this research to identify a time-varying physical driver loading pattern. An adaptive driver activity function is then developed to simulate the amount of support needed for the driver in the following stage. Additionally, this is combined with a monitored driver state that signals the driver's mental loading. A unique higher-order sliding mode controller is developed to maintain closed-loop stability in order to seamlessly switch control between different modes, based on the generated amounts of assistance. Additionally, the conflict is reduced by using a new sharing parameter that is proportionate to the torques originating from both the driver and the autonomous controller.

In summary, there is a huge potential for the application of CAV in areas of traffic perception, decision-making, and driving safety improvement. This Special Issue contributes to this by proposing solutions to the general problems of state estimation [1], objective detection [2,3], trajectory prediction [4], driving safety improvement [5,6], decision-making [7,8], path planning [9,10] and vehicle motion control [11], which largely represent the theoretical challenges and practical interest of this research topic. Finally, we wish to thank the authors, reviewers, and journal staff for their commitment and effort, without which we could not have completed this Special Issue on time. We hope that readers enjoy reading the articles and that the published works contribute to the progression of connected and automated vehicles.

Chao Huang, Yafei Wang, Peng Hang, Zhiqiang Zuo, and Bo Leng
Editors

sensors

MDPI

Editorial

Advanced Sensing and Safety Control for Connected and Automated Vehicles

Chao Huang [1,*], Yafei Wang [2], Peng Hang [3], Zhiqiang Zuo [4] and Bo Leng [5]

[1] Department of Industrial and Systems Engineering, The Hong Kong Polytechnic University, Hong Kong 999077, China
[2] School of Mechanical Engineering, Shanghai Jiao Tong University, Shanghai 200240, China
[3] Department of Traffic Engineering, Tongji University, Shanghai 201804, China
[4] School of Electrical and Information Engineering, Tianjin University, Tianjin 300072, China
[5] School of Automotive Studies, Tongji University, Shanghai 201804, China
* Correspondence: hchao.huang@polyu.edu.hk; Tel.: +86-852-2766-4226

The connected and automated vehicle (CAV) is a promising technology, anticipated to enhance the safety and effectiveness of mobility. Advanced sensing technologies and control algorithms, working to acquire environmental data, analyze data, and regulate vehicle movements, are key functional components of CAVs. In recent years, the creation of innovative sensing technologies for CAVs has gained substantial attention. CAVs can now interpret sensory data to more accurately detect impediments, locate their locations, navigate autonomously in a dynamic environment, and communicate with other nearby vehicles. This has been made possible by advances in sensing technology. Additionally, by utilizing computer vision and other sensing techniques, in-cabin persons' bodily movements, facial expressions, and even mental states can be identified.

This Special Issue on *Sensors* aims to report on some of the recent research efforts on this increasingly important topic. The 11 accepted papers in this Issue cover vehicle position estimation [1], 3D Object Detection [2], pedestrian state sense [3], trajectory prediction [4], criticality assessment [5], active fault-tolerant control for actuator failure [6], decision-making in the scenario of highway driving out of the ramp [7] and uncertain interactive traffic scenarios [8], RRT-based path planning algorithm [9], C/GMRES motion planning algorithm [10], and lane-keeping controller design [11]. In this introduction, a brief description of the content of each contribution forming the Special Issue is provided.

Localization, or a vehicle's capacity to ascertain its position and orientation, is one of the crucial capacities needed to accomplish fully autonomous driving. [1] The first article describes a general, modular image processing pipeline that improves the robustness and accuracy of feature-based VO techniques, allowing for better pose estimation. Contrast-limited adaptive histogram equalization (CLAHE), square covering (SSC), and angle-based outliers rejection (AOR) are the three stages of the proposed pipeline. Each level deals with a different problem related to posing estimate errors. In order to improve the performance of any feature-based algorithm, the proposed pipeline is designed to be generic and modular so that it can be incorporated into any existing method. The proposed pipeline offers a large improvement in VO accuracy and resilience, with just a slight increase in processing time, according to the quantitative and qualitative results.

Ref. [2] proposes an easy-to-use technique for up-sampling low-resolution point clouds, a process which improves the accuracy of 3D object detection by reconstructing objects rendered in sparse point cloud data into more dense data. First, 4-Chs is used to transform the 3D point cloud dataset into a 2D range image. In order to preserve the forms of the original object throughout the reconstruction, the interpolation on the empty space is calculated based on both the pixel distance and the range values of six neighbor points.

In [3], a novel pedestrian status-detecting approach, based on multi-sensor fusion, is developed for automated patrol vehicles. First, the 2D and 3D pedestrian data are acquired

Citation: Huang, C.; Wang, Y.; Hang, P.; Zuo, Z.; Leng, B. Advanced Sensing and Safety Control for Connected and Automated Vehicles. *Sensors* **2023**, *23*, 1037. https://doi.org/10.3390/s23021037

Received: 9 January 2023
Accepted: 11 January 2023
Published: 16 January 2023

through YOLO V4 and Euclidean clustering algorithms, respectively. A novel multi-sensor fusion method, based on the R-Tree algorithm, is then designed to detect the pedestrian and the body temperature is detected based on thermal imaging. The experimental results on an automated patrol vehicle indicated that the single-sensor algorithm's pedestrian detection results could be improved by more than 17% and that the multi-layer fusion method provided more understandable density estimation results.

A multi-head attention-based LSTM sequence-to-sequence model is designed in [4] to predict the future trajectories of the ego vehicle's surrounding vehicles. The social and temporal interaction are successfully extracted by the authors using a multi-head transformer method, and each one is then encoded into the input as hidden information using an LSTM encoder in the encoder module. The hidden information is decoded using the decoder module, which uses an additional multi-head attention layer to increase the accuracy of the subsequent decoding. The proposed model, according to ablative studies, resolves the cumulative error brought on by the transformer's autoregressive decoding behavior. The visualization outcomes demonstrate the model's robustness in challenging congested conditions.

In [5], a novel risk assessment method, based on the time-to-react measurement, is proposed to determine overall criticality. The novelty of this method lies in the introduction of variable threshold values which are converted into the time domain from minimal safety distance metrics and acceleration values for criticality level determination. The designed variable criticality threshold closes a research gap by accounting for the kinematic relationships between vehicles and, as a result, the variable nature of criticality. The proposed method is proven to have the advantages of simplifying the consideration of the surrounding ego vehicles and evaluating possible evasive actions.

In [6], an active fault-tolerant control (AFTC) system is designed for the longitudinal motion control of autonomous mobile robots (AMRs) in the condition of braking failure of actuators. Specifically, a velocity-tracking controller is designed, with integrated feedforward and feedback control for normal longitudinal driving control. Once a key actuator failure is detected, the driving/braking torques of the remaining normal actuators are then reallocated based on the weighted least-squares (WLS) method for failure compensation. The simulation results show that the proposed AFTC algorithm can deal with the braking failure of IWMs and realize braking energy recovery at the same time.

Ref. [7] proposes a generalized single-vehicle-based graph neural network reinforcement learning (SGRL) algorithm that incorporates interactive information between agents in the environment into the decision-making process of autonomous driving. The proposed SGRL algorithm uses the training approach for a single agent, constructs a clearer incentive reward function, and greatly increases the dimension of the action space over the conventional deep neural network (DQN) algorithm. The simulation results show that the proposed SGRL algorithm excels in terms of network convergence, decision-making impact, and training effectiveness.

Similar to [7], a reward function matrix for training different decision-making modes is provided in [8], with emphasis placed both on decision-making styles and, additionally, on rewards and penalties. A decision-weighted coefficient matrix, an incentive–punishment-weightedcoefficient matrix, and a reward–penalty function matrix are all included in the proposed reward function matrix. The simulation results demonstrate that the proposed reward function can significantly increase the algorithm's stability and speed of convergence.

Ref. [9] proposes using an improved heuristic Bi-RRT algorithm to obtain a smooth and asymptotically optimal path, with continuous curvature possessing high efficiency and accuracy in an uncertain dynamic environment. The consideration of the driver's driving habit and the obstacle-free direct connection mode of two trees, as well as the introduction of the greedy step size and the design of the path reorganization, can expand the node more effectively, make the path smooth, and ensure the ride comfort of the vehicle. The simulation results show that the proposed algorithm can generate the smoothest path and take the shortest time compared with the current studies.

In [10], a computationally efficient motion planning method that both considers traffic interactions and accelerates calculation is proposed. A nonlinear predictive controller is designed to dynamically generate a path by considering the predicted trajectories of other traffic participants. In addition, the C/GMRES algorithm is used to accelerate the calculation of trajectory generation. The simulation experiments show that the proposed path planning algorithm can enable greater rationality of movement planning.

Ref. [11] proposes a cooperative control strategy for lane-keeping by integrating driving monitoring, a variable level of assistance allocation, and human-in-the-loop control. The relationship between lateral acceleration, road curvature, and the observed maximum driver torque is used in the first stage of this research to identify a time-varying physical driver loading pattern. An adaptive driver activity function is then developed to simulate the amount of support needed for the driver in the following stage. Additionally, this is combined with a monitored driver state that signals the driver's mental loading. A unique higher-order sliding mode controller is developed to maintain closed-loop stability in order to seamlessly switch control between different modes, based on the generated amounts of assistance. Additionally, the conflict is reduced by using a new sharing parameter that is proportionate to the torques originating from both the driver and the autonomous controller.

In summary, there is a huge potential for the application of CAV in areas of traffic perception, decision-making, and driving safety improvement. This Special Issue contributes to this by proposing solutions to the general problems of state estimation [1], objective detection [2,3], trajectory prediction [4], driving safety improvement [5,6], decision-making [7,8], path planning [9,10] and vehicle motion control [11], which largely represent the theoretical challenges and practical interest of this research topic. Finally, we wish to thank the authors, reviewers, and journal staff for their commitment and effort, without which we could not have completed this Special Issue on time. We hope that readers enjoy reading the articles and that the published works contribute to the progression of connected and automated vehicles.

Funding: This work received funding from the PolyU (UGC), via grant A0040253 associated with grant A0039179.

Conflicts of Interest: The authors declare no conflict of interest.

References

1. Sabry, M.; Osman, M.; Hussein, A.; Mehrez, M.W.; Jeon, S.; Melek, W. A Generic Image Processing Pipeline for Enhancing Accuracy and Robustness of Visual Odometry. *Sensors* **2022**, *22*, 8967. [CrossRef] [PubMed]
2. You, J.; Kim, Y.-K. Up-Sampling Method for Low-Resolution LiDAR Point Cloud to Enhance 3D Object Detection in an Autonomous Driving Environment. *Sensors* **2022**, *23*, 322. [CrossRef] [PubMed]
3. Wang, P.; Liu, C.; Wang, Y.; Yu, H. Advanced Pedestrian State Sensing Method for Automated Patrol Vehicle Based on Multi-Sensor Fusion. *Sensors* **2022**, *22*, 4807. [CrossRef] [PubMed]
4. Hasan, F.; Huang, H. MALS-Net: A Multi-Head Attention-Based LSTM Sequence-to-Sequence Network for Socio-Temporal Interaction Modelling and Trajectory Prediction. *Sensors* **2023**, *23*, 530. [CrossRef] [PubMed]
5. Nalic, D.; Mihalj, T.; Orucevic, F.; Schabauer, M.; Lex, C.; Sinz, W.; Eichberger, A. Criticality Assessment Method for Automated Driving Systems by Introducing Fictive Vehicles and Variable Criticality Thresholds. *Sensors* **2022**, *22*, 8780. [CrossRef] [PubMed]
6. Hang, P.; Lou, B.; Lv, C. Nonlinear Predictive Motion Control for Autonomous Mobile Robots Considering Active Fault-Tolerant Control and Regenerative Braking. *Sensors* **2022**, *22*, 3939. [CrossRef] [PubMed]
7. Yang, F.; Li, X.; Liu, Q.; Li, Z.; Gao, X. Generalized Single-Vehicle-Based Graph Reinforcement Learning for Decision-Making in Autonomous Driving. *Sensors* **2022**, *22*, 4935. [CrossRef] [PubMed]
8. Gao, X.; Li, X.; Liu, Q.; Li, Z.; Yang, F.; Luan, T. Multi-Agent Decision-Making Modes in Uncertain Interactive Traffic Scenarios via Graph Convolution-Based Deep Reinforcement Learning. *Sensors* **2022**, *22*, 4586. [CrossRef] [PubMed]
9. Zhang, X.; Zhu, T.; Du, L.; Hu, Y.; Liu, H. Local Path Planning of Autonomous Vehicle Based on an Improved Heuristic Bi-RRT Algorithm in Dynamic Obstacle Avoidance Environment. *Sensors* **2022**, *22*, 7968. [CrossRef] [PubMed]

10. Zhang, Y.; Wang, J.; Lv, J.; Gao, B.; Chu, H.; Na, X. Computational Efficient Motion Planning Method for Automated Vehicles Considering Dynamic Obstacle Avoidance and Traffic Interaction. *Sensors* **2022**, *22*, 7397. [CrossRef]
11. Perozzi, G.; Oudainia, M.R.; Sentouh, C.; Popieul, J.-C.; Rath, J.J. Driver Assisted Lane Keeping with Conflict Management Using Robust Sliding Mode Controller. *Sensors* **2022**, *23*, 4. [CrossRef]

sensors

MDPI

Article

A Generic Image Processing Pipeline for Enhancing Accuracy and Robustness of Visual Odometry

Mohamed Sabry [1,*], Mostafa Osman [2], Ahmed Hussein [3], Mohamed W. Mehrez [2], Soo Jeon [2] and William Melek [2]

1 Autonomous Mobility and Perception Lab (AMPL), Universidad Carlos III de Madrid (UC3M), 28911 Leganes, Spain
2 Mechanical and Mechatronics Engineering, University of Waterloo, Waterloo, ON N2L 3G1, Canada
3 Intelligent Driving Function Department, IAV GmbH, 10587 Berlin, Germany
* Correspondence: mohamed.sabry@alumnos.uc3m.es

Abstract: The accuracy of pose estimation from feature-based Visual Odometry (VO) algorithms is affected by several factors such as lighting conditions and outliers in the matched features. In this paper, a generic image processing pipeline is proposed to enhance the accuracy and robustness of feature-based VO algorithms. The pipeline consists of three stages, each addressing a problem that affects the performance of VO algorithms. The first stage tackles the lighting condition problem, where a filter called Contrast Limited Adaptive Histogram Equalization (CLAHE) is applied to the images to overcome changes in lighting in the environment. The second stage uses the Suppression via Square Covering (SSC) algorithm to ensure the features are distributed properly over the images. The last stage proposes a novel outliers rejection approach called the Angle-based Outlier Rejection (AOR) algorithm to remove the outliers generated in the feature matching process. The proposed pipeline is generic and modular and can be integrated with any type of feature-based VO (monocular, RGB-D, or stereo). The efficiency of the proposed pipeline is validated using sequences from KITTI (for stereo VO) and TUM (for RGB-D VO) datasets, as well as experimental sequences using an omnidirectional mobile robot (for monocular VO). The obtained results showed the performance gained by enhancing the accuracy and robustness of the VO algorithms without compromising on the computational cost using the proposed pipeline. The results are substantially better as opposed to not using the pipeline.

Keywords: visual odometry; image processing pipeline; computer vision; robot operating system (ROS)

Citation: Sabry, M.; Osman, M.; Hussein, A.; Mehrez, M.W.; Jeon, S.; Melek, W. A Generic Image Processing Pipeline for Enhancing Accuracy and Robustness of Visual Odometry. *Sensors* **2022**, *22*, 8967. https://doi.org/10.3390/s22228967

Academic Editors: Yafei Wang, Chao Huang, Peng Hang, Zhiqiang Zuo and Bo Leng

Received: 27 October 2022
Accepted: 14 November 2022
Published: 19 November 2022

1. Introduction

Throughout the past few years, interest in autonomous robotic systems has increased drastically. One of the key modules required to achieve complete autonomy is localization, which is the ability of a mobile platform to determine its position and orientation. Currently, several sensors are used to achieve localization, including LiDAR [1], radar [2], Global Positioning System (GPS) [3], Inertial Measurement Unit (IMU) [4], wheel encoders [5], and cameras [6].

One of the common methods for localization using cameras is through Visual Odometry (VO) [7,8]. VO estimates the ego-motion of a camera by determining the incremental motion between the successive camera frames. Like other odometry methods (wheel encoders, LiDAR, and so on), VO relies on integrating the incremental motion between successive frames to compute the overall trajectory of the camera, leading to drift errors over long distances.

To avoid drift errors, VO is usually integrated into a Simultaneous Localization and Mapping (SLAM) system with a loop-closure module to correct the drift error [9–11]. Although using SLAM is beneficial for acquiring a map of the environment, it incurs unnecessary computational costs when the map is already known. A better solution could be using an accurate low-drift odometry module alongside localization in a pre-acquired map [12].

Several VO algorithms were developed in the literature aiming to increase pose estimation accuracy. Those algorithms are classified by the type of camera used in the motion estimation as monocular, stereo, and RGB-D VO. Alternately, they can also be classified by the method of motion estimation into feature-based and direct VO [13].

Monocular VO uses images captured by only one camera to determine its trajectory. This technique usually relies on Structure from Motion (SFM) [14–17]. With one camera, the motion of the robot can be captured up to an unobservable scale. This scale can then be determined through the use of an external velocity measurement from a wheel encoder or an IMU. Several researchers developed methods for estimating the scale of monocular VO without the use of external sensors, such as [18,19].

Lately, researchers have started developing deep-learning techniques for monocular VO [20,21]. Unlike the monocular VO, the stereo and the RGB-D VO estimate the full pose of the vehicle without any external measurements [22–24].

All three categories of VO can be either direct or feature-based approaches. On the one hand, the feature-based method relies on visual features for calculating the transformation between consecutive frames. For the detection of such features, a feature detector and descriptor are used such as Scale Invariant Features Transform (SIFT) [25], Speeded-up Robust Features (SURF) [26], or Oriented FAST and Rotated BRIEF (ORB) [27] (see [28] or [29] for a comparison between the different detectors and descriptors). On the other hand, direct approaches compute the relative transformation between frames based on the whole-image intensities [30,31].

Although such VO techniques are well-designed, VO algorithms have a drawback in that they rely on integration, which may suffer from drift errors due to, for example, false-matched features, bad lighting and illumination problems, random noise, or motion bias [32].

To overcome the above-mentioned issues, this paper proposes a generic modular image processing pipeline to enhance the accuracy and robustness of feature-based VO algorithms, independent of the camera type (monocular, RGB-D or stereo vision). The proposed pipeline includes additional filtration and pre-processing stages through Contrast Limited Adaptive Histogram Equalization (CLAHE) with adaptive thresholding to dynamically overcome variable lighting conditions. Additionally, the Suppression via Square Covering (SSC) is used to make the extracted features more equally distributed across the image to reduce the bias in the motion estimation [33]. Finally, a novel outlier rejection algorithm called the Angle-based Outliers Rejection (AOR) is proposed to reject false-matched features, as well as features captured on a moving object in the scene. The pipeline is integrated into a monocular, RGB-D, and stereo VO and validated using KITTI [34] and TUM datasets [35], as well as experimental sequences generated by an omnidirectional mobile robot.

The contribution of this paper is integrating the CLAHE filter, the SSC algorithm, and the proposed AOR algorithm into the VO. The results show that the three stages play an integral part in enhancing the performance of VO while overcoming the drawbacks of every individual stage.

The remainder of this paper is organized as follows, Section 2 discusses the related work, and Section 3 explains the proposed pipeline with the different filtration steps. The experimental work is presented in Section 4, which also includes the implementation details, along with the used datasets and the evaluation metrics. Section 5 presents the results and discussions. Finally, Section 6 provides concluding remarks and future works.

2. Related Works

2.1. Image Filtration

Several works have addressed the problem of noise in images for VO enhancement. The noise may be attributed to poor lighting conditions caused by light source flare, random visual sensor noise, or other noise sources [36].

In [37], a direct VO algorithm using binary descriptors was used to overcome poor lighting conditions. The authors showed that the algorithm performed in a robust and efficient way even under low lighting conditions. This was accomplished by the illumination invariance property of the binary descriptor within a direct alignment framework. The VO algorithm proposed therein is a direct method, which is usually more computationally expensive compared to feature-based VO.

In [38], a method for reducing drift in VO is introduced. The authors develop a new descriptor called the SYnthetic BAsis (SYBA) descriptor to reduce false-matched features. This is accomplished with the help of a sliding window approach. The features matching step is applied to features in a window instead of matching features between two consecutive frames. Although a sliding window approach can, in fact, increase the accuracy of the feature matching step, it will also significantly increase the computational cost of the matching task.

In [39], a robust feature-matching scheme was combined with an effective anti-blurring frame. The algorithm uses the singular value decomposition to mitigate the effect of blurring due to vibrations or other factors.

In [40], a stereo visual SLAM algorithm was proposed, which uses CLAHE to locally enhance the image contrast and obtain more feature details. The CLAHE-enhanced SLAM algorithm was compared to the results of a VO enhanced by a conventional histogram equalization and the results of ORB-SLAM2 [11]. The results showed a superior performance of the CLAHE-enhanced algorithm compared to the other algorithms. Furthermore, in [41], a robust VO for underwater environments was proposed. In order to overcome the turbid image quality of underwater imaging, the authors used CLAHE for contrast enhancement. The authors showed that the use of CLAHE resulted in brighter and larger visible regions. As a result, unclear structures were significantly reduced.

Therefore, in this paper, CLAHE is selected as a pre-processing stage for the camera frames to overcome the effect of poor lighting conditions.

2.2. Non-Maximal Suppression

Non-maximal suppression can be used to avoid poor distribution of features over the image, which leads to poor VO performance and motion bias. Several non-maximal suppression algorithms were used in VO [42,43]. In [44], a feature descriptor was proposed to facilitate fast feature matching processing while preserving matching reliability. The authors chose to use the FAST (Features from Accelerated Segment Test) detector [45] along with a non-maximal suppression algorithm.

In [46], a stereo/RGB-D VO was proposed for mobile robots. Therein, the authors used the adaptive non-maximal suppression introduced in [47] to enhance the performance of the feature detector algorithm BRIEF (Binary Robust Independent Elementary Features) [48] by ensuring uniform distribution of features over the image.

In [33], three new and efficient adaptive non-maximal suppression approaches were introduced, which included the SSC algorithm. The positive impact of the three algorithms on visual SLAM was demonstrated. Authors in [33] showed that the output of the three algorithms is visually and statistically similar; however, SSC showed lower computational costs, which suggests that it is more suitable for real-time applications such as VO.

Although the authors of [33] showed the effect of the SSC on the enhancement of the output of a visual SLAM algorithm, to the best of the authors' knowledge, the SSC algorithm was not used in any other VO or visual SLAM algorithm afterward. In this paper, SSC is selected as an additional stage for feature detection and matching to avoid the bias in the motion estimation due to poor distribution of the features.

2.3. Outliers Rejection

Feature-based VO relies on feature detection and matching for motion estimation. Commonly, the feature matching algorithms generate a considerable amount of false-matched features [7]. This is mainly due to the limitation of using local feature matching. These false-matched features lead to the increased error in motion estimation or the complete divergence of the VO output, as well as increased computational costs. Several works in the literature addressed this problem. In [49], an iterative outlier rejection scheme for stereo-based VO was proposed. The proposed algorithm was designed to improve the VO motion estimation for high-speed and large-scale depth environments.

In [50], a stereo VO was proposed that relies on using reference frames instead of all frames. This was accomplished by first selecting the stable features from a frame using quad-matching testing and grid-based motion statistics. Afterward, the features in this frame were matched to the features in a reference frame (instead of the previous frame), which contained the stable features found in the current frame.

A commonly used outlier rejection approach is the Random Sampling Consensus (RANSAC). RANSAC is an iterative outlier rejection algorithm, which relies on the computation of model hypotheses, from a randomly selected set of the matched features, followed by the verification of the hypotheses using the rest of the matched features [8]. In [51], a stereo VO algorithm was proposed, which uses a RANSAC-based outliers rejection along with an iterated sigma point Kalman filter to achieve robust frame-to-frame VO performance. Although RANSAC is effective in removing outliers, the iterative process sometimes results in poor performance due to a large number of iterations for convergence. Furthermore, if the number of outliers in the matched points is large, this may lead to wrong convergence entailing incorrect motion estimation.

Hence, in this paper, a non-iterative outliers rejection algorithm is proposed, which relies on the angular distance of the matched features in the matched frames. The algorithm can be incorporated before RANSAC in order to reduce the number of iterations required for convergence and thus increase the overall accuracy of motion estimation while reducing the computational cost of the algorithm.

3. Proposed Pipeline

In this section, the components of the proposed image processing pipeline are introduced. The flow chart of the pipeline is shown in Figure 1.

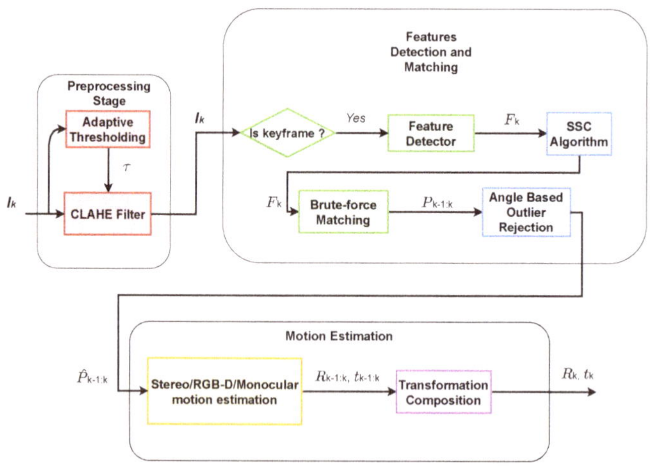

Figure 1. Proposed pipeline for the robust feature-based VO with the added filtration stages highlighted in red.

3.1. Image Pre-Processing

The pre-processing stage consists of applying a simple blurring filter to remove some noise in the image, followed by applying an Adaptive Histogram Equalization (AHE) technique, namely CLAHE [52]. CLAHE is applied to each input frame to enable the feature detector to find a sufficient number of features per frame. Although traditional AHE techniques tend to over-amplify the noise in the nearly constant regions in an image, the CLAHE filter prevents this over-amplification by limiting the histogram values. The effect of the CLAHE is shown in Figure 2.

Figure 2. The effect of CLAHE filter on the image. Top: the image before applying CLAHE, Bottom: the image after applying the CLAHE algorithm with the adaptive thresholding. The image was taken from the KITTI dataset.

Moreover, to ensure that the CLAHE filter adapts to different lighting conditions, the threshold value of the CLAHE is made adaptive to the ratio between the minimum, maximum, and the median of the intensity values in the frame, as presented in (1). The adaptation of the threshold value enables the CLAHE filter to adapt to different lighting conditions during the mobile platform operation and to avoid deterioration of the VO performance caused by too high or too low brightness in the images. Specifically, the contrast at which the CLAHE filter clips the histogram is computed as

$$\tau_k = \frac{\max(I_k) - \min(I_k)}{\text{median}(I_k)} \tag{1}$$

where τ_k and I_k are the contrast threshold for the CLAHE filter and the 2D image data at the k-th time step, respectively.

An example of the output of the CLAHE filter is shown in Figure 2. The effect of the sun can be seen in the original image, which leads to bright regions in the top middle of the image and dark regions on the left and the right of the image. Applying CLAHE results in decreasing the effect of the sunlight on the image and increasing the amount of extractable information, which can then be used by the feature detector algorithm to capture more stable features from the image.

3.2. Features Detection and Matching

After the image pre-processing, the current frame I_k is passed to a feature detector. The extracted set of features, denoted by $\mathcal{F}_k$, is then matched with the set of those from the reference frame $\mathcal{F}_{k-1}$. Then, the set of matched features $\mathcal{P}_{k-1:k}$ is used for estimating the incremental motion between the two frames I_{k-1} and I_k.

One of the causes of error in the motion estimation is the nonuniform distribution of features associated with the image [32,46]. Another cause of poor motion estimation is the presence of a high number of outliers in the detected and matched features. Therefore, in this paper, we added the SSC as well as the proposed AOR steps to the proposed pipeline.

3.2.1. Suppression via Square Covering

Before passing the feature set $\mathcal{F}_k$ to the feature matching algorithm, the features are first passed to the SSC [33] algorithm to make sure the captured features are homogeneously distributed over the whole captured image I_k.

The SSC algorithm is an approximation of the Suppression via Disk Covering (SDC). It relies on an approximate nearest neighbor algorithm, which uses a randomized search tree. In contrast, the SSC achieves comparable results with a single query operation per search range guess. Accordingly, the SSC has better efficiency and scalability than the SDC. In addition, SSC applies a square approximation for the SDC to avoid computing the Euclidean distance between a large number of features. This allows the SSC algorithm to execute in a runtime with lower complexity as the number of features increases.

The effect of using the SSC algorithm is shown in Figure 3. Figure 3a shows the original output of the SURF feature detector, where the feature density is higher in the top right region of the image. As shown in Figure 3b, after applying the SSC, the feature distribution across the image is almost the same (except for regions that did not contain any features).

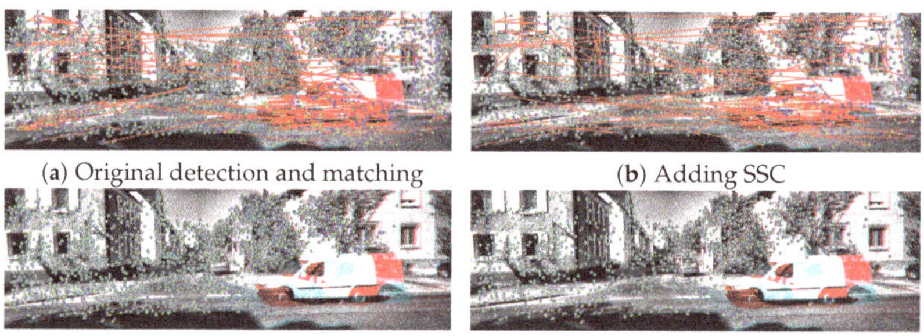

(**a**) Original detection and matching (**b**) Adding SSC (**c**) Adding AOR (**d**) Adding SSC and AOR

Figure 3. The effect of SSC and AOR on the features detection and matching. The previous frame is shown as the red component of I_{k-1}, and the current frame is shown as the blue component of I_k. The green crosses (+) and the blue circles (○) represent the features $\mathcal{F}_{k-1}$ and $\mathcal{F}_k$, respectively. Finally, the red lines represent the matched pairs in $\mathcal{P}_{k-1:k}$. Each of the images represents the following: (**a**) the original feature pairs through using the SURF detector and a brute-force matching algorithm, (**b**) shows the feature pairs after adding the SSC algorithm only. (**c**) shows the feature pairs after adding the AOR algorithm only, and finally (**d**) shows the feature pairs after adding both SSC and AOR. The majority of the outliers were removed due to the large difference between the angles, as illustrated in Figure 4.

Although several feature matching algorithms were introduced in the literature [53], those algorithms generate a considerable amount of false-matched points (as can be seen in Figure 3a,b. Motivated by this issue, in this paper, a novel outlier rejection algorithm is introduced and integrated into the overall VO pipeline to remove those false-matched points.

3.2.2. Angle-Based Outliers Rejection (AOR) for Feature Matching

To filter the produced matched points from outliers, a new outlier rejection algorithm called the AOR is proposed. In addition to removing false-matched features, the AOR can remove features that do not belong to the ego vehicle motion in the scene. This is applied before the motion estimation stage (RANSAC/LMEDS) in order to reduce the number of iterations required for convergence and thus increase the overall accuracy of motion estimation while reducing the computational cost of the algorithm.

Usually, during the motion of the camera, the further the feature from the vanishing point of the image, the more the feature moves. Although the amount of movement of a feature in the successive images is different depending on its position, it should be comparable to the motion of other features. False-matched points tend to show a larger amount of feature motion through the image, as shown in Figure 3a. The AOR uses the distance traveled by the feature in the successive images along with the actual position of the feature in those images to remove false-matched points, as illustrated in Figure 4.

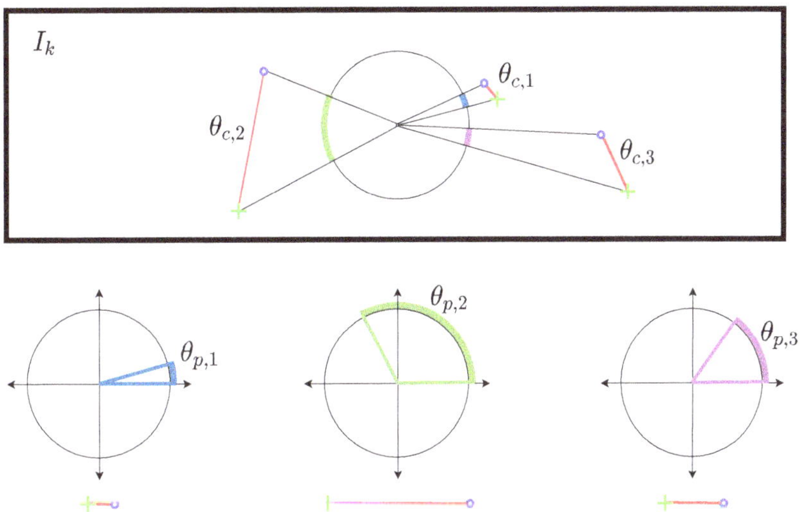

Figure 4. Visual illustration of θ_c and θ_p for three different feature pairs in an image.

Figure 4 illustrates the two metrics used by the proposed AOR. AOR can be divided into two steps. First, the angle θ_c between the lines drawn from the center of the image to the feature (shown in Figure 4) in I_{k-1} and I_k is simply calculated as [54].

$$\theta_{c,i} = \arccos \frac{x_{k-1,i} \cdot x_{k,i} + y_{k-1,i} \cdot y_{k,i}}{\sqrt{x_{k-1,i}^2 + y_{k-1,i}^2} \cdot \sqrt{x_{k,i}^2 + y_{k,i}^2}}, \tag{2}$$

where $(x_{k-1,i}, y_{k-1,i})$ and $(x_{k,i}, y_{k,i})$ are the coordinates of the i-th feature in the frames I_{k-1} and I_k, with respect to the center of the image, respectively. Notice that θ_c represents the amount of feature motion, irrespective of its position in the frame. θ_c for different feature positions is illustrated in the top plot of Figure 4.

Second, the Euclidean distance traveled by the feature projected on a reference circle, as shown in the bottom plot of Figure 4, and the corresponding angle θ_p is calculated as [55]:

$$E_i := \left\| \begin{bmatrix} x_{k,i} \\ y_{k,i} \end{bmatrix} - \begin{bmatrix} x_{k-1,i} \\ y_{k-1,i} \end{bmatrix} \right\|_2, \tag{3}$$

$$\theta_{p,i} = \frac{E_i}{R} \tag{4}$$

where R is the radius of the reference circle. Notice that R determines the sensitivity of the values of θ_p. The larger the radius, the smaller the angle for larger Euclidean distances. The radius can be calculated as,

$$R = \sqrt{\frac{c_x^2 + c_y^2}{\zeta}}, \tag{5}$$

where c_x and c_y are the centers of the image, and ζ is a parameter that controls the size of the radius. During the experimentation, the best results were obtained by $\zeta = 8$; however, different values may work better for different conditions. Notice that here we assume that the vanishing point is in the center of the image. Although this is not generally true, in the case of a ground vehicle motion, such an assumption did not affect the results acquired by the algorithm. Furthermore, through testing the algorithm with a vanishing point extraction algorithm [56], the output was similar in accuracy. Therefore, we chose to omit such a step to achieve faster VO pipeline.

Using the two angles θ_c and θ_p, a score S is computed for each feature as:

$$S_i = |\theta_{c,i}\theta_{p,i}(\theta_{c,i} - \theta_{p,i})|. \tag{6}$$

Notice that for every matched feature, the greater θ_p and θ_c values and the difference between them, the larger the score AOR yields.

Finally, a feature is selected as an inlier, if its AOR score value is less than a threshold η calculated as

$$\eta = c \cdot \text{median}\,(\mathcal{S}), \tag{7}$$

where $\mathcal{S}$ is the score set of the matched features, and $c > 1$ is a parameter, which is set in this paper to 2 (through tuning). The overall AOR algorithm is summarized in Algorithm 1.

Figure 3c shows the effect of AOR on the detected features. By using AOR, all the outliers, which are present in Figure 3a, are removed, and only the true features describing the motion of the camera remain. Furthermore, notice the effect of the AOR algorithm in removing the features detected on the moving vehicle present in the image since the motion of such features does not agree with the motion of the remaining features. Figure 3d shows the effect of both SSC and AOR on the image, where the inliers remaining in the image are better distributed due to the SSC effect.

Algorithm 1: AOR Algorithm

```
Set R and η
for p_i ∈ P_{k−1:k} do
    Calculate θ_{c,i} as in Equation (2).
    Calculate the euclidean distance of the feature motion E defined in (3).
    Calculate θ_{p,i} as in Equation (4).
    Calculate the feature AOR score S as in Equation (6).
    Push S_i → 𝒮.
end
Calculate η as in Equation (7).
for p_i ∈ P_{k−1:k} do
    if S_i < η then
        p_i → P̂_{k−1:k}
    end
end
where P̂_{k−1:k} is the set of matched features inliers.
```

The filtered matched feature set $\hat{\mathcal{P}}_{k-1:k}$ can then be passed to any VO algorithm to estimate the incremental motion of the camera and to compute the odometry.

4. Experimental Work

To show the generic aspect of the proposed pipeline, simple stereo, RGB-D, and monocular VO algorithms are implemented for validation. The motion estimation techniques used are the same as those described in [7].

The algorithms are implemented in Python using OpenCV library, SURF for feature tracking, and the extracted features are matched between consecutive frames by Brute-Force matching. All experiments and tests were conducted on a computer with an Intel i7-8850H 6-core processor running at 2.60 GHz using 16 GB of RAM, running Ubuntu 16.04. Furthermore, the algorithms were implemented with a Robot Operating System (ROS) wrapper node to be compatible with the ROS framework [57].

The VO algorithms were then used to estimate the motion of the camera using sequences from KITTI [34], TUM [35], as well as experimental sequences generated by Summit-XL Steel manufactured by Robotnik Inc. [58]. The performance of the pipeline is evaluated through several comparisons, which demonstrate the effect of the added stages to the VO pipeline.

4.1. Motion Estimation

4.1.1. Stereo/RGB-D Visual Odometry

The stereo and RGB-D VO algorithms used in this paper rely on solving the same 3D-to-2D correspondence problem. First, the features in the image I_{k-1} along with the disparity map or the depth image are used to produce the 3D features F_k in I_{k-1}. The motion of the camera is then estimated by solving the Perspective-n-Point (PnP) problem. The PnP problem is solved in a RANSAC scheme [8]. This is after utilizing the AOR to achieve better and more efficient motion estimation.

$$T_{k-1:k} = \underset{T_{k-1:k}}{\arg\min} \sum_{i=0}^{N_f} \left\| f_k^i - \hat{F}_{k-1}^i \right\|_2^2 \tag{8}$$

$T_{k-1:k} \in SE(3)$ is the transformation matrix describing the incremental motion between time-steps $k-1$ and k, f_k^i is i-th 2D feature in the current image, $\hat{F}_{k-1}^i$ is the same feature in 3D, reprojected from image I_{k-1} onto the current image I_k through $T_{k-1:k}$, and N_f is the total number of features in the image.

4.1.2. Monocular Visual Odometry

To estimate the motion from a single camera, the Epipolar constraint between frames is used as

$$\begin{bmatrix} x_{k-1,i} & y_{k-1,i} \end{bmatrix} \mathbf{E} \begin{bmatrix} x_{k,i} \\ y_{k,i} \end{bmatrix} = 0, \ \forall\, 0 \le i \le N_f \tag{9}$$

where $\mathbf{E} \in \mathbb{R}^{3\times 3}$ is the essential matrix for the calibrated camera [16].

The essential matrix $\mathbf{E}$ is estimated using the five-point algorithm proposed in [59]. After obtaining the essential matrix, it is decomposed into the translation and rotation of the camera as described in [16]. Furthermore, the motion estimation algorithm is executed with the Least Median Of Squares (LMEDS) [60] scheme in order to achieve better motion estimation.

Using a monocular camera, the ego-motion of the camera can be estimated up to a scale. To compensate for this scale, a velocity measurement of the vehicle needs to be available through the use of an external sensor such as wheel encoders, an IMU, a GPS, or through the CAN data from the vehicle's tachometer.

The implementation of the VO algorithms from scratch was intended for the ease of integration of the proposed pipeline. However, the pipeline, in general, can be integrated to any VO implementation.

4.2. Datasets

4.2.1. KITTI Vision Benchmark Dataset

KITTI Vision Benchmark Suite was selected as a publicly available dataset [34]. The dataset provides ground-truth ego-motion for 11 training sequences and 11 test sequences. The ground-truth is provided as a list of 6D poses for the training sequences, whereas for the test sequences, evaluation results are obtained by submitting them to the KITTI website. The dataset is sampled at 10 Hz at an average speed of 90 km/h, which creates a challenge in using the dataset for training and testing. Sequence 3 from the training subset is no longer available, as it was removed by KITTI for its similarities with the test sequences.

The dataset comprises the following information: raw synced and rectified color images from the left and right cameras and raw 3D GPS/IMU unfiltered data, along with the timestamps for all recordings. In order to convert the raw data to ROS bagfiles, the *kitti2bag* package was used [61]. The dataset also provides a tool for evaluating the performance of the VO and visual SLAM algorithms. This tool was used in the paper to evaluate the proposed pipeline in the case of the KITTI dataset.

4.2.2. TUM RGB-D Dataset

The TUM RGB-D dataset is a large dataset containing sequences captured by an RGB-D camera along with its ground-truth to establish a benchmark for evaluation of VO and visual SLAM algorithms [35]. The dataset contains the color and depth images taken by a Microsoft Kinect camera, while the ground truth was recorded using a high-accuracy motion capture system with eight high-speed tracking cameras (100 Hz). The data were recorded using a 30 Hz rate with a camera resolution of 640×460. The dataset also provides an online tool through which the results are submitted for evaluating the performance of VO and visual SLAM systems. In this paper, the TUM sequences are evaluated using the Relative Pose Error (RPE), which is recommended by the dataset for VO algorithms [62].

RPE is basically the error in relative motion between the pairs of the VO output. The evaluation tool by the TUM dataset computes the error between all pairs of the output and generates the evaluation metrics such as the Root Mean Square Error (RMSE), mean, max, etc. In this paper, the RMSE error for the translation and orientation is used for evaluation.

4.2.3. Images from Omnidirectional Robot

Summit XLS is a ground mobile robot with mecanum wheels, shown in Figure 5. The robot is equipped with an Astra RGB-D Camera (https://shop.orbbec3d.com/Astra, accessed on 27 October 2022), as well as wheel encoders. Several experiments were made using the robot to validate the proposed pipeline, while using the VICON motion capture system (https://www.vicon.com/, accessed on 27 October 2022) as a reference. The VICON system used consists of 12 cameras and the VICON bridge package was used to couple VICON with ROS [63]. Since the RGB-D VO case is tested and validated using the TUM dataset, the Summit-XL Steel sequences are used to validate the monocular VO while relying on the wheel encoders to obtain the speed for motion scaling. The evaluation is again conducted using the RMSE error for translation and orientation.

Three different sequences were executed using the robot in remote-control mode. In the first two sequences, the robot moved in semi-rectangular paths while, in the third sequence, the robot moved in a circular path. The total length of each of the paths was 12.5 m in the case of the rectangular paths and 6.5 m in the case of the circular path.

Figure 5. The omni-directional mobile robot used to validate the monocular VO algorithm with the proposed pipeline.

5. Results and Discussion

5.1. KITTI Vision Benchmark Dataset/Stereo VO

5.1.1. Pose Accuracy Comparison

In order to show the efficacy of the proposed algorithm, the pose estimation results from a stereo VO are reported with and without the proposed pipeline. Table 1 shows the accuracy comparison using the 10 sequences available from the KITTI dataset. The results shown are the translation and rotation RMSE values generated by the dataset evaluation tool. As can be seen in the table, the pipeline enhanced the pose estimation accuracy in almost all the sequences.

In sequence 2 (shown in Figure 6), note that the effect of the pipeline is very obvious since the presence of the pipeline significantly enhanced the pose estimation accuracy compared to the VO pose estimation without the pipeline. Notice that the reason for the divergence of the VO in the first case is due to the absence of enough features in the images, which made the VO unable to estimate the incremental motion for long durations in the sequence. On the other hand, using the CLAHE filter increases the number of stable features in the images, while using the AOR algorithm along with the RANSAC (which is present in both cases) ensures more accurate incremental motion estimation for all received images. This leads to a much better VO output, as shown in Figure 6.

In sequence 5, although the average translation RMSE of the VO without the pipeline is lower than that with the pipeline, the actual performance of the pose estimation for the VO with the pipeline is much better (as shown in Figure 7). The real performance of the VO odometry is not reflected in Table 1 because the drift in the orientation of the VO without the pipeline causes some estimated poses to look closer to the ground truth compared to the VO output with the pipeline. However, the overall path estimated by the VO with the pipeline is superior to that of the VO without the pipeline.

Finally, in the case of sequence 6, the performance of the VO without the pipeline outperformed that of the VO with the pipeline, as seen in Figure 8. This may be attributed to the fact that the amount of features available after applying the AOR is not enough for accurate motion estimation. This is further discussed in Section 5.1.2.

Table 1. KITTI dataset accuracy comparison.

Sequence Number	VO without Pipeline		VO with AOR		VO with Pipeline	
	Tr (%)	Rot (\deg/m)	Tr (%)	Rot (deg/m)	Tr (%)	Rot (deg/m)
0	1.85	0.011	1.80	0.011	**1.68**	**0.011**
1	5.14	0.013	5.33	0.013	**4.72**	**0.012**
2	6.99	0.036	**1.82**	**0.016**	1.89	0.017
4	3.91	0.007	2.75	**0.006**	**0.33**	0.007
5	**2.38**	0.011	2.75	0.012	2.43	**0.009**
6	**2.62**	0.010	2.74	0.010	2.88	**0.010**
7	2.11	0.016	**1.95**	**0.014**	2.07	0.015
8	2.57	0.012	2.65	**0.011**	**1.92**	0.012
9	2.70	0.019	2.86	0.019	**1.78**	**0.019**
10	2.65	0.020	2.82	0.02	**2.02**	**0.020**
Overall	3.29	0.015	2.71	0.013	**2.18**	**0.013**

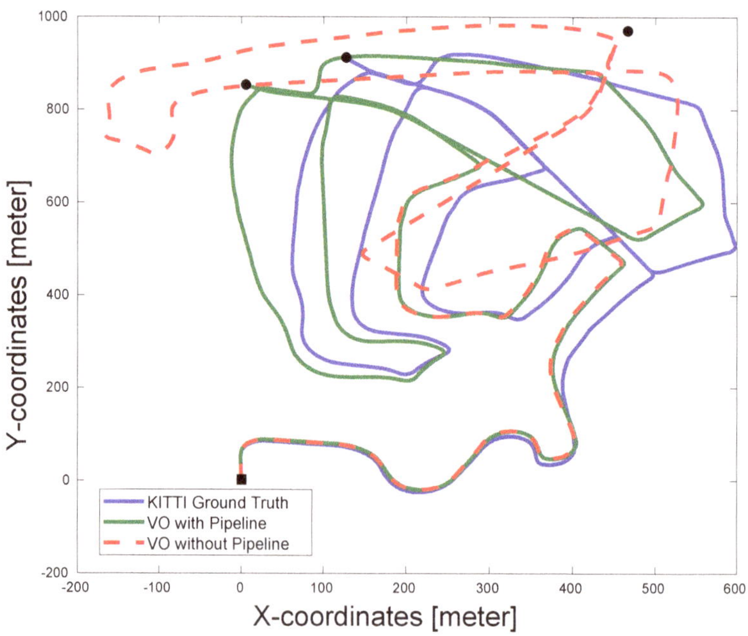

Figure 6. The pose estimation results of sequence 2 in the KITTI dataset with and without pipeline along with the dataset ground-truth. The proposed pipeline significantly enhances the output accuracy of the VO in this sequence. The VO output without the pipeline suffers from major drift errors and can even be considered to diverge. The black square represents the start of the paths and the black circles represent the ends of the paths.

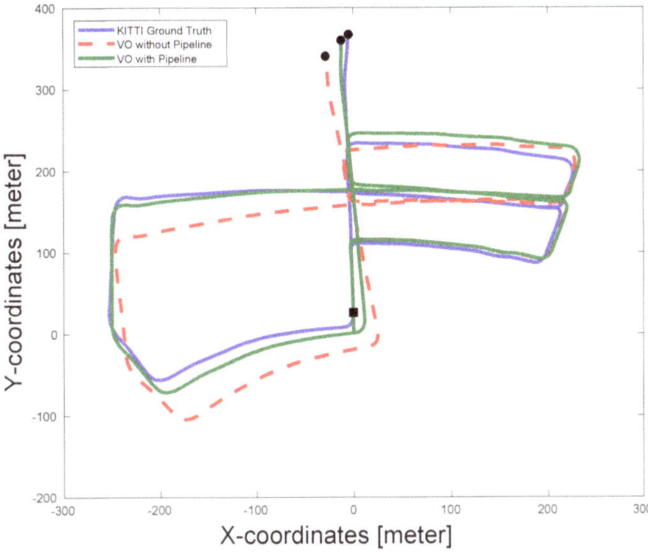

Figure 7. The pose estimation results of sequence 5 in the KITTI dataset with and without the pipeline, along with the dataset ground-truth. The performance of the VO with the pipeline enhances the performance of the VO in both translation and orientation estimation. This is not reflected in the average RMSE of the sequence due to the drift in the orientation of the VO without the pipeline, which causes some estimated poses to look close to the ground truth, although the overall estimated trajectory is much worse. The black square represents the start of the paths and the black circles represent the ends of the paths.

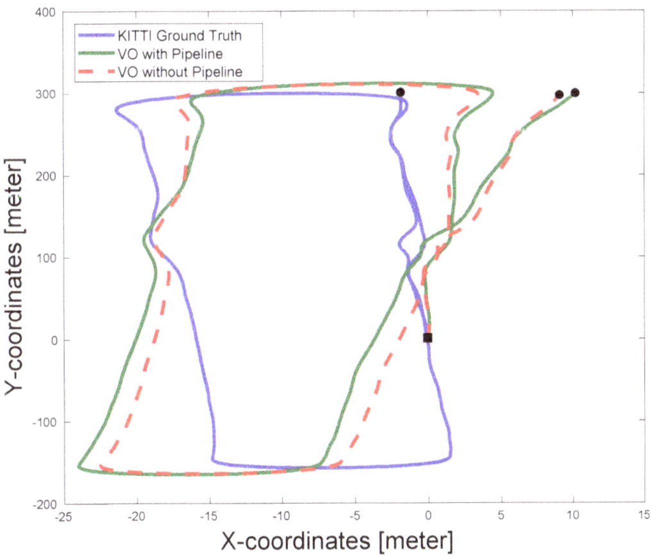

Figure 8. The pose estimation results of sequence 6 in the KITTI dataset with and without the pipeline along with the dataset ground-truth. The VO without the pipeline shows more accurate pose estimation compared to VO with the pipeline. However, the performance of both algorithms are very similar. The black square represents the start of the paths and the black circles represent the ends of the paths.

5.1.2. Effect of AOR

One of the contributions of the paper is the new outlier rejection algorithm named AOR. In this subsection, the effect of AOR alone on the performance of the VO is studied. To this end, the translation and orientation RMSE results for the VO with AOR are also reported in Table 1.

As shown in Table 2, the AOR significantly contributed to the enhancement of some of the sequences. For example, the table shows that the use of AOR was responsible for the enhancement of sequence 2, which diverged without the use of it, as shown in Figure 6. Furthermore, the use of AOR also resulted in better results for sequences 0, 4, and 7.

Table 2. AOR effect on pose estimation.

Seq. No.	VO without AOR		VO with AOR	
	Tr (%)	Rot (deg/m)	Tr (%)	Rot (deg/m)
0	1.85	0.011	**1.80**	0.011
1	**5.14**	0.013	5.33	**0.013**
2	6.99	0.036	**1.82**	**0.0158**
4	3.91	0.007	**2.75**	**0.0056**
5	**2.38**	**0.011**	2.75	0.012
6	**2.62**	0.010	2.74	0.010
7	2.11	0.016	**1.95**	**0.014**
8	**2.57**	0.012	2.65	**0.011**
9	**2.70**	0.019	2.86	0.019
10	**2.65**	0.020	2.82	0.020
Overall	3.29	0.015	**2.71**	**0.013**

The use of AOR resulted in worse results in the case of the other results. The reason for this effect is the absence of enough features for motion estimation after applying the AOR algorithm, which results in worse motion estimation due to the limited amount of information available. This problem can be addressed by increasing the threshold of the AOR η to increase the number of inliers.

As can be seen in the results, the AOR algorithm is an aggressive outlier rejection method. In other words, the AOR might result in the removal of matched features with slight deviations. This means that the use of AOR on an image requires the presence of a sufficient amount of stable features for motion estimation. Otherwise, the AOR will lead to the deterioration of the motion estimation due to decreasing the amount of matched features. The threshold weight can be altered to obtain a balanced amount of inliers to minimal outliers. However, this will improve the output of some sequences, and other situations might deteriorate. This is why the integration of AOR with CLAHE and SSC is a very good combination because each of the three stages plays a role in enhancing the motion estimation while complementing the negative effects of the other ones.

Although the use of AOR can lead to the removal of many matched features, the presence of CLAHE increases the number of stable features in the images. Moreover, despite the increase in features due to CLAHE, this might lead to a concentration of features in a certain region of the image. However, the use of SSC prevents such poor distribution of the features. Although the use of CLAHE, while adding more stable features, can also add more outliers in the features, the AOR acts to remove these outliers. In conclusion, the presence of the three stages of the pipeline acts as a desirable combination to overcome the drawbacks of each of the stages while utilizing their advantages.

It is worth mentioning that the AOR resulted in worse results for some sequences; however, the overall performance of the VO with the pipeline (2.71%) was better than that of the VO without the pipeline (3.29%). Furthermore, the use of the complete pipeline still resulted in better accuracy for almost all sequences compared to the VO without the pipeline, as shown in Table 1.

The VO with pipeline results in better performance because the use of CLAHE increased the overall amount of features while SSC uniformly distributed those features helping to achieve better results along with the AOR outlier removal.

5.1.3. Computational Cost

It is expected that adding processing stages to the VO algorithm will lead to an increase in computational time. This can indeed be seen in Table 3, where the average computation time for the VO is reported after adding each of the proposed stages of the pipeline. However, the significant benefits in the accuracy of the pose estimation outweigh the increased computational cost.

In Table 3, the computation time of the VO algorithm with the AOR algorithm only is reported. Notice that the computation time of the algorithm after adding the AOR to the VO is less than that of the VO without AOR. As discussed in Section 3, this is due to the fact that the AOR removes the false-matched features. Accordingly, this simplifies the mission of the RANSAC, which results in a faster motion estimation convergence and a faster performance, as shown in Table 3. This also explains why the full pipeline computational time is less than the case of adding CLAHE only or CLAHE and SSC.

Table 3. Computation time comparison for KITTI sequences.

VO Type	Comp. Time (mean $\pm$ std [ms])	Tr RMSE (%)	Rot RMSE (deg/m)
Vanilla	116 ± 31	3.29	0.0154
CLAHE	176 ± 36	2.88	0.0136
CLAHE and SSC	181 ± 34	2.86	0.0136
AOR Only	**106 ± 24**	2.71	0.0134
Full Pipeline	160 ± 35	**2.18**	**0.0133**

The mentioned average computational time shows that the algorithm is capable of working in real-time while receiving up to 6 fps. The performance of the algorithm, as well as the accuracy of the pose estimation, can be further enhanced through the use of a Graphical Processing Unit (GPU) and multithreading processing.

5.2. TUM RGB-D Dataset/RGB-D VO

5.2.1. Pose Accuracy Comparison

Table 4 shows the translation and orientation RMSE for nine sequences from the TUM RGB-D dataset. As shown in the table, the RGB-D VO with the proposed pipeline shows better pose estimation accuracy for all sequences. This enhancement varies from one sequence to another based on the lighting and the number of features available in each sequence.

Figure 9 shows an example of the pose estimation output from the VO algorithm with and without the pipeline. It can be seen that the VO without the proposed pipeline suffers from large motion estimation errors at the beginning of the path, which in turn results in large drift errors over the rest of the path. As for the VO with the pipeline, although there is still an error in the estimated path, the error is significantly smaller than that of the VO without the pipeline, especially at the beginning of the path, causing a much better estimation for the rest of the path. The better performance of the VO with the pipeline algorithm is attributed to the increased amount of detected features at the beginning of the path compared to that of the VO without the pipeline.

Table 4. TUM accuracy comparison.

Seq. Name	VO without Pipeline		VO with Pipeline	
	Tr (m)	Rot (deg)	Tr (m)	Rot (deg)
fr1/xyz	0.24	8.80	**0.04**	**2.08**
fr1/desk	0.43	19.5	**0.07**	**3.55**
fr1/desk2	0.46	23.8	**0.09**	**6.74**
fr1/room	0.32	23.5	**0.12**	**5.73**
fr2/pioneer_360	0.18	6.50	**0.10**	**3.58**
fr2/pioneer_slam	0.10	3.60	**0.09**	**2.38**
fr2/pioneer_slam2	0.07	2.82	**0.07**	**2.00**
fr2/pioneer_slam3	0.07	2.03	**0.05**	**1.44**
fr2/desk	0.19	5.20	**0.02**	**0.64**
Overall	0.23	10.6	**0.07**	**3.12**

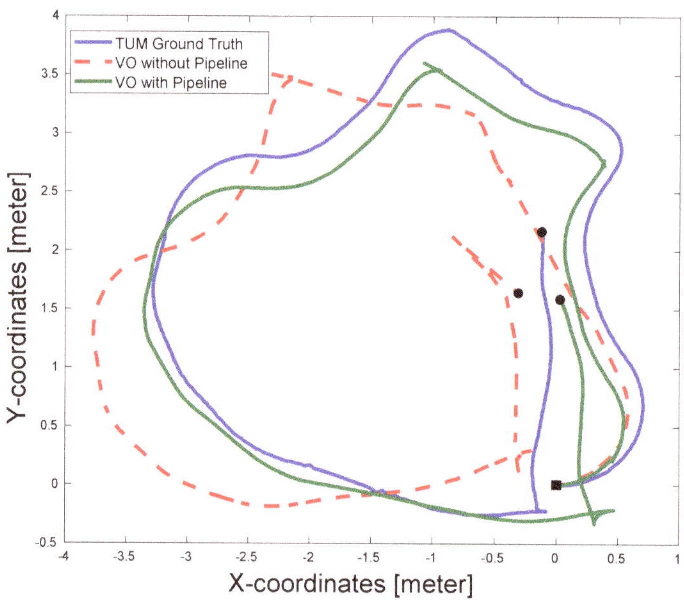

Figure 9. The pose estimation results for the sequence fr1/pioneer_360 in the TUM dataset with and without the pipeline along with the dataset ground-truth. The pose estimation accuracy of the VO with the pipeline is superior, while the VO without the pipeline suffers from both poor translation and orientation estimation. The black square represents the start of the paths and the black circles represent the ends of the paths.

Since the TUM sequences are recorded indoors, adding the CLAHE stage results in a significant increase in the detected features (some examples are shown in Figure 10). Note that for these sequences, an RGB-D VO algorithm is used, which means that only the features with an observable depth by the depth sensor can be used for motion estimation. In other words, the far features in the images cannot be used. Through the SSC algorithm, the features detected in the images are well distributed, and thus, the number of close features with observable depth increases (see Figure 10).

The results shown in Figure 9 and Table 4 confirm the efficacy of the proposed pipeline for VO algorithms, even for indoor scenarios.

Figure 10. Three examples from TUM sequences showing the effect of CLAHE and SSC in indoor scenarios. The images in the top row show the images with the detected features using SURF, detected without the use of CLAHE and SSC algorithms. The images in the bottom row show the features detected by SURF after adding the CLAHE and SSC stages.

5.2.2. Computational Cost

Table 5 shows the average computational time for the VO with the different added stages. Notice that, in this case, the results are slightly different from those of the computation cost analysis shown for KITTI sequences. As expected, the computational cost increases for the different stages of the proposed pipeline. However, in KITTI sequences, adding the AOR algorithm to the pipeline resulted in a faster performance compared to the VO without any stages. This is not the case for TUM sequences, where the computational cost still increases with the AOR algorithm. It is postulated that the amount of features detected in the case of TUM is larger or the number of outliers in the matched features set is larger, which is mainly based on qualitative and quantitative analysis. Nevertheless, adding the AOR to the pipeline results in lowering the computation cost of the VO algorithm. As can be seen in Table 5, the average computational cost of the VO algorithm when adding CLAHE and SSC is larger than that of the overall pipeline computational cost.

Table 5. Computation time comparison for TUM sequences.

VO Type	Comp. Time (mean ± std (ms))	Tr RMSE (m)	Rot RMSE (deg)
Vanilla	**118 ± 26**	0.229	10.6
CLAHE	179 ± 48	0.213	10.7
CLAHE and SSC	185 ± 53	0.210	9.98
AOR Only	169 ± 36	0.135	10.4
Full Pipeline	176 ± 40	**0.073**	**3.13**

5.3. *Summit XL Steel Sequences/Monocular VO*

5.3.1. Pose Accuracy Comparison

The translation and orientation RMSE are reported in Table 6 for three different scenarios. The results show the accuracy enhancement in the case of the VO with the pipeline. For the Summit XL Steel robot sequences, a monocular VO algorithm was used for validating the proposed processing pipeline. Since the scale is unobservable, the robot's wheel encoders are used to calculate its speed and the scale of the odometry. This means that a component of the error is due to the error in the velocity measurement taken by the encoders. However, since the same data are used for both cases, this effect is the same for both cases and will not make any bias in the comparison.

Table 6. Summit accuracy comparison.

Seq. Name	VO without Pipeline		VO with Pipeline	
	Tr (m)	Rot (deg)	Tr (m)	Rot (deg)
Rectangle 1	0.49	1.67	**0.34**	**1.51**
Rectangle 2	0.25	1.69	**0.25**	**1.52**
Circle	0.50	2.09	**0.48**	**2.07**
Overall	0.31	1.75	**0.27**	**1.67**

Figure 11 shows the estimation results for VO with and without the pipeline. As can be seen in the figure, the accuracy is superior in the case of the VO with the proposed pipeline. This is especially true at the end of the sequence, at which the VO without the pipeline suffered from a large drift error. The presence of the AOR resulted in removing many matched outliers, which would have caused bad motion estimation and a significant amount of drift errors in the case of VO without the proposed pipeline.

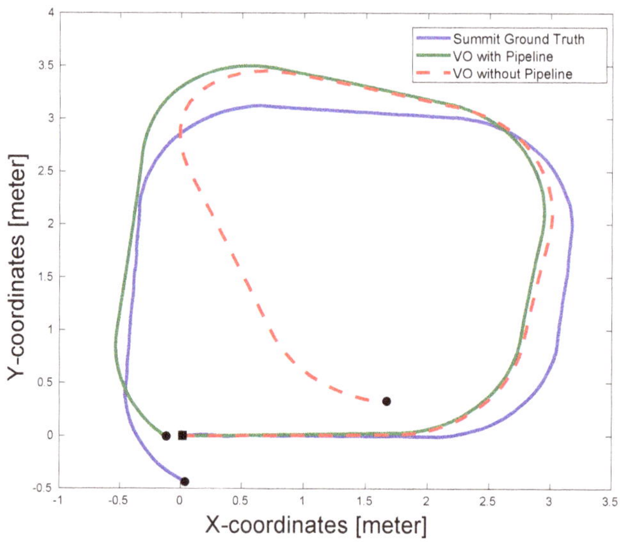

Figure 11. The pose estimation results for a rectangular sequence executed by Summit XLS steel. The figure shows the output of the VO with and without the proposed pipeline. The VO with the pipeline shows better pose accuracy especially at the end of the sequence while the VO without the pipeline suffers from significant amount of drift error. The black square represents the start of the paths and the black circles represent the ends of the paths.

5.3.2. Computational Cost

Table 7 shows the computational time analysis of several combinations of VO algorithms, including the proposed VO algorithm. As was illustrated before, the best computational performance was for the VO with the AOR algorithm. This is a direct result of the better outliers removal of the AOR, which results in a better and faster convergence of the motion estimation algorithm. The table also shows the translation and orientation RMSE for each of the different VO combinations. The results confirm that the proposed VO results in the best performance.

Table 7. Computation time comparison for summit XLS steel sequences.

VO Type	Comp. Time (mean ± std (ms))	Tr RMSE (m)	Rot RMSE (deg)
Vanilla	148 ± 28	0.315	1.68
CLAHE	165 ± 47	0.350	1.68
CLAHE and SSC	168 ± 42	0.302	1.74
AOR Only	**132 ± 19**	0.287	1.68
Full Pipeline	158 ± 34	**0.275**	**1.67**

5.4. Discussion

For all the scenarios, and for all VO types used in this paper, the proposed pipeline showed better performance compared to the VO without the pipeline. Specifically, adding the three additional stages to the actual VO algorithm enhanced the accuracy by an average of 37% for the considered datasets. These additional three stages can be added to any feature-based VO algorithm to enhance its accuracy and robustness.

In Tables 3, 5 and 7, the results for different combinations of the three proposed stages were reported. In the three cases, the VO with the full proposed pipeline showed better pose estimation accuracy. This shows that the three stages proposed in this paper are integral. Each one of the three serves its own purpose and contributes to the overall enhancement. However, as expected, this came with an increase in the computational cost of the algorithm. Notice that the increase in the computational cost is still not significant (an average of 37 ms) and does not result in a large reduction in the number of frames per second. In most applications, this increase in the computational cost can be accepted to improve the pose estimation accuracy in return (as shown in Table 8).

Table 8. The pipeline effect on VO.

Dataset	Tr (%)	Rot (%)	Comp. Time (ms)
KITTI	−33%	−13%	+44
TUM	−68%	−70%	+58
Summit	−12%	−0.5%	+10

6. Conclusions and Future Works

In this paper, an image processing pipeline was introduced to enhance the accuracy and robustness of VO algorithms. The proposed pipeline consists of three stages, CLAHE, SSC, and AOR. Each stage addresses a separate issue associated with pose estimation error.

The proposed pipeline is intended to be generic and modular, which can be embedded in any feature-based algorithm in order to enhance its performance. In order to validate the proposed pipeline, sequences from KITTI and TUM datasets, as well as experimental sequences generated by a commercial omnidirectional mobile robot, were used. For each dataset, one type of VO was used for validation, namely stereo, RGB-D and monocular. The quantitative and qualitative results show that the proposed pipeline has a significant enhancement in the VO accuracy and robustness, with a minor increase in the computational time.

As mentioned earlier, a VO algorithm relies on integration and, consequently, can suffer from large amounts of errors or the overall divergence of the pose estimation through operation. This can occur due to several causes, such as poor lighting conditions, false-matched features or motion bias. Throughout this paper, the three aforementioned causes of error were addressed by designing a generic pipeline that can be integrated into any visual odometry algorithm to enhance its accuracy.

As a future work, the proposed pipeline is planned to be integrated into visual SLAM algorithms, and the effect of the pipeline will be studied. Furthermore, comparisons with deep learning approaches will also be conducted to see which approach works best with which conditions. Several additional filtration steps will also be investigated to further enhance the performance of VO algorithms. Meanwhile, the computational cost of the algorithm is expected to be reduced through the use of GPUs and parallel computing techniques.

Author Contributions: Conceptualization, M.S., M.O. and A.H.; methodology, M.S., M.O. and A.H.; software, M.S., M.O. and A.H.; validation, M.S., M.O. and A.H.; formal analysis, M.S., M.O. and A.H.; writing—original draft preparation, M.S., M.O. and A.H.; writing—review and editing, M.S., M.O., A.H., M.W.M., S.J. and W.M.; supervision, A.H., M.W.M., S.J. and W.M.; project administration, A.H., M.W.M., S.J. and W.M.; funding acquisition, A.H. All authors have read and agreed to the published version of the manuscript.

Funding: This research received no external funding.

Data Availability Statement: Publicly available datasets were analyzed in this study. This data can be found here: [https://vision.in.tum.de/data/datasets/rgbd-dataset, www.cvlibs.net/datasets/kitti, accessed on 27 October 2022].

Acknowledgments: The work in this paper is in part supported by Natural Sciences and Engineering Research Council (NSERC) of Canada under the Strategic Partnership Grant for Projects (STPGP-506987-2017) and under Postdoctoral Fellowship grant (PDF-532957-2019).

Conflicts of Interest: The authors declare no conflict of interest.

Abbreviations

The following abbreviations are used in this manuscript:

MDPI	Multidisciplinary Digital Publishing Institute
DOAJ	Directory of Open Access Journals
TLA	Three Letter Acronym
LD	Linear Dichroism

References

1. Zhang, J.; Singh, S. LOAM: Lidar Odometry and Mapping in Real-time. In Proceedings of the Robotics: Science and Systems Conference, Rome, Italy, 13–15 July 2014 ; Volume 2.
2. Quist, E.B.; Niedfeldt, P.C.; Beard, R.W. Radar odometry with recursive-RANSAC. *IEEE Trans. Aerosp. Electron. Syst.* **2016**, *52*, 1618–1630. [CrossRef]
3. Drawil, N.M.; Amar, H.M.; Basir, O.A. GPS localization accuracy classification: A context-based approach. *IEEE Trans. Intell. Transp. Syst.* **2012**, *14*, 262–273. [CrossRef]
4. Yi, J.; Zhang, J.; Song, D.; Jayasuriya, S. IMU-based localization and slip estimation for skid-steered mobile robots. In Proceedings of the IEEE/RSJ International Conference on Intelligent Robots and Systems, San Diego, CA, USA, 29 October–2 November 2007; pp. 2845–2850.
5. Lee, S.; Song, J.B. Robust mobile robot localization using optical flow sensors and encoders. In Proceedings of the IEEE International Conference on Robotics and Automation (ICRA), New Orleans, LA, USA, 26 April–1 May 2004; Volume 1; pp. 1039–1044.
6. Shim, J.H.; Cho, Y.I. A mobile robot localization via indoor fixed remote surveillance cameras. *Sensors* **2016**, *16*, 195. [CrossRef]
7. Scaramuzza, D.; Fraundorfer, F. Visual odometry [tutorial]. *IEEE Robot. Autom. Mag.* **2011**, *18*, 80–92. [CrossRef]
8. Fraundorfer, F.; Scaramuzza, D. Visual odometry: Part ii: Matching, robustness, optimization, and applications. *IEEE Robot. Autom. Mag.* **2012**, *19*, 78–90. [CrossRef]
9. Strasdat, H.; Montiel, J.; Davison, A.J. Scale drift-aware large scale monocular SLAM. *Robot. Sci. Syst. VI* **2010**, *2*, 7.

10. Engel, J.; Schops, T.; Cremers, D. LSD-SLAM Large-scale direct monocular SLAM. In *Proceedings of the European Conference on Computer Vision*; Springer: Berlin/Heidelberg, Germany, 2014; pp. 834–849.
11. Mur-Artal, R.; Montiel, J.M.M.; Tardos, J.D. ORB-SLAM a versatile and accurate monocular SLAM system. *IEEE Trans. Robotics* **2015**, *31*, 1147–1163. [CrossRef]
12. Qu, X.; Soheilian, B.; Paparoditis, N. Landmark based localization in urban environment. *ISPRS J. Photogramm. Remote. Sens.* **2018**, *140*, 90–103. [CrossRef]
13. Yang, N.; Wang, R.; Cremers, D. Feature-based or direct An evaluation of monocular visual odometry. *arXiv* **2017**, arXiv:1705.04300.
14. Tomono, M. 3-D localization and mapping using a single camera based on structure-from-motion with automatic baseline selection. In Proceedings of the IEEE International Conference on Robotics and Automation, Barcelona, Spain, 18–22 April 2005; pp. 3342–3347.
15. Furukawa, Y.; Hernández, C.; et al. Multi-view stereo: A tutorial. *Found. Trends Comput. Graph. Vis.* **2015**, *9*, 1–148. [CrossRef]
16. Hartley, R.; Zisserman, A. *Multiple View Geometry in Computer Vision*; Cambridge University Press: Cambridge, UK, 2003.
17. Sabry, M.; Al-Kaff, A.; Hussein, A.; Abdennadher, S. Ground Vehicle Monocular Visual Odometry. In Proceedings of the IEEE Intelligent Transportation Systems Conf. (ITSC), Auckland, NZ, USA, 27–30 October 2019; pp. 3587–3592.
18. Zhou, D.; Dai, Y.; Li, H. Ground-Plane-Based Absolute Scale Estimation for Monocular Visual Odometry. *IEEE Trans. Intell. Transp. Syst.* **2019**, *21*, 791–802. [CrossRef]
19. Zhou, D.; Dai, Y.; Li, H. Reliable scale estimation and correction for monocular visual odometry. In Proceedings of the IEEE Intelligent Vehicles Symposium (IV), Gotenburg, Sweden, 19–22 June 2016; IEEE: Piscataway, NJ, USA, 2016; pp. 490–495.
20. Li, R.; Wang, S.; Long, Z.; Gu, D. Undeepvo Monocular visual odometry through unsupervised deep learning. In Proceedings of the IEEE International Conference on Robotics and Automation (ICRA), Brisbane, Australia, 21–25 May 2018; pp. 7286–7291.
21. Zhan, H.; Garg, R.; Saroj Weerasekera, C.; Li, K.; Agarwal, H.; Reid, I. Unsupervised learning of monocular depth estimation and visual odometry with deep feature reconstruction. In Proceedings of the IEEE Conference on Computer Vision and Pattern Recognition, Salt Lake City, UT, USA, 18–22 June 2018; pp. 340–349.
22. Steinbrücker, F.; Sturm, J.; Cremers, D. Real-time visual odometry from dense RGB-D images. In Proceedings of the IEEE International Conference on Computer Vision Workshops (ICCV Workshops), Barcelona, Spain, 7 November 2011; pp. 719–722.
23. Henry, P.; Krainin, M.; Herbst, E.; Ren, X.; Fox, D. RGB-D mapping: Using Kinect-style depth cameras for dense 3D modeling of indoor environments. *Int. J. Robot. Res.* **2012**, *31*, 647–663. [CrossRef]
24. Li, X.; Zhang, C. Robust RGB-D Visual Odometry Based on the Line Intersection Structure Feature in Low-Textured Scenes. In Proceedings of the 5th IEEE International Conference on Cloud Computing and Intelligence Systems (CCIS), Nanjing, China, 23–25 November 2018; pp. 390–394.
25. Lowe, D.G. Distinctive image features from scale-invariant keypoints. *Int. J. Comput. Vis.* **2004**, *60*, 91–110. [CrossRef]
26. Bay, H.; Tuytelaars, T.; Van Gool, L. Surf Speeded up robust features. In Proceedings of the European Conference on Computer Vision, Graz, Austria, 7–13 May 2006; Springer: Berlin/Heidelberg, Germany, 2006; pp. 404–417.
27. Rublee, E.; Rabaud, V.; Konolige, K.; Bradski, G. ORB: An efficient alternative to SIFT or SURF. In Proceedings of the IEEE International Conference on Computer Vision, Washington, DC, USA, 6–13 November 2011; pp. 2564–2571.
28. BAYRAKTAR, E.; Boyraz, P. Analysis of feature detector and descriptor combinations with a localization experiment for various performance metrics. *Turk. J. Electr. Eng. Comput. Sci.* **2017**, *25*, 2444–2454. [CrossRef]
29. Dong, X.; Dong, X.; Dong, J.; Zhou, H. Monocular visual-IMU odometry: A comparative evaluation of detector-descriptor-based methods. *IEEE Trans. Intell. Transp. Syst.* **2019**, *21*, 2471–2484. [CrossRef]
30. Alismail, H.; Kaess, M.; Browning, B.; Lucey, S. Direct visual odometry in low light using binary descriptors. *IEEE Robot. Autom. Lett.* **2016**, *2*, 444–451. [CrossRef]
31. Engel, J.; Sturm, J.; Cremers, D. Semi-dense visual odometry for a monocular camera. In Proceedings of the IEEE International Conference on Computer Vision, Sydney, Australia, 1–8 December 2013; pp. 1449–1456.
32. Yang, N.; Wang, R.; Gao, X.; Cremers, D. Challenges in monocular visual odometry Photometric calibration, motion bias, and rolling shutter effect. *IEEE Robot. Autom. Lett.* **2018**, *3*, 2878–2885. [CrossRef]
33. Bailo, O.; Rameau, F.; Joo, K.; Park, J.; Bogdan, O.; Kweon, I.S. Efficient adaptive non-maximal suppression algorithms for homogeneous spatial keypoint distribution. *Pattern Recognit. Lett.* **2018**, *106*, 53–60. [CrossRef]
34. Geiger, A.; Lenz, P.; Urtasun, R. Are we ready for autonomous driving? the kitti vision benchmark suite. In Proceedings of the IEEE Conference on Computer Vision and Pattern Recognition, Providence, RI, USA, 16–21 June 2012; pp. 3354–3361.
35. Sturm, J.; Engelhard, N.; Endres, F.; Burgard, W.; Cremers, D. A benchmark for the evaluation of RGB-D SLAM systems. In Proceedings of the IEEE/RSJ International Conference on Intelligent Robots and Systems, Algarve, Portugal, 7–12 October 2012; pp. 573–580.
36. Ramanath, R.; Snyder, W.E.; Yoo, Y.; Drew, M.S. Color image processing pipeline. *IEEE Signal Process. Mag.* **2005**, *22*, 34–43. [CrossRef]
37. Campbell, J.; Sukthankar, R.; Nourbakhsh, I.; Pahwa, A. A robust visual odometry and precipice detection system using consumer-grade monocular vision. In Proceedings of the IEEE International Conference on Robotics and Automation, Barcelona, Spain, 18–22 April 2005; pp. 3421–3427.

38. Desai, A.; Lee, D.J. Visual odometry drift reduction using SYBA descriptor and feature transformation. *IEEE Trans. Intell. Transp. Syst.* **2016**, *17*, 1839–1851. [CrossRef]
39. Zhao, X.; Min, H.; Xu, Z.; Wu, X.; Li, X.; Sun, P. Image antiblurring and statistic filter of feature space displacement: application to visual odometry for outdoor ground vehicle. *J. Sens.* **2018**, *2018*, 2987819. [CrossRef]
40. Yang, W.; Zhai, X. Contrast Limited Adaptive Histogram Equalization for an Advanced Stereo Visual SLAM System. In Proceedings of the IEEE International Conference on Cyber-Enabled Distributed Computing and Knowledge Discovery (CyberC), Guilin, China, 17–19 October 2019; pp. 131–134.
41. Zhang, J.; Ila, V.; Kneip, L. Robust visual odometry in underwater environment. In Proceedings of the IEEE OCEANS-MTS/IEEE Kobe Techno-Oceans (OTO), Kobe, Japan, 28–31 May 2018; pp. 1–9.
42. Wu, M.; Lam, S.K.; Srikanthan, T. A framework for fast and robust visual odometry. *IEEE Trans. Intell. Transp. Syst.* **2017**, *18*, 3433–3448. [CrossRef]
43. Jiang, Y.; Xu, Y.; Liu, Y. Performance evaluation of feature detection and matching in stereo visual odometry. *Neurocomputing* **2013**, *120*, 380–390. [CrossRef]
44. Mou, W.; Wang, H.; Seet, G. Efficient visual odometry estimation using stereo camera. In Proceedings of the 11th IEEE International Conference on Control & Automation (ICCA), Sapporo, Japan, 6–9 July 2014; pp. 1399–1403.
45. Rosten, E.; Drummond, T. Machine learning for high-speed corner detection. In Proceedings of the European Conference on Computer Vision, Tel Aviv, Israel, 23–27 October 2006; Springer: Berlin/Heidelberg, Germany, 2006; pp. 430–443.
46. Aladem, M.; Rawashdeh, S.A. Lightweight visual odometry for autonomous mobile robots. *Sensors* **2018**, *18*, 2837. [CrossRef]
47. Brown, M.; Szeliski, R.; Winder, S. Multi-image matching using multi-scale oriented patches. In Proceedings of the IEEE Computer Society Conference on Computer Vision and Pattern Recognition (CVPR'05), San Diego, CA, USA, 20–25 June 2005; Volume 1, pp. 510–517.
48. Calonder, M.; Lepetit, V.; Ozuysal, M.; Trzcinski, T.; Strecha, C.; Fua, P. BRIEF: Computing a local binary descriptor very fast. *IEEE Trans. Pattern Anal. Mach. Intell.* **2011**, *34*, 1281–1298. [CrossRef] [PubMed]
49. Buczko, M.; Willert, V. How to distinguish inliers from outliers in visual odometry for high-speed automotive applications. In Proceedings of the IEEE Intelligent Vehicles Symposium (IV), Gotenburg, Sweden, 19–22 June 2016; pp. 478–483.
50. Fan, H.; Zhang, S. Stereo Odometry Based on Careful Frame Selection. In Proceedings of the 10th IEEE International Symposium on Computational Intelligence and Design (ISCID), Hangzhou, China, 9–10 December 2017; Volume 2, pp. 177–180.
51. Kitt, B.; Geiger, A.; Lategahn, H. Visual odometry based on stereo image sequences with ransac-based outlier rejection scheme. In Proceedings of the IEEE Intelligent Vehicles Symposium (IV), La Jolla, CA, USA, 21–24 June 2010; pp. 486–492.
52. Zuiderveld, K. Contrast limited adaptive histogram equalization. In Proceedings of the Graphics Gems IV ; Academic Press Professional, Inc.: Cambridge, MA, USA, 1994; pp. 474–485.
53. Mount, D.M.; Netanyahu, N.S.; Le Moigne, J. Efficient algorithms for robust feature matching. *Pattern Recognit.* **1999**, *32*, 17–38. [CrossRef]
54. Weisstein, E.W. Law of Cosines from MathWorld—A Wolfram Web Resource. Available online: https://mathworld.wolfram.com/LawofCosines.html (accessed on 10 June 2020).
55. Weisstein, E.W. Angular Distance from MathWorld—A Wolfram Web Resource. Available online: https://mathworld.wolfram.com/AngularDistance.html (accessed on 10 June 2020).
56. Lu, X.; Yaoy, J.; Li, H.; Liu, Y.; Zhang, X. 2-line exhaustive searching for real-time vanishing point estimation in Manhattan world. In Proceedings of the Winter Conference on Applications of Computer Vision (WACV), Santa Rosa, CA, USA, 24–31 March 2017; IEEE: Piscataway, NJ, USA, 2017; pp. 345–353.
57. Quigley, M.; Conley, K.; Gerkey, B.; Faust, J.; Foote, T.; Leibs, J.; Wheeler, R.; Ng, A.Y. ROS an open-source Robot Operating System. *ICRA Workshop Open Source Softw.* **2009**, *3*, 1–5.
58. Robotnik. SUMMIT- XL STEEL Datasheet Available online: www.robotnik.eu (accessed on 10 June 2020).
59. Nister, D. An efficient solution to the five-point relative pose problem. *IEEE Trans. Pattern Anal. Mach. Intell.* **2004**, *26*, 756–770. [CrossRef]
60. Rousseeuw, P.J. Least median of squares regression. *J. Am. Stat. Assoc.* **1984**, *79*, 871–880. [CrossRef]
61. Krejci, T. kitti2bag. Available online: https://github.com/tomas789/kitti2bag/ (accessed on 1 February 2020).
62. TUM. Submission form for automatic evaluation of RGB-D SLAM results. Available online: https://vision.in.tum.de/data/datasets/rgbd-dataset/online_evaluation (accessed on 10 June 2020).
63. Achtelik, M. vicon bridge. GitHub Repository. Available online: https://github.com/ethz-asl/vicon_bridge (accessed on 1 October 2020).

sensors

MDPI

Article

Up-Sampling Method for Low-Resolution LiDAR Point Cloud to Enhance 3D Object Detection in an Autonomous Driving Environment

Jihwan You and Young-Keun Kim *

School of Mechanical and Control Engineering, Handong Global University, Pohang 37554, Republic of Korea
* Correspondence: ykkim@handong.edu

Abstract: Automobile datasets for 3D object detection are typically obtained using expensive high-resolution rotating LiDAR with 64 or more channels (Chs). However, the research budget may be limited such that only a low-resolution LiDAR of 32-Ch or lower can be used. The lower the resolution of the point cloud, the lower the detection accuracy. This study proposes a simple and effective method to up-sample low-resolution point cloud input that enhances the 3D object detection output by reconstructing objects in the sparse point cloud data to produce more dense data. First, the 3D point cloud dataset is converted into a 2D range image with four channels: x, y, z, and intensity. The interpolation on the empty space is calculated based on both the pixel distance and range values of six neighbor points to conserve the shapes of the original object during the reconstruction process. This method solves the over-smoothing problem faced by the conventional interpolation methods, and improves the operational speed and object detection performance when compared to the recent deep-learning-based super-resolution methods. Furthermore, the effectiveness of the up-sampling method on the 3D detection was validated by applying it to baseline 32-Ch point cloud data, which were then selected as the input to a point-pillar detection model. The 3D object detection result on the KITTI dataset demonstrates that the proposed method could increase the mAP (mean average precision) of pedestrians, cyclists, and cars by 9.2%p, 6.3%p, and 5.9%p, respectively, when compared to the baseline of the low-resolution 32-Ch LiDAR input. In future works, various dataset environments apart from autonomous driving will be analyzed.

Keywords: 3D upsampling; LiDAR super-resolution; interpolation; 3D object detection

Citation: You, J.; Kim, Y.-K. Up-Sampling Method for Low-Resolution LiDAR Point Cloud to Enhance 3D Object Detection in an Autonomous Driving Environment. *Sensors* **2023**, *23*, 322. https://doi.org/10.3390/s23010322

Academic Editors: Yafei Wang, Chao Huang, Peng Hang, Zhiqiang Zuo and Bo Leng

Received: 31 October 2022
Revised: 20 December 2022
Accepted: 23 December 2022
Published: 28 December 2022

1. Introduction

Light detection and ranging (LiDAR) is a core mapping and detection sensor technology that ensures the safe and autonomous navigation of robots, drones, and vehicles. LiDAR is typically implemented in autonomous driving applications for 3D object detection, and particularly the detection of pedestrians, cyclists, and vehicles from a few to hundreds of meters. 3D LiDAR with high vertical resolution, such as 64-channel (Ch) LiDAR, can achieve highly accurate 3D detection of relatively small objects within a large volume area. Consequently, it is used in autonomous vehicles to obtain a more vertical dense point cloud and to achieve better object segmentation and a safer driving experience.

Furthermore, the sparsity of the point cloud increases with the increase in the distance of the object owing to the fixed vertical angles of the laser diodes of a 3D LiDAR. A low-resolution LiDAR, such as 16-Ch, may not be able to effectively capture the points required to distinguish a pedestrian from the background, causing major safety issues.

Accordingly, the representative open datasets of LiDAR 3D object detection, such as the KITTI [1] and Waymo Open [2] datasets, obtain the data point clouds of the driving environment by using 64-Ch LiDARs.

A high-resolution LiDAR, although preferable for accurate detection and safe driving, is expensive. A 64-Ch LiDAR is approximately 10–20 times more expensive than a low-resolution 16-Ch LiDAR. The high cost of high-resolution LiDAR considerably restricts its widespread implementation in commercial vehicles. This cost can be reduced through the super-resolution of the sparse point cloud of a low-resolution LiDAR while achieving a high 3D object detection accuracy.

The conventional 3D data up-sampling methods include bilinear or bicubic interpolations [3]. Although they present a high processing speed, they overly smoothen the object edges. Most existing deep-learning super-resolution models are designed for 2D images, and only a few models can be implemented for 3D point clouds. PU-Net [4] proposed a method for up-sampling 3D point cloud models; however, they primarily focused on obtaining the resolution of small objects. Shan et al. [5] proposed a deep-learning model to increase the LiDAR resolution, wherein the super-resolution effects of the model on the simulation data were compared with those of other conventional methods. However, object detection enhancement due to the super-resolution was not analyzed and validated. Contrary to such methods, we propose a simple and efficient method for the up-sampling of the point cloud based on a non-deep-learning model. This method enhances the 3D object detection accuracy by predicting and filling the sparse space in the point cloud data obtained by a low-resolution LiDAR. The proposed method also solves the over-smoothing problem of the conventional interpolation methods while achieving higher operational speed and better object detection performance when compared to the aforementioned deep-learning 3D super-resolution method [4,5]. The proposed method first converts the 3D point cloud data into 2D range images using four channels. It then implements a weighted interpolation with a decaying factor and excludes the zero value outliers. Subsequently, the up-sampled point cloud is passed to the widely used point-pillar [6] 3D object detection model to evaluate the improvement in the detection accuracy.

The contributions of this paper can be summarized as follows:

- This study presents a simple and efficient up-sampling method for the sparse point cloud of a low-resolution LiDAR,
- The proposed up-sampling method enhances the 3D object detection of a low-resolution LiDAR,
- A comparative analysis is performed using the conventional interpolation methods and recent deep-learning super-resolution models on point clouds for 3D object detection.

2. Related Work

2.1. 2D Image Up-Sampling

Previous studies conducted on super-resolution were primarily focused on the enhancement of a single 2D image [7]; in this process, a low-resolution image with coarse details is converted to a corresponding high-resolution image with refined details. Other similar terminologies for super-resolution in the research community include image scaling, interpolation, and up-sampling. In the current study, we use the term up-sampling to refer to super-resolution.

The conventional image up-sampling methods employ interpolation methods that fill an empty space with data points using a low-order polynomial or other general kernels. The most commonly used interpolation approaches on 2D images include the nearest-neighbor, bilinear, bicubic, and lanczos [7] schemes; these methods are simple and present a low computational overhead. However, the visual quality obtained is low due to the heavy smoothing performed on the edges.

Deep-learning methods, which employ highly non-linear kernels, present better up-sampling performance for complex scenarios. The super-resolution convolutional neural network (SR-CNN) [8] is a pioneer end-to-end up-sampling model that uses several convolution layers stacked on top of each other. The efficient sub-pixel convolutional neural network (ESPCN) [9] is a more complex and faster up-sampling method that extracts several features in the low-resolution feature map that are aggregated to reconstruct a high-resolution image output. Additionally, various approaches of deep-learning have been presented for image up-sampling, such as residual networks methods [10,11], recursive network methods [12,13], progressive reconstruction designs [14,15], densely connected networks [16,17], multi-branch designs [18,19], attention-based networks [20,21], multiple degradation handling networks [22,23], and GAN (generative adversarial networks)-based models [24,25].

2.2. 3D Point Cloud Up-Sampling

The 3D point cloud up-sampling process requires interpolation in a higher-dimension space, including the depth map, when compared to the 2D image up-sampling process. Typically, the low-resolution point cloud data is a 3D or higher-dimension space comprising an empty space with sparse, unordered data points, which increases the complexity of the interpolation.

2.2.1. 3D Up-Sampling with RGB Image

Most existing studies on depth map up-sampling have combined corresponding 2D RGB images. The existing image-guided up-sampling methods use fusion filters such as Markov random field (MRF) [26,27] bilateral filtering with pixel color information [28]. The depth interpolation is estimated using a filter based on the information of the neighboring pixels such as the geometric distance, intensity, and anisotropic diffusion [29]. Some recent studies have introduced deep-learning models for image-guided 3D up-sampling. Shan et al. [5] projected the 3D point cloud as a 2D range image and up-sampled the cloud to a denser map using an encoder–decoder architecture with residual connections similar to REDNet [11] and UNet [30]. Although the reconstruction loss is lower when compared to conventional interpolation methods, the simulation-based method [5] did not demonstrate the effectiveness of reconstruction in improving 3D object detection accuracy.

2.2.2. 3D Up-Sampling with Point Cloud

For the applications where only LiDAR is available, the 3D up-sampling of the sparse point cloud must be performed without using any other high-resolution guides. Only a few studies have been conducted on LiDAR up-sampling methods that map the sparse 3D points into a higher resolution using deep-learning models.

PU-Net [4] is a recently developed data-driven up-sampling method on the 3D point cloud data; it learns the multi-level features of each point and implicitly expands the point sets via multi-branch convolution units in feature spaces. This model performs the up-sampling of small sized objects in indoor environments. The reconstruction loss involved is lower when compared to conventional interpolation methods; however, the effectiveness of the reconstruction in enhancing the 3D object detection accuracy was not demonstrated.

The proposed up-sampling method for low-resolution LiDAR point clouds directly enhances the 3D object detection. The proposed method is not based on deep learning, contrary to the recent methods; hence, it achieves more efficient calculations. The reconstructed 3D point clouds are then fed to the 3D object detection model to evaluate the effectiveness of the object detection. Furthermore, we validated the up-sampling effectiveness of the proposed method by performing a comparative analysis of our method with the recently developed deep-learning method [5] and the conventional interpolation methods in terms of the object detection performance on the KITTI dataset.

3. Comparison Methods

3.1. Test Data Preparation

This section presents the KITTI dataset [1] of the bird's-eye view used for 3D object detection. The dataset was acquired using the Velodyne HDL 64-Ch LiDAR and a total of 7481 frames were used for training. High-resolution LiDAR range images were obtained by converting the 64-Ch LiDAR point cloud data into range. The low-resolution range images were obtained by extracting some rows out of the 64-Ch LiDAR range images. In the following section, we present a comparison of the up-sampling performance between the low-resolution (16-Ch) LiDAR point cloud and the high-resolution point cloud (64-Ch) of the conventional and proposed methods.

3.2. Conventional Up-Sampling

The representative conventional up-sampling methods were selected for the performance comparison with the proposed up-sampling method. The conventional up-sampling methods include the interpolation-based and deep-learning-based methods, as explained in Section 2.

Figure 1 presents the comparisons of the basic representative interpolation methods, which include the nearest-neighborhood, bilinear, and bicubic interpolations. In Figure 1, the nearest-neighborhood interpolation method cannot describe complex shapes such as curves of objects. Several noisy points are generated around the object in the bilinear and bicubic interpolation. Considering the sparsity characteristic of a range image, it is essential to perform interpolation for all of the points apart from the zero value point in a range image.

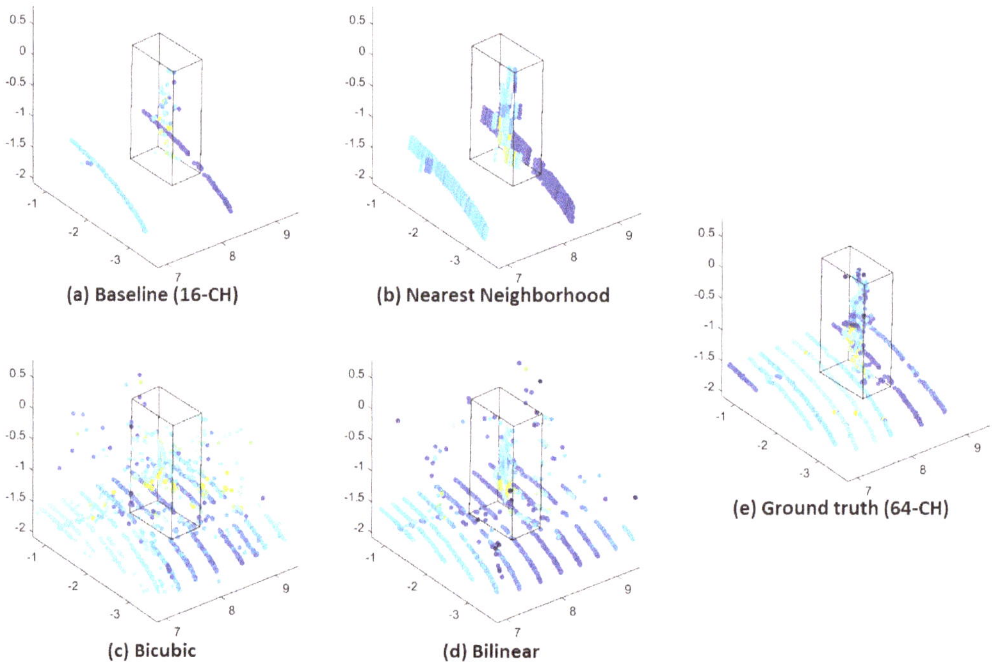

Figure 1. Up−sampled point cloud from 16-Ch (**a**) to 64-Ch (**e**) using classical interpolations (pedestrian). Comparison of (**b**) nearest-neighborhood, (**c**) bilinear, and (**d**) bicubic interpolations.

3.3. Deep Learning Based Up-Sampling

Among the deep-learning-based up-sampling methods, we compared the ESPCN [9] network and the method proposed by Shan et al. [5] as a reference. When the ESPCN [9] model was implemented, the distribution of points did not sufficiently describe the shape of an object, as shown in Figure 2. The simulation-based model [5] presented relatively good concurrence with the shape of the 64-Ch LiDAR; however, there were still some noisy points around the object.

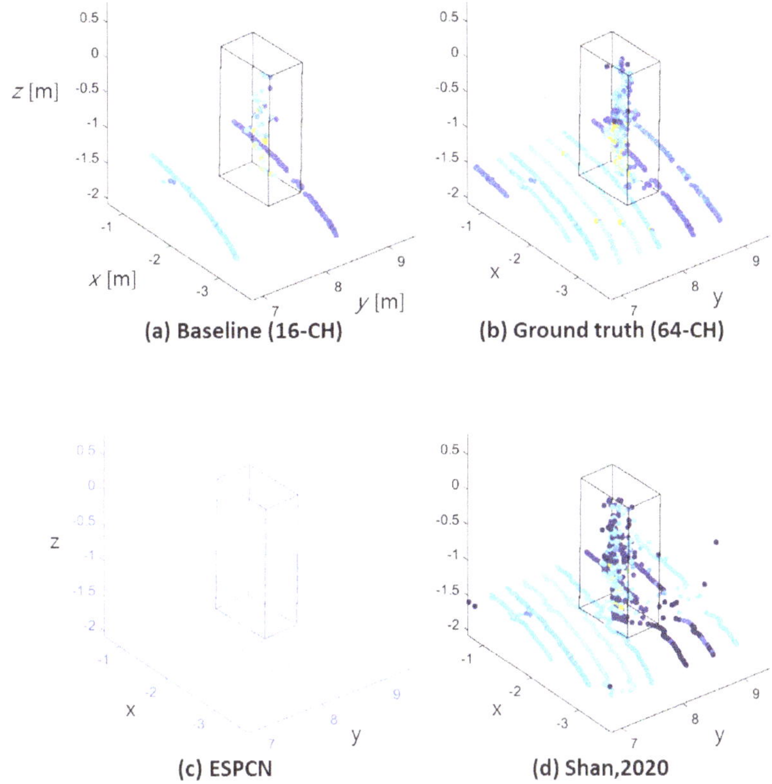

Figure 2. Up−sampled point cloud using deep-learning-based methods (pedestrian). The increase of the vertical angle resolution increases the density of the point cloud along all of the Cartesian axis directions.

4. Proposed 3D Up-Sampling Method

Figure 3 presents the outline of the research strategy used for the proposed method. First, the unordered 3D point cloud data are projected onto ordered range map images with multiple channels. The proposed up-sampling method reconstructs the sparse range image input into a dense 3D point cloud. It is then fed to the 3D object detection model as the input data and the detection performance is evaluated on the KITTI dataset.

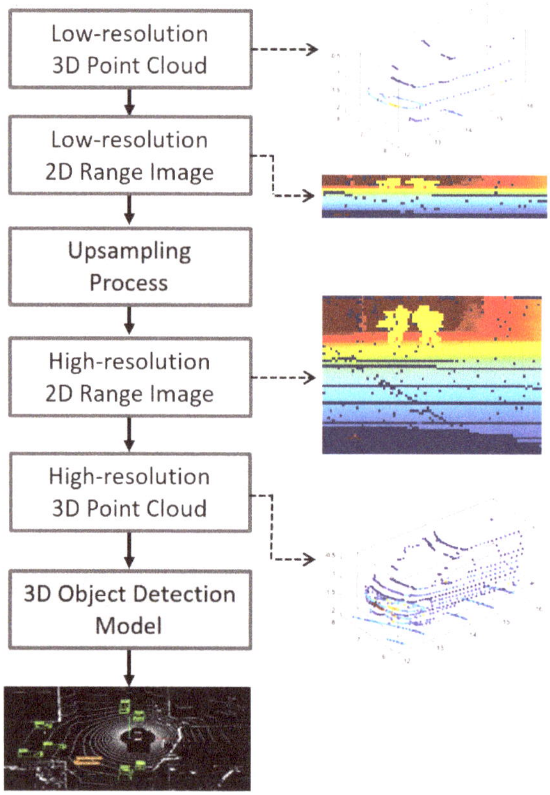

Figure 3. Research strategy for the proposed up-sampling method for 3D detection.

4.1. Projection to Range Image

The number of 3D points obtained through a single scan of a rotating LiDAR is determined by the vertical channel numbers, vertical resolution (V_{res}), horizontal FOV, and horizontal angular resolution (H_{res}).

The point cloud data is structured as a list of 3D positions and the intensity value of the laser beam points in the order of laser diode scanning, which may vary based on the LiDAR model. The dimension of the point cloud is expressed as N by D, where N represents the total number of beam points and D represents the feature numbers of the beam point, such as the 3D Cartesian coordinates and intensity value. Therefore, the data in the point cloud are in an unordered form in terms of the spatial coordinates, contrary to the 2D image pixels.

The complexity of handling such unordered data can be reduced by projecting the point cloud into a range map, which is spatially ordered using the 2D coordinates of the horizontal and vertical scanning angles, as shown in Figure 4. The dimensions of the range map are H by V by C, where H denotes the horizontal azimuth (α) angles, V denotes the number of LiDAR vertical channels, and C denotes the channel of range measurement and intensity. The up-sampling techniques for 2D images can be implemented easily since the range map is an ordered structure.

The projection of the 3D point cloud data to the range map in the azimuth and vertical angles can be processed using the transformation equation in the spherical coordinates, which is expressed as follows:

$$
\begin{aligned}
R &= \sqrt{x^2 + y^2 + z^2} \\
\alpha &= \arctan(y, x) \\
\omega &= \arctan\left(z, \sqrt{x^2 + y^2}\right)
\end{aligned} \tag{1}
$$

The dimension of the 3D point cloud from Velodyne LiDAR (HDL-64E) is 2048 × 64 × 4 Ch (X-Ch, Y-Ch, Z-Ch, Intensity-Ch), and it can be converted to a range map with the dimension of 2048 × 64 × 2 Ch (α-Ch, ω-Ch).

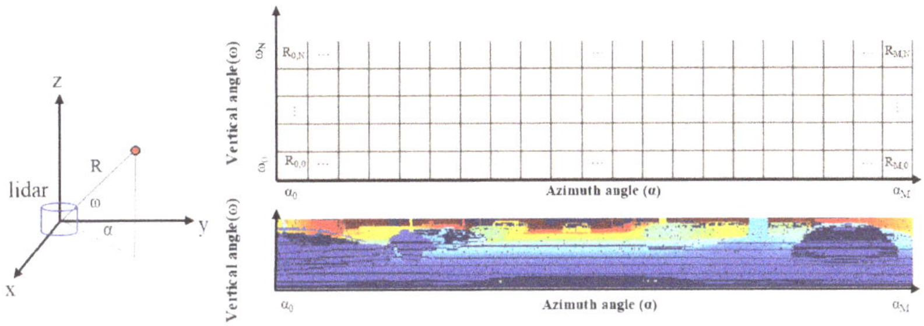

Figure 4. 2D Range image from 3D point cloud.

4.2. Proposed Up-Sampling Technique

The existing up-sampling methods do not adequately represent the shape of high-resolution LiDAR points when there are meaningless zero values within the pixels of the range image or when estimating the boundary of an object. In this study, we present a novel interpolation technique that minimizes the influence of the outlier points. The robustness of the proposed interpolation method is improved by implementing the following strategies.

4.2.1. Pixel-Distance Weighted Interpolation

The up-sampling is applied in the direction of elevation to increase the density of the vertical angle of the LiDAR channel. The proposed method interpolates the empty spaces of the LiDAR range image using the six surrounding neighbor pixels with weights based on their relative distances from the anchor point, similar to the four-point depth interpolation method proposed by Lim et al. [31]. The six neighborhood pixels for the interpolations are labeled from P_1 to P_6, starting from the top-left pixel, as shown in Figure 5. The interpolated value, P', is obtained from the weighted sum of the neighbors with the decaying factor against the distance as follows:

$$W_i = e^{-0.5d_i} \tag{2}$$

$$P' = \frac{\sum_{i=1}^{6} W_i P_i}{\sum_{i=1}^{6} W_i}, \tag{3}$$

where P_i, W_i, and d_i represent the range values, the weight ratio, and the pixel distance between P_i and P', respectively.

The interpolated output in the 2D range image is converted to a 3D point cloud to visualize and analyze the interpolation effect, as shown in Figure 6.

The distance-weighted interpolation result presented in Figure 6c depicts a denser 3D output with amplified noisy data owing to the fact that the interpolation with the neighboring points that have zero or maximum range values can be considered as the outlier when compared to the surrounding data. The negative effect of the zero-valued pixels in the interpolation process can be reduced by removing the outlier neighbor pixels. The zero or maximum-valued pixel in the 2D range image refers to the LiDAR beam point that does not project on the object in the measurable range.

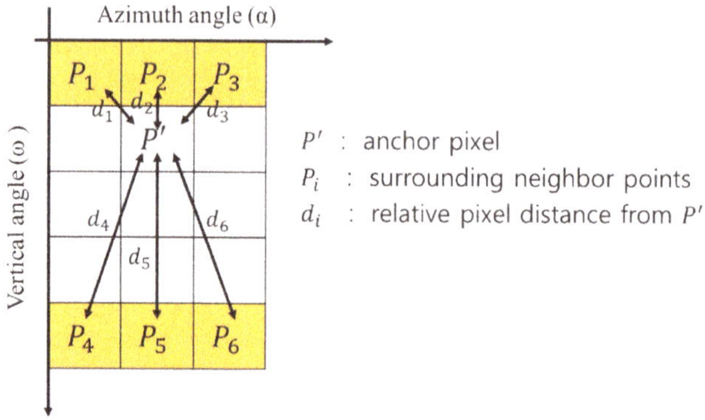

Figure 5. Interpolation with the weighted sum of neighbor pixels. The blank grid represents empty space where there are no LiDAR points.

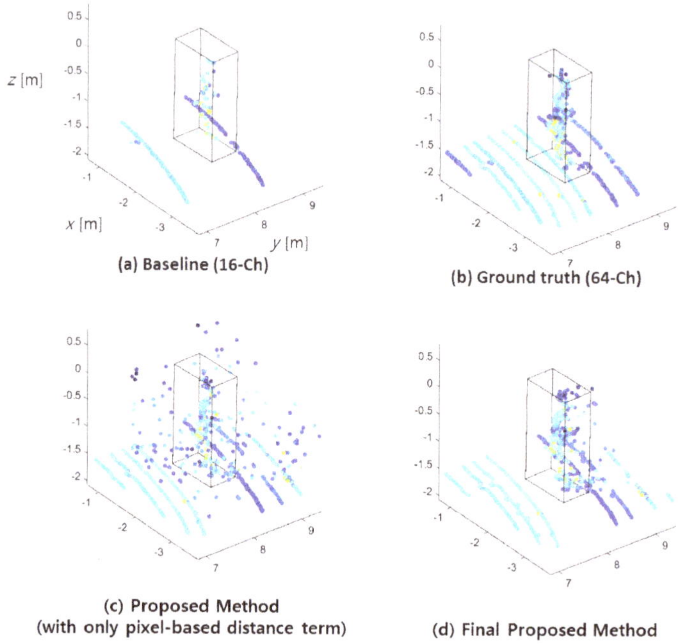

Figure 6. Up-sampled point cloud using proposed methods (pedestrian). Comparison of: (**a**) raw data of 16-Ch LiDAR; (**b**) ground truth 64-Ch LiDAR; (**c**) our proposed method with only pixel-based distance terms; and (**d**) the final proposed method.

4.2.2. Pixel-Distance and Range Weighted Interpolation

These outlier points can be skipped in the weighted sum process using the coefficient, s_i, which presents a value of 0 for outliers; otherwise, it presents a value of 1.

Furthermore, the up-sampling can be enhanced by readjusting each weight, W_i, based on the relative depth range among the neighbor pixels. The closer the depth of the pixel is, the more weight is given in the interpolation, which is similar to assuming that the closer range neighbor pixels are likely to be projected on the same object. Then, the modified interpolation equation can be expressed as:

$$P' = \frac{\sum_{i=1}^{6} s_i W_i P_i}{\sum_{i=1}^{6} s_i W_i} \tag{4}$$

with:

$$W_i = e^{-0.5 d_i} \cdot \left(\frac{2}{1 + e^{(R_i - R_{\min})}} \right), \tag{5}$$

where R_i is the range value of each neighbor data and $R_{\min}$ is the minimum range value among them. This interpolation result, as shown in Figure 6d, clearly removes the noisy interpolated data for up-sampling enhancement.

4.2.3. Increasing Channel Dimension of 2D Range-Image

The interpolation performance can be further improved by increasing the dimensions of the 2D range image. The 1-Ch 2D range image can be converted to a higher-dimension 2D range image with 4-Ch of the Cartesian coordinates (x, y, z) and intensity values. The values of x, y, and z can easily be obtained from the azimuth, elevation, and range values as:

$$\begin{aligned} x &= r \cos\omega \cos\alpha \\ y &= r \cos\omega \sin\alpha \\ z &= r \sin\omega, \end{aligned} \tag{6}$$

where azimuth (α) and elevation (ω) represent the coordinates of the 2D range image, and r denotes the range value of the pixel.

Figure 6 presents the up-sampling result of the proposed method on a pedestrian dataset. The result demonstrates that this method closely resembles the 64-Ch ground truth. Additionally, the proposed interpolation method presents the highest resemblance to the reference interpolated output when compared to the conventional methods presented in Figures 1 and 2.

5. Up-Sampling Effect on 3D Object Detection

5.1. Experiment Overview

The effectiveness of the proposed LiDAR up-sampling method was demonstrated by feeding it as the input to the selected 3D object detection model and evaluating the detection accuracy. The point-pillar model [6], which is an effective and widely used 3D detection model with LiDAR data, was used as the 3D model. This model was trained using the dataset up-sampled via the proposed method from the baseline point cloud data.

The baseline dataset comprises the point cloud data of a low-resolution (32-Ch) LiDAR, which was generated by down-sampling the KITTI [1] 3D dataset of a 64-Ch LiDAR.

The numbers of data points in the training, validation, and test datasets were 7481, 1856, and 3769, respectively. The overall performances of the 3D detection task were evaluated based on the mAP (mean average precision) of easy, moderate, and hard difficulty for each class of cars, pedestrians, and cyclists, which are the most common obstacles in an autonomous driving environment.

5.2. Algorithm Design (Ablation Studies)

An ablation study was conducted to analyze the effect of the following methods: (A) pixel-distance weighted interpolation, (B) pixel and range weighted interpolation, (C) additional channel on the 2D range image, and (D) ground removal on the 3D object detection. The baseline dataset of the 32-Ch point cloud was up-sampled to a 64-Ch point cloud using the aforementioned methods and the accuracy of the 3D object detection was evaluated for comparison.

5.2.1. (A) Effect of Pixel Distance Weight Interpolation

The baseline dataset was up-sampled to a 64-Ch point cloud using only the six-point pixel distance weighted interpolation based on Equations (2) and (3).

This method presented inferior detection results for all of the classes when compared to the detection output with the baseline dataset of the 32-Ch point cloud, as described in Table 1. The interpolation performed using only the pixel distances of the neighbor points, similar to the method reported in [31], raised the 3D detection accuracy from 31.32% to 43.85%.

Table 1. 3D object detection performance (mAP) of the proposed methods on the test dataset.

Class	Baseline (32-Ch)			(A) Pixel Weighted			(B) Pixel and Depth Weighted		
	Easy	Mod.	Hard	Easy	Mod.	Hard	Easy	Mod.	Hard
Pedestrian	30.01	25.98	23.95	27.59	24.94	21.86	36.21	32.31	29.61
Cyclist	58.13	36.88	35.58	52.18	33.02	31.55	56.31	36.59	35.00
Car	65.82	51.88	46.25	51.78	38.15	33.44	72.39	53.41	47.84
Overall	51.32	38.25	35.26	43.85	32.04	28.95	54.97	40.77	37.48
Class	**(C) Ground-Removal**			**(D) 2D Range Image of 4-Ch**			**Ground-Truth (64-Ch)**		
	Easy	**Mod.**	**Hard**	**Easy**	**Mod.**	**Hard**	**Easy**	**Mod.**	**Hard**
Pedestrian	35.38	31.27	28.31	39.96	35.18	31.71	52.03	46.4	42.48
Cyclist	62.45	40.92	38.19	64.95	43.19	39.95	78.72	59.95	57.25
Car	70.18	55.13	51.60	75.06	57.72	54.19	85.41	73.98	67.76
Overall	56.00	42.44	39.36	59.99	45.36	41.95	72.05	60.11	55.83

5.2.2. (B) Effect of Pixel Distance and Range Weight Interpolation

The up-sampling method was enhanced by neglecting the zero-value range points to avoid the outlier effect on the interpolation. Additionally, the range difference between the anchor point and neighbors was considered to apply a higher weight on the points that had a higher chance to be on the same object surface. The data obtained from the interpolation performed using Equations (4) and (5) were applied to the baseline dataset and used as the training data for the 3D detection model.

The mAP of the overall classes for all of the difficulty levels increased by up to 3.6% when compared to the baseline result, as presented in Table 1. The mAP of the car detection task increased to 72.39%, when compared to the mAP of 65.82% achieved by the baseline. However, the mAP of the cyclist detection task decreased slightly.

5.2.3. (C) Effect on Increasing Range Image Channels

In the previous sections, the point clouds were up-sampled to a 2D range image using 2-Ch of the range and intensity values. The axes of the image are the azimuth and vertical angles, as shown in Figure 4.

The 2D image with a higher number of channels was generated using the Cartesian 3D coordinates of each point with Equation (1). Therefore, the input image used for the interpolation was a 2D image with 4-Ch of the x, y, and z-axis position values and the intensity value.

The detection achieved by the up-sampled image of the 4-Ch presented superior performance when compared to the baseline. The overall mAP for all the classes was enhanced by as much as 8%. The mAP of the easy level cyclist detection increased from 58.1% of the baseline to 64.95%. The overall mAP was enhanced by 5.01%p, 4.59%p, and 4.47%p for the easy, moderate, and hard levels, respectively, when compared to the proposed interpolation method with the 2-Ch image.

Figures 7 and 8 show the 3D object detection results of the selected scene for pedestrians and vehicles, respectively. The up-sampled point cloud and the detection results closely resemble the ground truth. This helped in accurately reconstructing the objects for locating the pedestrians and vehicles.

(a) RGB Image Scene

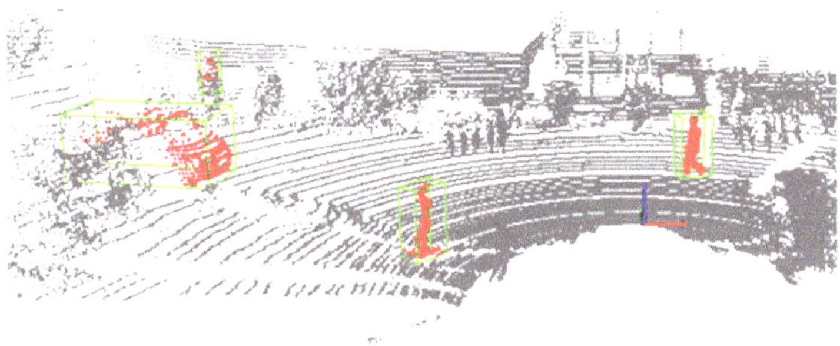

(b) 3D Detection with 64 ch Point Cloud (Ground Truth)

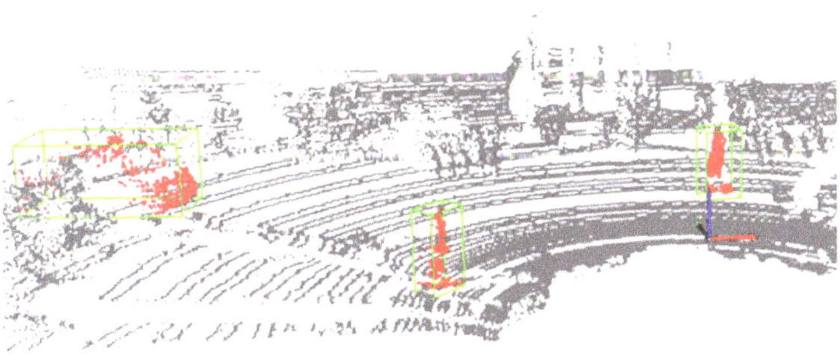

(c) 3D Detection with Proposed Upsampling Method

Figure 7. An example of 3D object detection performed using the proposed up-sampling method for pedestrian detection

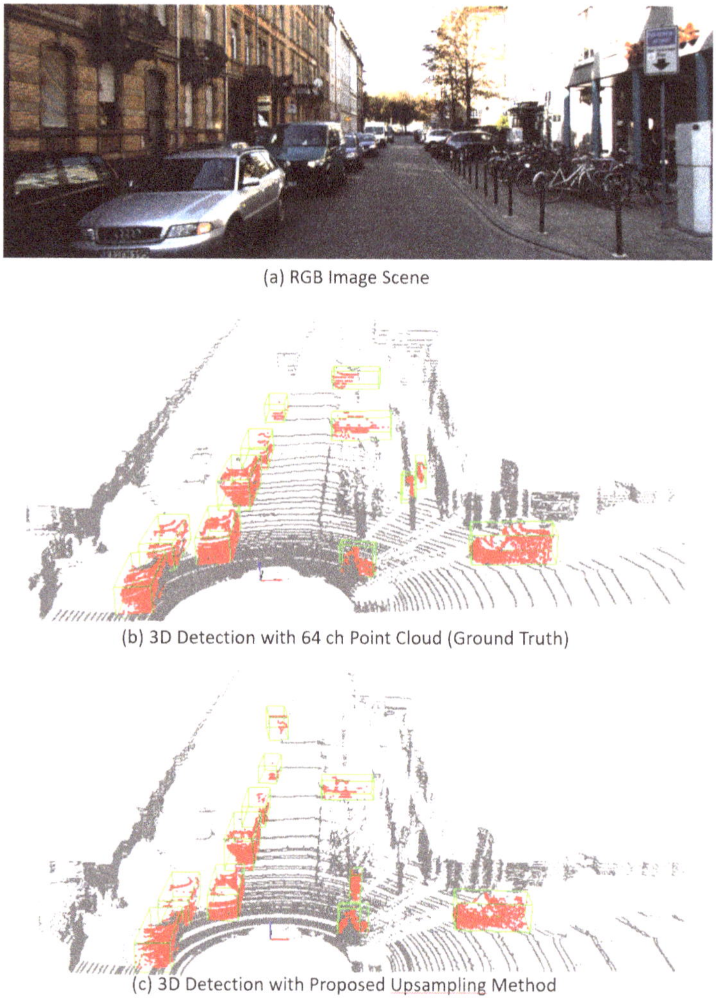

Figure 8. An example of 3D object detection performed using the proposed up-sampling pre-processing for vehicles detection

5.2.4. (D) Effect on Ground-Removal Pre-Processing

A high proportion of the point cloud data included the points on the ground. The effect on the 3D detection accuracy due to the ground-removal pre-processing performed before the up-sampling was analyzed.

The ground removal process was simplified by discarding all of the points located at a lower elevation than the ego-vehicle. Although this approach can work effectively in most driving scenarios, it may fail if the relative angle between the ground and the ego-vehicle is not flat. In this study, we applied the ground removal algorithm presented by [32], which is based on the inclination angle of LiDAR beams corresponding to the horizontal axis. The up-sampling of the 4-Ch image was conducted after the ground removal processing.

The 3D detection results presented in Table 1 demonstrated that the ground removal decreased the detection accuracy, indicating that the ground points have important roles in the detection model that could not be ignored.

5.3. Comparison with Previous Methods

The performance enhancement of the 3D object detection task owing to the use of the proposed up-sampling method was compared with the widely used previous methods and the SOTA (state-of-the-art) method for 3D up-sampling. The same baseline datasets were used for the up-sampling process, whereas the point-pillar model was used for the detection.

Table 2 presents the values of the detection mAP for the classes of pedestrians, cyclists, and cars that were compared to analyze the detection enhancement presented by different up-sampling methods. The strict intersection over union (IoU) of 0.7 was applied for the evaluation. The abbreviations E, M, and H represent the difficulty levels of easy, moderate, and hard, respectively.

The CNN-based up-sampling methods of ESPCN [9] and the method proposed by Shan et al. [5] exhibited inferior 3D detection performances. These methods were able to reconstruct a denser point cloud with a low overall loss score. However, they could not accurately reconstruct the object shapes, thereby leading to inferior detection performance.

Table 2 presents a comparison between the moderate-level class detections. It can be clearly observed that the proposed solution (Case C) exhibited the best performance in each class detection when compared to the existing up-sampling methods. The mAP of the overall class (level M) was 45.4%, which is approximately 7% higher than that of the baseline dataset.

Table 2. Comparison table of object detection performance using pretrained point-pillars model (up-sampled from 32-Ch to 64-Ch). The strict IoU of 0.7 was applied.

		Ped	Cyc	Car	Overall
Baseline (32-Ch)	E	30.01	58.13	65.82	51.32
	M	25.98	36.88	51.88	38.25
	H	23.95	35.58	46.25	35.26
NN Interpolation	E	27.39	49.09	40.60	39.03
	M	24.22	30.95	30.13	28.43
	H	22.00	29.75	26.99	26.25
Bilinear Interpolation	E	20.67	38.31	30.97	29.98
	M	18.13	24.51	21.12	21.25
	H	16.99	22.95	17.72	19.22
ESPCN [9]	E	2.45	2.90	6.67	4.01
	M	2.57	2.55	5.77	3.63
	H	2.61	2.56	4.55	3.24
Shan,2020 [5]	E	9.29	5.72	5.20	6.74
	M	9.09	9.34	4.55	7.66
	H	9.09	9.29	4.55	7.64
Proposed (Case C)	E	**39.96**	**64.95**	**75.06**	**59.99**
	M	**35.18**	**43.19**	**57.72**	**45.36**
	H	**31.71**	**39.95**	**54.19**	**41.95**

6. Conclusions

This study proposed an up-sampling method for a low-resolution LiDAR to enhance 3D object detection by reconstructing the objects in sparse point cloud data into data with a higher vertical angle resolution.

The existing 3D up-sampling methods based on deep neural networks focus on decreasing the overall loss in the up-sampling reconstruction process instead of enhancing the object detection performance. Our proposed sampling method is designed as a pre-processing stage to implement an enhanced 3D object detection model. First, we converted the 3D point cloud dataset into a 2D range image with 4-Ch. The interpolation on the empty space was calculated based on both the pixel-distance and range values of six neighbor points to conserve the shapes of the original object during the reconstruction. This approach solved the over-smoothing problem faced by the conventional interpolation methods while presenting higher operational speed and better object detection performance when compared to the aforementioned deep-learning super-resolution methods.

The effectiveness of our proposed up-sampling method on the 3D detection was demonstrated by applying it to baseline 32-Ch point cloud data and then feeding them as the input to a point-pillar detection model. The 3D object detection result on the KITTI dataset demonstrates that the proposed method could increase the mAP (strict IoU with 0.7) of pedestrians, cyclists, and cars by 9.2%p, 6.3%p, and 5.9%p, respectively, when compared to the baseline with a low-resolution 32-Ch LiDAR input.

The scope of this study was limited to the reconstruction of 32-ch LiDAR to 64-ch LiDAR on the KITTI dataset. In future works, various dataset environments apart from autonomous driving will be analyzed. Additionally, further analysis is required to integrate the proposed up-sampling method with the 3D object detection model based on 2D range images such as the range-sparse network (RSN) [33], for the construction of an end-to-end network.

Author Contributions: Conceptualization, J.Y. and Y.-K.K.; methodology, J.Y. and Y.-K.K.; writing—original draft, J.Y.; writing—review and editing, Y.-K.K.; project administration, Y.-K.K. All authors have read and agreed to the published version of the manuscript.

Funding: This research received no external funding.

Acknowledgments: We thank Hyungseok Song for his assistance in the experiment.

Conflicts of Interest: We declare that we have no financial and personal relationships with other people or organizations that can inappropriately influence our work. There is no professional or other personal interest of any nature or kind in any product, service, and/or company that could be construed as influencing the position presented in the review of this manuscript.

References

1. Geiger, A.; Lenz, P.; Stiller, C.; Urtasun, R. Vision meets robotics: The kitti dataset. *Int. J. Robot. Res.* **2013**, *32*, 1231–1237. [CrossRef]
2. Sun, P.; Kretzschmar, H.; Dotiwalla, X.; Chouard, A.; Patnaik, V.; Tsui, P.; Guo, J.; Zhou, Y.; Chai, Y.; Caine, B.; et al. Scalability in perception for autonomous driving: Waymo open dataset. In Proceedings of the IEEE/CVF Conference on Computer Vision and Pattern Recognition, Seattle, WA, USA, 13–19 June 2020; pp. 2446–2454.
3. Siu, W.C.; Hung, K.W. Review of image interpolation and super-resolution. In Proceedings of the 2012 Asia Pacific Signal and Information Processing Association Annual Summit and Conference. Hollywood, CA, USA 3–6 December 2012; pp. 1–10.
4. Yu, L.; Li, X.; Fu, C.W.; Cohen-Or, D.; Heng, P.A. Pu-net: Point cloud upsampling network. In Proceedings of the IEEE Conference on Computer Vision and Pattern Recognition, Salt Lake City, UT, USA, 18–23 June 2018; pp. 2790–2799.
5. Shan, T.; Wang, J.; Chen, F.; Szenher, P.; Englot, B. Simulation-based lidar super-resolution for ground vehicles. *Robot. Auton. Syst.* **2020**, *134*, 103647. [CrossRef]
6. Lang, A.H.; Vora, S.; Caesar, H.; Zhou, L.; Yang, J.; Beijbom, O. Pointpillars: Fast encoders for object detection from point clouds. In Proceedings of the IEEE/CVF Conference on Computer Vision and Pattern Recognition, Long Beach, CA, USA, 15–20 June 2019; pp. 12697–12705.
7. Fadnavis, S. Image interpolation techniques in digital image processing: An overview. *Int. J. Eng. Res. Appl.* **2014**, *4*, 70–73.
8. Dong, C.; Loy, C.C.; He, K.; Tang, X. Image super-resolution using deep convolutional networks. *IEEE Trans. Pattern Anal. Mach. Intell.* **2015**, *38*, 295–307. [CrossRef] [PubMed]
9. Shi, W.; Caballero, J.; Huszár, F.; Totz, J.; Aitken, A.P.; Bishop, R.; Rueckert, D.; Wang, Z. Real-time single image and video super-resolution using an efficient sub-pixel convolutional neural network. In Proceedings of the IEEE Conference on Computer Vision and Pattern Recognition, Las Vegas, NV, USA, 27–30 June 2016; pp. 1874–1883.
10. Lim, B.; Son, S.; Kim, H.; Nah, S.; Mu Lee, K. Enhanced deep residual networks for single image super-resolution. In Proceedings of the IEEE Conference on Computer Vision and Pattern Recognition Workshops, Honolulu, HI, USA, 21–26 July 2017; pp. 136–144.
11. Mao, X.; Shen, C.; Yang, Y.B. Image restoration using very deep convolutional encoder-decoder networks with symmetric skip connections. *Adv. Neural Inf. Process. Syst.* **2016**, *29*, 2802–2810.
12. Kim, J.; Lee, J.K.; Lee, K.M. Deeply-recursive convolutional network for image super-resolution. In Proceedings of the IEEE Conference on Computer Vision and Pattern Recognition, Las Vegas, NV, USA, 27–30 June 2016; pp. 1637–1645.
13. Tai, Y.; Yang, J.; Liu, X.; Xu, C. Memnet: A persistent memory network for image restoration. In Proceedings of the IEEE International Conference on Computer Vision, Venice, Italy, 22–29 October 2017; pp. 4539–4547.

14. Wang, Z.; Liu, D.; Yang, J.; Han, W.; Huang, T. Deep networks for image super-resolution with sparse prior. In Proceedings of the IEEE International Conference on Computer Vision, Santiago, Chile, 7–13 December 2015; pp. 370–378.
15. Lai, W.S.; Huang, J.B.; Ahuja, N.; Yang, M.H. Deep laplacian pyramid networks for fast and accurate super-resolution. In Proceedings of the IEEE Conference on Computer Vision and Pattern Recognition, Seattle, WA, USA, 21–23 June 2017; pp. 624–632.
16. Tong, T.; Li, G.; Liu, X.; Gao, Q. Image super-resolution using dense skip connections. In Proceedings of the IEEE International Conference on Computer Vision, Venice, Italy, 22–29 October 2017; pp. 4799–4807.
17. Haris, M.; Shakhnarovich, G.; Ukita, N. Deep back-projection networks for super-resolution. In Proceedings of the IEEE Conference on Computer Vision and Pattern Recognition, Salt Lake City, UT, USA, 18–23 June 2018; pp. 1664–1673.
18. Ren, H.; El-Khamy, M.; Lee, J. Image super resolution based on fusing multiple convolution neural networks. In Proceedings of the IEEE Conference on Computer Vision and Pattern Recognition Workshops, Honolulu, HI, USA, 21–26 July 2017; pp. 54–61.
19. Hui, Z.; Wang, X.; Gao, X. Fast and accurate single image super-resolution via information distillation network. In Proceedings of the IEEE Conference on Computer Vision and Pattern Recognition, Seattle, WA, USA, 21–23 June 2018; pp. 723–731.
20. Choi, J.S.; Kim, M. A deep convolutional neural network with selection units for super-resolution. In Proceedings of the IEEE Conference on Computer Vision and Pattern Recognition Workshops, Honolulu, HI, USA, 21–26 July 2017; pp. 154–160.
21. Anwar, S.; Barnes, N. Densely residual laplacian super-resolution. *IEEE Trans. Pattern Anal. Mach. Intell.* **2020**. *44*, 1192–1204. [CrossRef]
22. Shocher, A.; Cohen, N.; Irani, M. "zero-shot" super-resolution using deep internal learning. In Proceedings of the IEEE Conference on Computer Vision and Pattern Recognition, Salt Lake City, UT, USA, 18–23 June 2018; pp. 3118–3126.
23. Zhang, K.; Zuo, W.; Zhang, L. Learning a single convolutional super-resolution network for multiple degradations. In Proceedings of the IEEE Conference on Computer Vision and Pattern Recognition, Salt Lake City, UT, USA, 18–23 June 2018; pp. 3262–3271.
24. Ledig, C.; Theis, L.; Huszár, F.; Caballero, J.; Cunningham, A.; Acosta, A.; Aitken, A.; Tejani, A.; Totz, J.; Wang, Z.; et al. Photo-realistic single image super-resolution using a generative adversarial network. In Proceedings of the IEEE Conference on Computer Vision and Pattern Recognition, Honolulu, HI, USA, 21–26 July 2017; pp. 4681–4690.
25. Wang, X.; Yu, K.; Wu, S.; Gu, J.; Liu, Y.; Dong, C.; Qiao, Y.; Change Loy, C. Esrgan: Enhanced super-resolution generative adversarial networks. In Proceedings of the European Conference on Computer Vision (ECCV) Workshops, Munich, Germany, 8–14 September 2018.
26. Diebel, J.; Thrun, S. An application of markov random fields to range sensing. In Proceedings of the 18th International Conference on Neural Information Processing Systems, 5 December 2005; Volume 18. Available online: https://dl.acm.org/doi/10.5555/2976248.2976285 (accessed on 22 December 2022).
27. Wirges, S.; Roxin, B.; Rehder, E.; Kühner, T.; Lauer, M. Guided depth upsampling for precise mapping of urban environments. In Proceedings of the 2017 IEEE Intelligent Vehicles Symposium (IV), Los Angeles, CA, USA, 11–14 June 2017; pp. 1140–1145.
28. Dolson, J.; Baek, J.; Plagemann, C.; Thrun, S. Upsampling range data in dynamic environments. In Proceedings of the 2010 IEEE Computer Society Conference on Computer Vision and Pattern Recognition, San Francisco, CA, USA, 13–18 June 2010; pp. 1141–1148.
29. He, Y.; Chen, L.; Li, M. Sparse depth map upsampling with rgb image and anisotropic diffusion tensor. In Proceedings of the 2015 IEEE Intelligent Vehicles Symposium (IV), Seoul, Republic of Korea, 28 June–1 July 2015; pp. 205–210.
30. Ronneberger, O.; Fischer, P.; Brox, T. U-net: Convolutional networks for biomedical image segmentation. In Proceedings of the International Conference on Medical Image Computing and Computer-Assisted Intervention, Munich, Germany, 5–9 October 2015; pp. 234–241.
31. Lim, H.b.; Kim, E.s.; Rathnayaka, P.; Park, S.Y. Low-Resolution LiDAR Upsampling Using Weighted Median Filter. In *Advances in Computer Science and Ubiquitous Computing*; Springer: Berlin, Germany, 2021; pp. 213–220.
32. Bogoslavskyi, I.; Stachniss, C. Efficient online segmentation for sparse 3D laser scans. *PFG—J. Photogramm. Remote Sens. Geoinf. Sci.* **2017**, *85*, 41–52. [CrossRef]
33. Sun, P.; Wang, W.; Chai, Y.; Elsayed, G.; Bewley, A.; Zhang, X.; Sminchisescu, C.; Anguelov, D. RSN: Range Sparse Net for Efficient, Accurate LiDAR 3D Object Detection. In Proceedings of the IEEE/CVF Conference on Computer Vision and Pattern Recognition, Nashville, TN, USA, 20–25 June 2021; pp. 5725–5734.

sensors

MDPI

Article

Advanced Pedestrian State Sensing Method for Automated Patrol Vehicle Based on Multi-Sensor Fusion

Pangwei Wang [1,*], Cheng Liu [1], Yunfeng Wang [1,2] and Hongsheng Yu [1]

[1] Beijing Key Lab of Urban Intelligent Traffic Control Technology, North China University of Technology, Beijing 100144, China; liucheng1130@126.com (C.L.); wuyu13522777537@126.com (Y.W.); yuhongsheng110@126.com (H.Y.)

[2] Key Laboratory of Operation Safety Technology on Transport Vehicles, Research Institute of Highway, Ministry of Transport, Beijing 100088, China

* Correspondence: wpw@ncut.edu.cn

Abstract: At present, the COVID-19 pandemic still presents with outbreaks occasionally, and pedestrians in public areas are at risk of being infected by the viruses. In order to reduce the risk of cross-infection, an advanced pedestrian state sensing method for automated patrol vehicles based on multi-sensor fusion is proposed to sense pedestrian state. Firstly, the pedestrian data output by the Euclidean clustering algorithm and the YOLO V4 network are obtained, and a decision-level fusion method is adopted to improve the accuracy of pedestrian detection. Then, combined with the pedestrian detection results, we calculate the crowd density distribution based on multi-layer fusion and estimate the crowd density in the scenario according to the density distribution. In addition, once the crowd aggregates, the body temperature of the aggregated crowd is detected by a thermal infrared camera. Finally, based on the proposed method, an experiment with an automated patrol vehicle is designed to verify the accuracy and feasibility. The experimental results have shown that the mean accuracy of pedestrian detection is increased by 17.1% compared with using a single sensor. The area of crowd aggregation is divided, and the mean error of the crowd density estimation is 3.74%. The maximum error between the body temperature detection results and thermometer measurement results is less than 0.8°, and the abnormal temperature targets can be determined in the scenario, which can provide an efficient advanced pedestrian state sensing technique for the prevention and control area of an epidemic.

Keywords: advanced sensing technology; multi-sensor fusion; crowd density estimation; movable temperature detection; automated patrol vehicle

Citation: Wang, P.; Liu, C.; Wang, Y.; Yu, H. Advanced Pedestrian State Sensing Method for Automated Patrol Vehicle Based on Multi-Sensor Fusion. *Sensors* **2022**, *22*, 4807. https://doi.org/10.3390/s22134807

Academic Editors: Chao Huang, Yafei Wang, Peng Hang, Zhiqiang Zuo and Bo Leng

Received: 28 April 2022
Accepted: 23 June 2022
Published: 25 June 2022

1. Introduction

At present, in order to curb the COVID-19 pandemic, governments have divided the current areas into disease control areas and normal areas according to the different status of the epidemic. The source of infection, which may break out in the disease control area, has an important impact on the daily life of the general public. The crowd aggregation and contact temperature measurement method based on hand-held thermometers increases the risk of cross-infection. Therefore, how to reduce the risk of infection for medical practitioners and pedestrians in disease control areas and ensure the effective implementation of epidemic prevention measures have become important issues. In order to solve the above problems, it is necessary to detect crowd aggregation and body temperature rapidly based on the advanced sensing technology. Due to the improvement in sensor accuracy and detection algorithms in recent years, sensing technology and application technology are developing rapidly. A connected and automated vehicle (CAV), which is equipped with sensors, such as a camera, radar and LiDAR, has been widely used in short-distance logistics distribution and security patrols to improve transportation efficiency and driving safety [1]. As one of the CAVs, the automated patrol vehicle can be used as a new solution for epidemic prevention and control. Therefore, how to improve the accuracy of sensing and apply high-precision sensing technology to the automated patrol vehicle have become key research directions.

However, there are still some problems in the above directions, which can be classified in the following conditions: (1) pedestrians in public areas, such as in parks, museums and shopping malls, do not have a fixed route, and the difference in crowd density distribution in adjacent areas is tiny, which makes it difficult to detect dense areas effectively; (2) there are technique and evaluation difficulties in the combination of high-precision crowd density estimation and body temperature detection. In order to solve the above problems, an advanced pedestrian state sensing method for automated patrol vehicles based on multi-sensor fusion is proposed. Considering the current pandemic situation and the limitations of the experimental platform, a large and enclosed area with less traffic flow is selected as the work scenario of the automated patrol vehicle. Firstly, through fusing the data of the camera and LiDAR, the multi-dimension data of pedestrians are obtained as the basis of density estimation and face detection. Secondly, the centroids of the clustered pedestrian objects are calculated as the cluster center point. Combined with the positional relationship between pedestrians, crowd targets are detected based on the foreground image. Then, the whole detection area is divided into multiple layers by various scales, and each sub-area of each layer is constructed as the basic unit of the multi-layer fusion method. Through weighted mean theory, the crowd density in the scenario is estimated based on the crowd density distribution matrix. Finally, because of the precision limitation of the sensor, the automated patrol vehicle needs to maintain a fixed distance from crowd targets, and a thermal infrared camera is used to detect the temperature, which can reduce the risk of cross-infection.

2. Related Works

The advanced sensing technologies lead to the rapid development of CAV, providing a platform foundation for mobile detection. Meanwhile, the gradual development of deep learning [2], pedestrian detection, crowd density estimation and pedestrian body temperature detection techniques have been hot topics.

Pedestrian detection is one of the important parts of automated vehicle sensing systems and has been researched for many years. It is mainly achieved through the neural network [3–5] and open-source computer vision library (OpenCV) [6]. Among them, the convolutional neural network (CNN) proposed by Lecun [7] was representative. Local pixels of the image were perceived by the network and combined as global features in higher layers. This meant that the pooling layer would pay more attention to local features and ignore the overall target. Multi-sensor data fusion methods [8–10] have been paid more and more attention in recent years. Multi-sensor data fusion is a method to combine data with multi-level and multi-spatial information and output a consistent interpretation of the targets. There are three levels of fusion, including: (1) data-level fusion; (2) feature-level fusion and (3) decision-level fusion. In order to reduce the false positive (FP) rate of pedestrians in a complex environment, Han et al. [11] integrated image data and point cloud data deeply through decision-level fusion to improve detection accuracy. Combining the object detection algorithm of you only look once (YOLO) and a light field camera, Zhang et al. [12] proposed an indoor obstacle detection algorithm. This algorithm was suitable for detection in the indoor environment but had not been tested in outdoor environments. In order to detect heavily occluded pedestrians, Seong et al. [13] proposed a fusion algorithm of LiDAR and radar, which avoided the strict requirements regarding the conditions of lighting for the camera, and it was suitable for indoor and outdoor environments.

Crowd density estimation has been paid more and more attention due to the requirements of abnormal event sources localization and the policy under the pandemic situation in recent years. With the advancement in sensor techniques, the current mainstream crowd density estimation is achieved through neural networks [14–18]. Huang et al. [19] proposed a multi-scale fusion conditional generative adversarial network based on a bidirectional fusion module. The network could solve the problem of scale variation effectively. In order to reduce crowd counting errors caused by large viewing angle changes and severe occlusions, Sajid et al. [20] proposed an end-to-end crowd counting network based on patch rescaling module (PRM) and multi-resolution fusion. In order to make deep features recognizable by CNN, Nguyen et al. [21] fused depth information into a density estimation algorithm. Through a multi-task learning method, the scale variation problem was solved and crowd counting was achieved accurately. Through applying an edge computing method, Wang et al. [22] achieved crowd density estimation efficiently to solve the problem of high network latency caused by the deployment of the density estimation platform in the server; Zhang et al. [23] proposed the adaptive multi-scale context aggregation network (MSCANet) to obtain the full-scale information of the crowd. After the information of different scales was extracted by the network, this information was fused by the network adaptively, which was suitable for crowd counting.

With the gradual development of automated vehicle sensing techniques and body temperature detection techniques, body temperature detection is gradually being applied in the automated vehicle [24,25]. At present, body temperature detection is mainly achieved by the method of thermal imaging [26,27]. With the continuous development of image processing techniques, thermal imaging technology combined with image recognition has become the mainstream method. Based on image processing, Muhammad et al. [28] proposed a pedestrian body temperature detection method to fuse infrared and visible images. The target and the background could be distinguished and the infection problem that might be caused by the gun-type thermometer was avoided by this method. Considering the difficulty of detecting body temperature without distance limitation, this method could measure temperature accurately by strictly fixing the temperature measurement distance and controlling environmental variables. Based on the micro control system 51 series (MCS 51) equipped with infrared temperature measurement function (GY-MLX90614 module), Zhu et al. [29] designed a non-contact infrared temperature measurement system and applied it to the access control platform in the community. As the distances between pedestrians and detective equipment are fixed, the influence of distance on temperature measurement accuracy is reduced. In order to ensure social safety in public areas, Rosepreet et al. [30] proposed an infrared body temperature detection method based on CNN. Non-contact pedestrian body temperature detection was achieved by this method, and the system was applied to the tunnel scenario, improving the detection accuracy by 1.4%.

In the above research achievements, many pedestrian detection methods rely on the high quality of pedestrian datasets. Meanwhile, the data fusion algorithm will be limited by the fixed scenarios. Crowd density estimation is still mainly based on the bird's-eye-view data, which are obtained by surveillance cameras. However, automated-vehicle-based estimation methods are still to be researched. Additionally, there are difficulties in detecting body temperature at the unrestricted distance, and environmental temperature has a large impact on the accuracy of body temperature detection.

Therefore, we propose a novel method to solve the above problems, and the main contributions in pedestrian detection, crowd density estimation and body temperature detection are summarized as follows:

1. A multi-sensor data fusion method is proposed to improve the accuracy of large, medium and small pedestrian target detection, which is suitable for more scenarios.
2. Based on two-dimensional (2D) foreground image data and three-dimensional (3D) point cloud data, a multi-layer fusion method is designed for crowd density estimation, which can improve the accuracy and intuitiveness of estimation.
3. Based on the foreground image, crowd targets are detected to fix the distance between the sensor and the crowd target. Furthermore, an empirical method, external bold body method and outlier filtering method are proposed to correct the temperature detection results.

3. Pedestrian State Sensing Method

Based on intelligent automated vehicles and information fusion technology, an advanced pedestrian state sensing method for automated patrol vehicle based on multi-sensor fusion is proposed to reduce the risk of cross-infection in the epidemic prevention and control areas. Pedestrian state includes pedestrian target, crowd aggregation and pedestrian body temperature. Hence, our method is designed in three parts, including pedestrian data acquisition, crowd density estimation and pedestrian body temperature detection. The method includes sensing equipment, decision equipment and control equipment. LiDAR, camera and infrared are mounted on the body of the automated patrol vehicle as the sensing devices; the industrial personal computer (IPC) with the robot operating system (ROS) is used as the decision-making device; the STM32 Microcontroller is used as the control device. The detection of pedestrian and crowd objects, the estimation of crowd density and the detection of pedestrian body temperature are achieved through the detection method proposed in this paper. The structure of proposed method is shown in Figure 1.

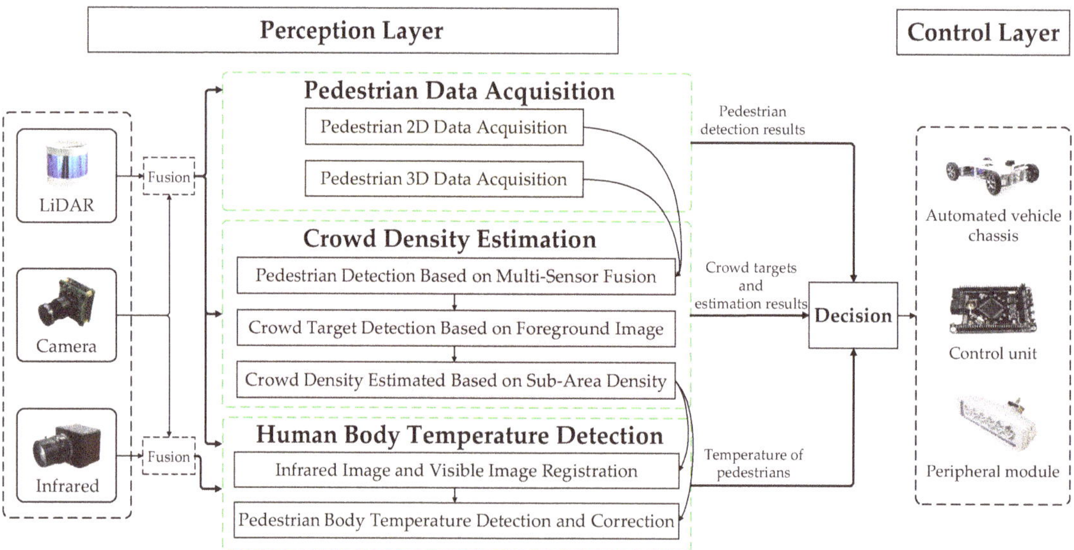

Figure 1. Structure of advanced pedestrian state sensing method for automated patrol vehicle based on multi-sensor fusion.

3.1. *Pedestrian Data Acquisition*

How to acquire the pedestrian data is the basis of the proposed method in this paper, which consists of two parts: (1) 2D pedestrian data acquisition based on YOLO V4 algorithm; (2) 3D pedestrian data acquisition based on Euclidean clustering algorithm. The obtained data can be used as the data support for crowd density estimation and temperature detection algorithm.

3.1.1. Pedestrian 2D Data Acquisition

The 2D pedestrian data include pedestrian detection box and face detection box based on visual algorithm. To implement the pedestrian detection with strong timeliness and high precision in a specific enclosed area, the Darknet framework based on the YOLO algorithm is used to obtain 2D image data of pedestrians with different occlusion. Since the spatial pyramid pooling (SPP) is adopted in YOLO V4, the network and loss function are optimized, and the intersection over union (IoU) [31] can reflect the degree of intersection between two boxes accurately. Therefore, YOLO V4 is selected as the visual model of proposed pedestrian detection and face detection algorithm. The structure of pedestrian detection based on YOLO V4 is shown in Figure 2.

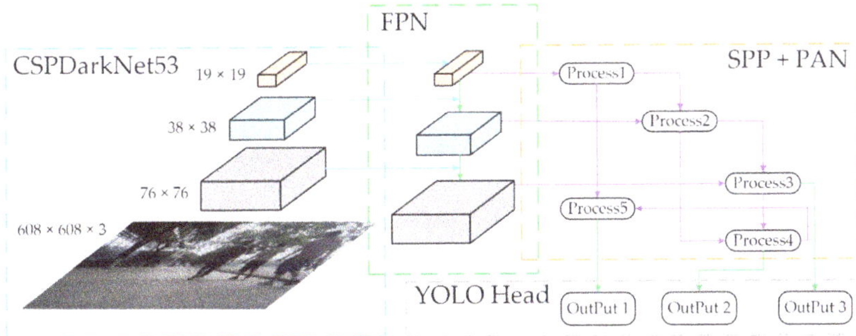

Figure 2. Structure of pedestrian detection based on YOLO V4.

In order to obtain 2D pedestrian detection box, the pedestrian dataset of the Massachusetts Institute of Technology (MIT) and the data of pedestrians in the campus environment collected by us are adopted as the input dataset of YOLO network training. Through adjusting the brightness and saturation of the pedestrian image, the learning ability and environmental resistance of pedestrian detection model can be improved. In order to solve the mesoscale problem of target detection, feature pyramid network (FPN) is used to construct a pyramid on the feature graph. Similarly, to reduce the amount of calculation and ensure the accuracy, a cross-stage partial (CSP) module is added to each Darknet53 in YOLO network, pedestrian features are extracted by the first 52 layers and fusion features are output by the last layer. In order to prevent overfitting, dropblock regularization is used to randomly drop features in the same pedestrian module. The SPP module and path aggregation network (PANet) module are added between the feature extraction network and the final output layer to enhance pedestrian feature level and complete multi-scale fusion. Finally, the large, medium and small pedestrian targets are predicted, and pedestrian detection boxes are obtained.

In order to obtain face detection boxes, it is necessary to train the YOLO V4 network with a dataset containing facial features. The MIT pedestrian dataset does not contain facial feature information. Therefore, the dataset with face features needs to be added. As Celeba (an open-source dataset of pedestrian faces) [32] does not take up too many resources and the accuracy of face detection box acquisition satisfies the requirements of this paper, Celeba is selected to add to the dataset. The flow chart of training YOLO V4 network and achieving face detection is shown in Figure 3. During training, the adjusted brightness and saturation can improve the robustness of the algorithm. Firstly, we modify Celeba to a dataset format, which can be recognized by YOLO V4. Then, the configuration files, such as "transform.py", "train.txt" and "val.txt", are modified. The number of classifications is adjusted to three types, and the learning rate is adjusted to 0.001. Finally, face detection boxes can be obtained through the YOLO V4 network trained by Celeba.

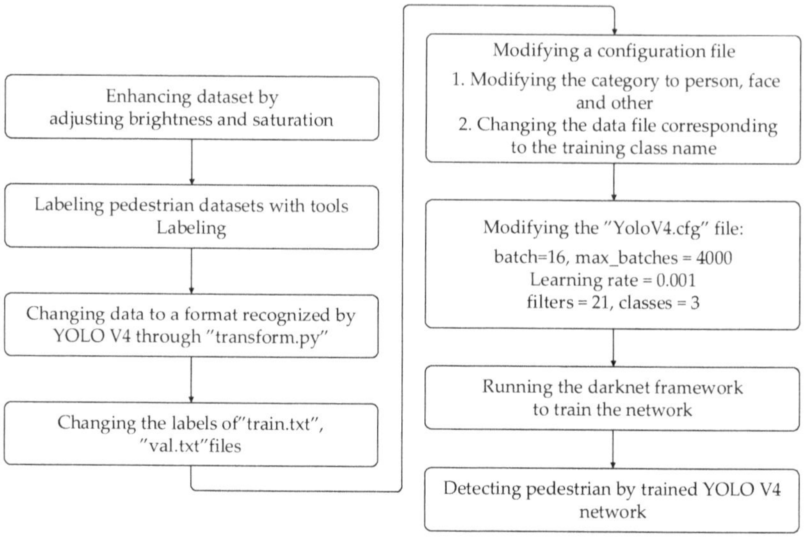

Figure 3. Flow chart of training YOLO V4 network and face detection.

3.1.2. Pedestrian 3D Data Acquisition

Considering the insufficient accuracy of the YOLO V4 algorithm for long-distance detection, LiDAR is used to obtain the 3D pedestrian data to increase the dimension of data. Euclidean clustering algorithm is applied to obtain 3D pedestrian detection boxes, which contain three primary parameters in this algorithm. It includes the threshold of clustering radius R, the minimum number of clustering points $P_{\min}$ and the maximum number of clustering points $P_{\max}$. The flow chart of pedestrian point cloud detection based on Euclidean clustering algorithm is shown in Figure 4.

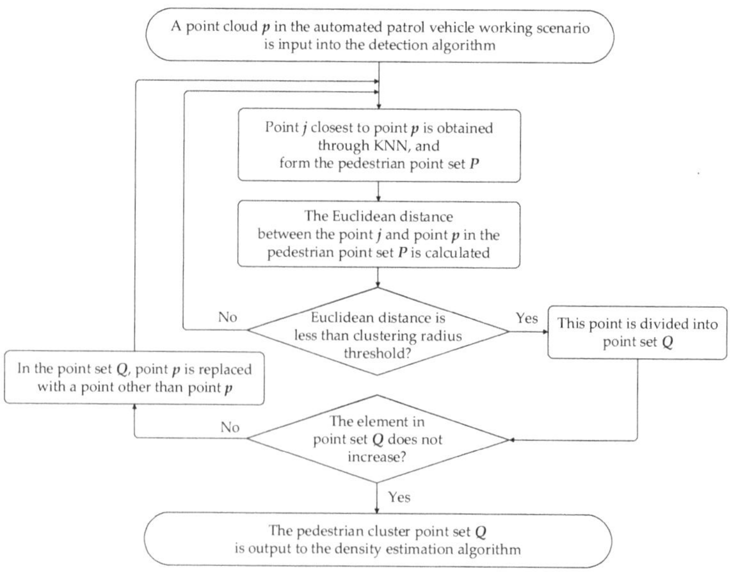

Figure 4. Flow chart of pedestrian point cloud detection based on Euclidean clustering algorithm.

In order to obtain 3D pedestrian data through Euclidean clustering, the threshold needs to be set first. Considering the sparse characteristic of cloud point data, different thresholds of clustering radius R are set according to the sparsity of point cloud at different distances. Details of the relationship between pedestrian distance L_{peds} and cluster radius threshold R are listed in Table 1.

Table 1. Corresponding relationship between pedestrian distance and clustering radius threshold.

Distance between Pedestrian and the Automated Patrol Vehicle $L_{peds}/(m)$	Threshold of Clustering Radius $R/(m)$
(0,10)	0.2
(10,20)	0.5
(20,30)	1.0
(30,40)	1.5
(40,50)	2

Then, the set of points to be searched in the point cloud space is defined as P. According to the features of P, we create a 3D k-dimensional tree (KD-Tree) object. A feature point $P_i = (x_i, y_i, z_i)^T$ is selected from P, its nearest neighbor points $P_n = \{P_j, j = 1, 2, \ldots, n\}$ are searched and the Euclidean distance D_{ij} between points P_i and P_j is calculated through K-nearest neighbor (KNN) algorithm. The Euclidean distance calculation function is represented as Equation (1):

$$D_{ij} = \sqrt{(x_i - x_j)^2 + (y_i - y_j)^2 + (z_i - z_j)^2} \tag{1}$$

where x_i, y_i and z_i indicate the coordinates of the i-th point; x_j, y_j and z_j indicate the coordinates of the j-th point. Once the Euclidean distance D_{ij} is less than the threshold of clustering radius R, its corresponding point P_j is classified as a clustered point cloud set Q. According to the relationship between its Euclidean distance and the threshold, determine whether the point belongs to the clustered point cloud set Q.

Finally, according to the preset minimum number of clustering points $P_{\min}$ and maximum number of clustering points $P_{\max}$, we judge whether the clustered point cloud set Q can be used as a valid single pedestrian target. If the number of point clouds in Q is greater than $P_{\min}$ and less than $P_{\max}$, the cluster point cloud set Q is valid. Meanwhile, a pedestrian clustering result is generated and the 3D pedestrian data are obtained.

3.2. Crowd Density Estimation

At present, the crowd aggregation is an important factor causing cross-infection. Crowd aggregation is related to the crowd density of a scenario. In order to locate crowded scenarios accurately, the 2D and 3D pedestrian data obtained in the previous section are combined to improve the accuracy of the estimation algorithm. A multi-layer fusion method is designed for crowd density estimation, which consists of three parts: (1) pedestrian detection based on multi-sensor fusion, (2) crowd target detection based on foreground image and (3) crowd density estimation based on sub-area density.

3.2.1. Pedestrian Detection Based on Multi-Sensor Fusion

After acquiring the 2D and 3D pedestrian data, multi-source data of pedestrians are fused based on the confidence optimization. The confidence of the pedestrians, applied in the YOLO V4 algorithm, has a great influence on the algorithm's detection results for pedestrians. The value of confidence is related to the constraints of pedestrian features in the YOLO V4 algorithm. The increase in confidence indicates that the constraints of the current pedestrian are more stringent, which will reduce FP rate. However, due to strict constraints, pedestrians with no obvious features may be missed by the algorithm. Therefore, in order to achieve accurate detection of long-distance and occluded pedestrians, the pedestrian confidence probability of the YOLO V4 algorithm is corrected by the Euclidean clustering algorithm. The flow chart of multi-sensor data fusion based on confidence optimization is shown in Figure 5.

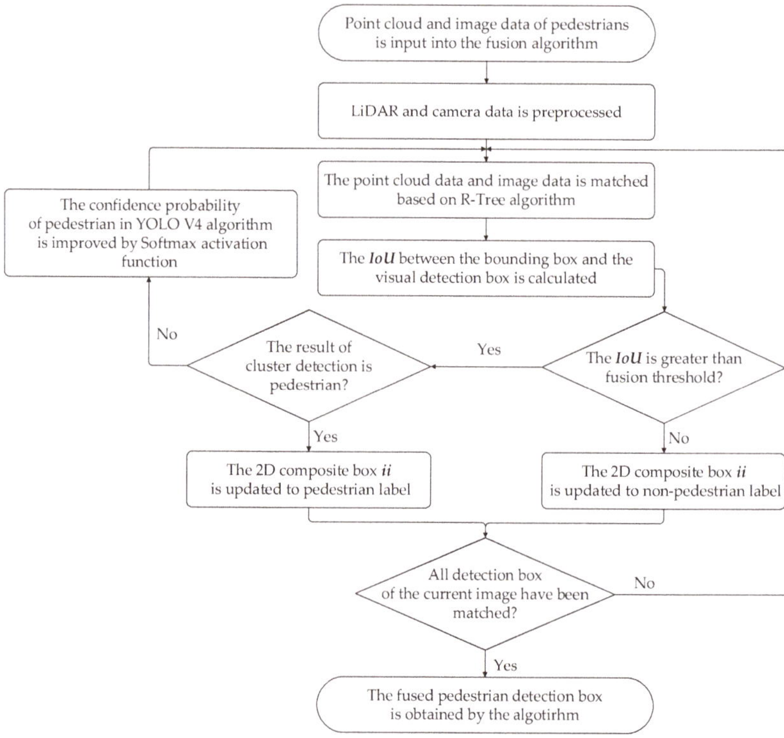

Figure 5. Flow chart of multi-sensor data fusion based on confidence optimization.

Firstly, the preprocessing of the fusion algorithm is achieved based on Zhang calibration [33] method and timestamp matching; secondly, the point cloud with the pixel data is matched through the R-Tree algorithm and the *IoU* between the bounding box and the 2D image detection box is calculated to determine the numerical relationship between the value of the *IoU* and the threshold. Finally, through the activation function Softmax(), the confidence probability of pedestrians in the image detection algorithm is corrected, which realizes the fusion of multi-sensor data. The fusion results can be used as a data source for crowd density estimation.

(1) Multi-sensor data preprocessing

The 2D and 3D detection boxes are input into the preprocessing part once the fusion algorithm is starting. Through Zhang calibration method, we obtain the relationship between the pixel coordinate system and the point cloud coordinate system. The coordinate system calibration of the multi-source sensors is shown in Figure 6. The four dots in different colors represent the vertices of the calibration board, and the lines in different colors represent the internal parameters of the camera and the degree of distortion. Through the calculation of the rotation vector and the translation vector, the matching of the image and the point cloud is achieved. The matching of timestamp is resolved by the Time Synchronizer mechanism. Since the point cloud sampling rate is 10 Hz, the image sampling rate is 25 Hz. Therefore, we add timestamps to each frame of data while collecting data from LiDAR and camera and call back the data with the same timestamp.

(2) Point cloud data and pixel data matching

Figure 6. The coordinate system calibration of the multi-source sensors.

The R-Tree algorithm is used to associate and match the detection box of the point cloud and the image to project the point cloud data to the pixel plane. According to the IoU of the image detection result and the projection result, the specific category information of the detected target in the point cloud space can be determined since there are several detection boxes, respectively, in each frame of image and point cloud data. We select a separate frame for matching each time, and the steps of the fusion algorithm are as follows:

Step 1: A frame of image data and point cloud data, including m 3D detection boxes and n 2D detection boxes, are input into the algorithm;

Step 2: The i-th 3D detection box and the j-th 2D detection box are selected. We set the initial values of parameters i and j to 1;

Step 3: The points in the i-th 3D detection box are projected onto the image to generate a 2D bounding box ii;

Step 4: The IoU of the 2D bounding box ii and the j-th 2D detection box are calculated through the algorithm. Through comparing the value of IoU of the two boxes and the threshold, we can determine the category information of the detection result.

IoU value is calculated as shown in Equations (2)–(4):

$$intersection = \left(\min\left(x_{ii}^{a_2}, x_j^{b_2}\right) - \max\left(x_{ii}^{a_1}, x_j^{b_1}\right)\right) \times \left(\min\left(y_{ii}^{a_2}, y_j^{b_2}\right) - \max\left(y_{ii}^{a_1}, y_j^{b_1}\right)\right) \quad (2)$$

$$union = S_{ii} + S_j - intersetion \quad (3)$$

$$IoU = \frac{intersection}{union} \quad (4)$$

where $intersection$ indicates the intersection between the bounding box and the 2D detection box, $(x_{ii}^{a_1}, y_{ii}^{a_1})$ indicates the coordinates of the upper left corner of the bounding box, $(x_{ii}^{a_2}, y_{ii}^{a_2})$ indicates the coordinates of the lower right corner, $(x_j^{b_1}, y_j^{b_1})$ indicates the coordinates of the upper left corner of the 2D detection box, $(x_j^{b_2}, y_j^{b_2})$ indicates the coordinates of the lower right corner, $union$ indicates the union between two 2D detection boxes, S_{ii} indicates the area of the point cloud detection box after mapping and S_j indicates the area of the 2D detection box.

(3) Pedestrian confidence correction

The fusion threshold of the intersection ratio is set to 0.75 [34], and the value of the IoU is compared with the fusion threshold to judge whether the current bounding box represents the pedestrian target. If the IoU is less than the fusion threshold, the label of the 2D bounding box is updated to be non-pedestrian; if the current IoU is greater than the fusion threshold and the clustering detection result is non-pedestrian, the confidence probability of the target is increased based on activation function Softmax(). Through adjusting the pixel and point cloud weight values, the correction effect is achieved. The confidence probability correction function is represented as Equation (5):

$$P = \max(Softmax(W_{LiDAR}, W_{vision}) = \frac{e^{W_{LiDAR}}}{\sum_{i=1}^{n} e^{W_i}}) \tag{5}$$

where P indicates the value of probability correction, W_i indicates the value of i-th sensor weight, where i includes vision and LiDAR, and n indicates the number of sensors.

3.2.2. Crowd Target Detection Based on Foreground Image

Crowd aggregation is one of the sources of virus spread, and current body temperature detection without distance restriction is difficult. Therefore, it is necessary to detect the aggregated crowd first and then detect the temperature of pedestrians in real time. Camera mounted on automated patrol vehicle can only acquire foreground image data. Due to occlusion and lack of depth information, the accuracy of detecting crowd targets based on foreground images cannot satisfy the patrol requirements. The 3D point cloud data has advantages in spatial sensing of the environment. Since the crowd target consists of multiple single pedestrians, we select the Euclidean clustering algorithm to detect the crowd target. Therefore, based on LiDAR and camera, perspective transformation algorithm is designed. The structure of perspective transformation is presented in Figure 7.

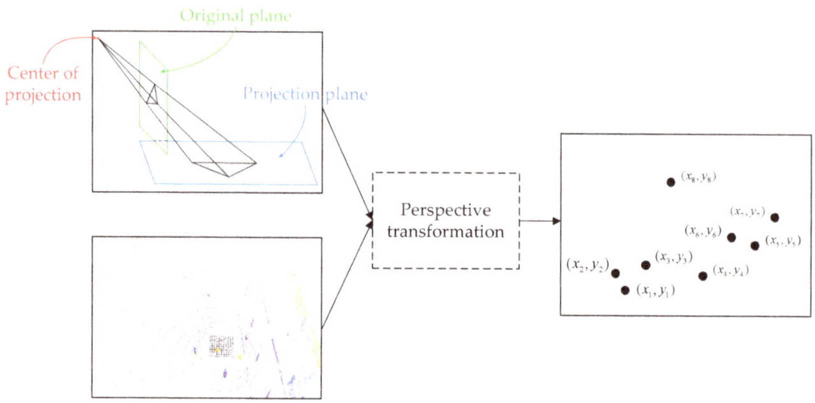

Figure 7. Structure of perspective transformation.

Through perspective transformation, the pedestrian detection box in the 2D image is mapped to a point in the top view. The perspective transformation function is represented as Equation (6):

$$\begin{bmatrix} Y_i \\ 1 \end{bmatrix}_{(i+1)\times 1} = \begin{bmatrix} A_{2\times 2} & t_{2\times 1} \\ 0_{1\times 2} & E_{1\times 1} \end{bmatrix} \begin{bmatrix} X_i \\ 1 \end{bmatrix}_{(i+1)\times i} \tag{6}$$

where Y_i takes the corner coordinate values of the three groups of detection boxes after transformation, X_i takes the corner coordinate values of the three groups of detection boxes before the transformation and A represents the matrix of the transformation.

Combined with the point cloud data, we assign the coordinate information of each target to the mapped pedestrian target points. The point set is clustered by the Euclidean clustering algorithm. The KNN algorithm based on KD-Tree object is an important pre-processing method of Euclidean clustering algorithm. Due to the fact that the errors in the parameter estimation process can be avoided by this algorithm and this algorithm has high classification accuracy, KD-Tree is used to traverse the adjacent pedestrian targets and obtain the minimum distance between pedestrians. The algorithm flow of the minimum distance calculation between pedestrians in 3D space is shown as follows:

Step 1: The pedestrian target (3D target recognition box, confidence degree) obtained by the data fusion algorithm is used as the input of the algorithm;

Step 2: The centroid coordinates of each pedestrian target $(X_{Peds}, Y_{Peds}, Z_{Peds})$ are extracted by the algorithm, and the dimension of the current pedestrian centroid data n is calculated as the set;

Step 3: The dimension d with the largest variance and the median m of all data items on dimension d are found. The dataset is divided according to the median m, which is divided into two parts, and the two data subsets are recorded as D_l and D_r, respectively;

Step 4: A pedestrian centroid KD-Tree node is established to store the division dimension d and the median m. Step 2 is performed again on the data subsets D_l and D_r divided by the Step 3, and the newly generated centroid KD-Tree node is set as the left and right child nodes;

Step 5: The dataset is continuously divided until the pedestrian centroid data subset contains less than two pieces of data after the current division. The corresponding centroid data are saved to the last leaf node, which is the KD-Tree object of the pedestrian centroid;

Step 6: Each pedestrian target is traversed by the KNN algorithm to find the current pedestrian and the nearest pedestrian. The Euclidean distance between the current pedestrian and the closest pedestrian is calculated through KNN algorithm;

Step 7: Pedestrian minimum distance sequence is obtained by minimum distance calculation algorithm.

The clustering threshold is one of the most important parameters in Euclidean clustering algorithm, and it is necessary to be set reasonably. Combined with the minimum distance sequence of pedestrians, which are obtained by the above algorithm, the distance features of pedestrians are distinguished. Due to the change in the objective of clustering, the thresholds are set according to the following two clustering conditions existing in the work scenario of the automated patrol vehicle;

Condition 1: There is a long distance between the center point and each clustering point, as shown in Figure 8a;

Figure 8. Detection results of crowd target. (**a**) Euclidean cluster in Condition 1; (**b**) Euclidean cluster in Condition 2.

Condition 2: The center of the cluster point is in the scenario with high crowd density, as shown in Figure 8b.

For Condition 1, the centroid point cloud of each pedestrian is used by the clustering algorithm to represent the current pedestrian target, and the centroid point cloud of all pedestrians is classified into a new clustering space. Through the experiments in Section 4, the clustering threshold is selected as 0.5 m and the pedestrian centroid point cloud is clustered. The requirements of crowd target detection are satisfied. For Condition 2, areas with high density are regarded as complete crowd targets by the clustering algorithm to achieve crowd detection.

As shown in Figure 8, a single black point in the scenario represents the pedestrian target detected by the algorithm, and the red point represents the non-pedestrian target. Firstly, the non-pedestrian target is eliminated through the algorithm. Then, two adjacent pedestrian targets are clustered based on the KNN algorithm. The clustered target is used as a new individual to cluster again until there is no single pedestrian target in the scenario. Finally, crowd targets are obtained as the algorithm output.

3.2.3. Crowd Density Estimated Based on Sub-Area Density

In order to improve the effect of density estimation, a crowd density estimation algorithm based on sub-area density is proposed to expand the difference in the distribution characteristics of crowd density. The layer is divided into equal-area units by the algorithm, and the density values of the units are selected as input of the algorithm. Through fusing crowd density data from multiple layers, the crowd density distribution is calculated. Based on the weighted mean theory, the crowd density of the current scenario is estimated.

According to the number of pedestrian targets detected by the automated patrol vehicle in the current scenario, the layer with a side length of 15 m is divided into 32 × 32 units by algorithm. Each of these units covers an area of 0.22 square meters, which fits the minimum footprint of an adult. The density value of unit is taken from the ratio of the number of pedestrians in each unit to its area of unit, whose calculation is shown as Equation (7):

$$\left\{\rho_{(i,j)} \middle| \rho_{(i,j)} = \frac{N_{ij}}{S}, i, j \in 50\right\} \tag{7}$$

where $\rho_{(i,j)}$ indicates the density value of unit, N_{ij} indicates the number of pedestrians in the unit of the i-th row and the j-th column and S indicates the area of a single unit within the layer. The calculation algorithm of N_{ij} is illustrated in Algorithm 1.

Algorithm 1: N_{ij} calculation algorithm in 3D space.

Initialize pedestrian detection results, target detection boxes and confidence level.
Create a List $L_{targets}$ of every unit, a centroid variable C and a pedestrian count variable N_{count}.
While ($i_{\min} < i < i_{\max}$ && $j_{\min} < j < j_{\max}$) **do**
  If (target detection boxes) **then**
    $L_{targets} = L_{targets} + 1$
  for each $L_{targets}$ **do**
    If (C is under the range of the corner coordinates of $L_{targets}$) **then**
      $N_{count} = N_{count} + 1$
    end if
  end for
  $N_{ij} = N_{count}$
End while
Return N_{ij}

The initial crowd density distribution matrix M_{old} is composed of the density values of unit $\rho_{(i,j)}$. The numerical values in each row and column of the matrix are the density values of unit. The initial crowd density distribution matrix is represented as Equation (8):

$$M_{old} = \begin{bmatrix} \rho(1,1) & \rho(1,2) & \cdots & \rho(1,j) \\ \rho(2,1) & \ddots & \rho(2,j-1) & \rho(2,j) \\ \vdots & \rho(i-1,2) & \ddots & \rho(i-1,j) \\ \rho(i,1) & \rho(i,2) & \cdots & \rho(i,j) \end{bmatrix}, i, j \in 50 \tag{8}$$

The initial crowd density distribution matrix is represented by the current density of each individual point. However, density is one of the attributes of the crowd; the arrangement of density values of individual points is not representative. Therefore, the density features of adjacent units in the layer are correlated by this algorithm. After the corresponding unit density values of different layers are linearly combined, the unit density values containing the features of adjacent units are obtained. The matrix M_{new}, which is composed of each unit density value containing adjacent unit features in the layer, is the final crowd density distribution matrix. The structure of crowd density distribution detection based on the multi-layer fusion algorithm is shown in Figure 9.

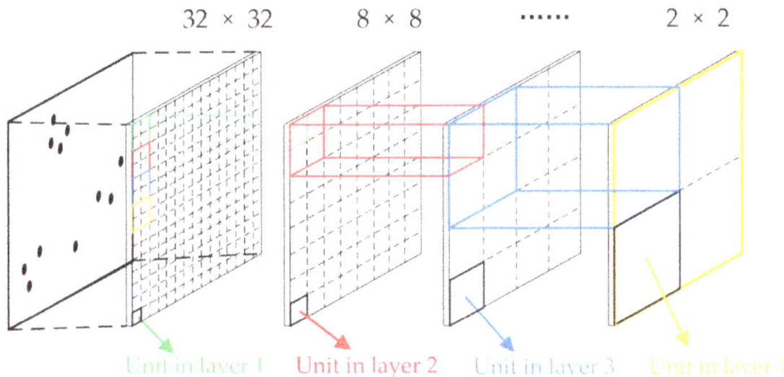

Figure 9. Structure of crowd density distribution detection based on the multi-layer fusion algorithm.

A 2D coordinate system is established according to the initial position of the automated patrol vehicle in Section 3.2.1. Along the initial direction of the automated patrol vehicle, the 2×2 units are sequentially taken to form a new unit in the next layer, a new layer of 16×16. The side length of any 2×2 unit in the current layer is 0.9 m. The pedestrian density value in the 2×2 unit is counted and superimposed with the density value of the unit in upper layer. The superposition algorithm is represented as Equation (9):

$$\rho^k_{(i,j)} = \sum_{i=1,j=1}^{2} \rho^{k-1}_{(i,j)} \tag{9}$$

where $\rho^k_{(i,j)}$ indicates the density value of the unit in the i-th row and the j-th column after the k-th division, and $\rho^{k-1}_{(i,j)}$ indicates the density value of the unit of the i-th row and the j-th column after the $(k-1)$-th division.

It is worth mentioning that the total area of the single layer does not change after the layer is divided by the above algorithm. The 2×2 units are finally iterated by the algorithm. We count the density values of the units after each division, and, based on the fusion of multi-layer by weighting, the density distribution matrix M_{new} is obtained as the output of the algorithm. The weighted superposition is represented as Equation (10):

$$M_{new} = \begin{bmatrix} \sum_{k=0}^{n} w_k \cdot \rho^k_{(1,1)} & \sum_{k=0}^{n} w_k \cdot \rho^k_{(1,2)} & \cdots & \sum_{k=0}^{n} w_k \cdot \rho^k_{(1,j)} \\ \sum_{k=0}^{n} w_k \cdot \rho^k_{(2,1)} & \ddots & \sum_{k=0}^{n} w_k \cdot \rho^k_{(2,j-1)} & \sum_{k=0}^{n} w_k \cdot \rho^k_{(2,j)} \\ \vdots & \sum_{k=0}^{n} w_k \cdot \rho^k_{(i-1,2)} & \ddots & \sum_{k=0}^{n} w_k \cdot \rho^k_{(i-1,j)} \\ \sum_{k=0}^{n} w_k \cdot \rho^k_{(i,1)} & \sum_{k=0}^{n} w_k \cdot \rho^k_{(i,2)} & \cdots & \sum_{k=0}^{n} w_k \cdot \rho^k_{(i,j)} \end{bmatrix}, i,j \in 50 \tag{10}$$

where w_k indicates weight of the k-th division. In order to improve the detection efficiency, the harmonic series is used as the weight update function to magnify the difference between the crowd density in different units. The weight update function is represented as Equation (11):

$$w_k = \frac{1}{k+1}, k \in [0, N] \tag{11}$$

where k indicates the current superposition times and N indicates the total number of divisions.

Real-time detection in disease control areas is very important. Due to the multi-dimensional nature of crowd features, the weighted mean theory used in the algorithm should have high real-time performance and the density estimation should satisfy the requirements of patrolling. Therefore, we combine the comparative requirements of the crowd density of different layers, and, based on the weighted mean theory, estimate the crowd density in the current scenario. The estimation equation is represented as Equation (12):

$$\rho_{crowd} = \frac{\sum_i \sum_j \sum_{k=0}^{n} w_k \cdot \rho^k_{(i,j)}}{\iint d\sigma}, i, j \in 50 \tag{12}$$

where ρ_{crowd} indicates the crowd density and σ indicates the overall area of the layer.

3.3. Pedestrian Body Temperature Detection

At present, more than 40% of the patients that test positive for the coronavirus have high temperature features in their clinical manifestations [35], and the risk of current epidemic spreading can be decided by temperature. Therefore, a body temperature detection algorithm based on thermal imaging is designed after the density estimation to ensure the risk of cross-infection can be reduced by temperature detection with high crowd density. Since the accuracy limitations of current temperature sensors, it is difficult to detect temperature at any distance accurately. Therefore, in order to reduce the cost of temperature detection and improve the accuracy, the distance from the sensor to the target to be detected is fixed at 3 m, and an external bold body, an empirical method and outlier filtering are used to correct the detection results. The flow chart of pedestrian body temperature detection is shown in Figure 10.

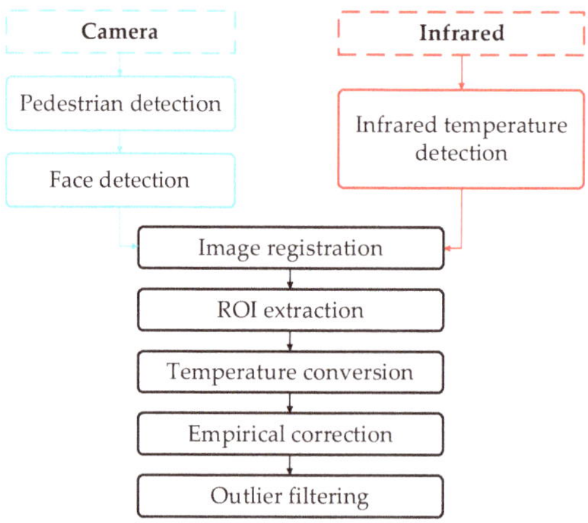

Figure 10. Flow chart of pedestrian body temperature detection.

3.3.1. Infrared Image and Visible Image Registration

In order to combine image data after fusion and infrared data, the registration method is used to fuse multi-sensor data. Color image data is output by the infrared camera in real time, in which each pixel represents the temperature value. Through detecting the temperature value of the face area, we calculate the current pedestrian's body temperature. However, since the angle and orientation of the output images of infrared and visible images to the same target are different, it is necessary to register the infrared image and the visible image. Several matching points in the infrared and visible images are selected, and we register the two images by means of affine transformation. Affine transformation function is shown in Equation (13):

$$\begin{bmatrix} x'_i \\ y_i' \end{bmatrix} = A \begin{bmatrix} x_i \\ y_i \end{bmatrix} + BI \tag{13}$$

where A and B indicates the parameters of affine transformation equation, and (x_i, y_i) and (x_i', y_i') indicate a set of corresponding coordinates in visible and infrared images. The equation of obtaining parameters A and B of affine transformation equation is shown in Equation (14):

$$\begin{cases} A = \begin{bmatrix} a_{11} & a_{12} \\ a_{21} & a_{22} \end{bmatrix} \\ B = \begin{bmatrix} b_1 \\ b_2 \end{bmatrix} \end{cases} \tag{14}$$

In order to determine the two parameters in the process of image registration in the above equation, different corners of the detection screen are obtained by the automated patrol vehicle in real time. When the number of extracted corners is more than three groups, the optimal parameters of affine transformation equation can be fitted by the least square method or gradient descent method. Since the least square method can be used to find the global solution instead of the local optimal solution in the gradient descent method. Therefore, multiple matched corners are selected to fitting the optimal solution based on the least square method. Infrared image and visible image are registered by the equation after obtaining the parameters of affine transformation equation.

3.3.2. Human Body Temperature Detection and Correction

The corresponding point can be found both in the infrared image and the visible image after the registration, which can establish the mapping relationship to match the face area. Considering the accuracy of the infrared sensor, we set the distance of the temperature detection between 3–8 m. However, there are temperature interference factors such as hats, glasses and hair in the actual scenario, which leads to its temperature lower than the real value. Therefore, through the 5 × 5 template in the region of interest (ROI), the mode index of temperature of each template in the current scenario is calculated by the algorithm. The template is averaged as the pedestrian's body temperature. The preset template is used to traverse the face area after the face target is detected in the dense area. And temperature data are extracted for each pixel within the template. The temperature of pedestrian calculation method is represented as Equation (15):

$$T_i = \frac{\sum\limits_{j=1}^{n} M(j)}{n} \tag{15}$$

where T_i indicates the temperature of i-th pedestrian, n indicates the number of face detection box pixels and $M(j)$ indicates the mode of infrared detection temperature in a single 5 × 5 template.

Meanwhile, hardware errors are still inevitable, and the algorithm detection results are quite different from the values detected by the thermometer. Considering that the most accurate body temperature detection is the thermometer, it is reasonable to use it as the standard. Therefore, we combined the algorithm detection results with the thermometer detection results to correct the body temperature detection algorithm by empirical method. The temperature deviation caused by the occlusion of glasses and other facial decoration can be reduced effectively. And the local abnormal temperature caused by the sensor distance accuracy can be reduced by taking the mean value. The correction algorithm is shown as Equation (16):

$$T_p = T_{f\_d} + \sqrt{\frac{T_{f\_d}^2}{T_{f\_d}^2 + (T_{f\_d} - T_{env})^2}} \times (T_t - \frac{\sum_{i=1}^{n} T_i}{n}) \quad (16)$$

where T_p indicates the detection temperature obtained after correction, T_{f_d} indicates the result of infrared temperature measurement, T_{env} indicates the current environmental temperature, T_t indicates the current pedestrian's body temperature measured by a thermometer and n indicates the number of selected measurements.

In addition, errors in the temperature detection are mainly due to temperature drift caused by environmental changes and long-term operation. The probe of an infrared detector is exposed in the infrared radiation, including air temperature and radiation generated by the surrounding environment. Due to changes in the environment, the state of infrared radiation is changed, which will reduce the accuracy of temperature detection. The long-term work of automated patrol vehicles can also cause temperature drift. Therefore, we adopt external bold body correction and outlier filtering to further improve the accuracy of temperature detection. By setting the external bold body in front of the infrared camera, the error of temperature detection can be reduced and the stability of the system can be improved. The external bold body correction coefficient function is represented as Equation (17):

$$K = \frac{T_0}{T_a} \quad (17)$$

where K indicates the temperature correction coefficient, T_0 indicates the absolute temperature set by the external bold body and T_a indicates the absolute temperature detected by the external bold body. The external bold body is a constant temperature source, whose temperature T_0 is fixed at 37°.

Moreover, outliers are usually caused by pedestrian occlusion, face detection errors, environmental temperature changes and other factors in the process of body temperature detection. In order to ensure the accuracy of temperature detection, temperature detection values of less than 35° and greater than 42° are removed with reference to the normal body temperature of pedestrians.

4. Experimental Results and Discussion

In this section, a detection platform and scenario are constructed based on the automated patrol vehicle. Considering the current situation of the pandemic, a straight road in the campus is chosen for the experiment. There are entrances for pedestrians on the north and south sides of the straight road. There are convenience stores and other factors at the end of the east side that cause pedestrian aggregation, as shown in Figure 11a. For this scenario, we design an automated patrol vehicle applied with a detection method. The automated patrol vehicle is equipped with a mechanical 16-line LiDAR, a 720P USB industrial camera, a thermal infrared camera, an Ackerman vehicle chassis, an industrial computer and a monitor screen, as shown in Figure 11b. The measuring range of the mechanical LiDAR does not exceed 150 m, and the minimum measuring range is 0.5 m. It is mounted perpendicular to the vehicle at 0.5 m. The camera is mounted perpendicular to the LiDAR at 0.4 m directly below. The thermal infrared camera is mounted perpendicular to the camera at 0.6 m directly below. The maximum velocity of the automated patrol vehicle is 4 m per second, which satisfies the security requirements of patrolling on campus.

Figure 11. Experimental platform based on automated patrol vehicle. (**a**) Experimental scenario. (**b**) Automated patrol vehicle.

Then, the feasibility and accuracy of pedestrian data acquisition, crowd density estimation and real-time body temperature detection are analyzed and discussed.

The experiments of pedestrian data acquisition consist of two parts: (1) feasibility of pedestrian data acquisition; (2) pedestrian detection accuracy verification. The 2D pedestrian data acquisition results using the YOLO V4 algorithm are shown in Figure 12a, and the rectangular boxes are the detection results of pedestrians. The 3D pedestrian data acquisition results using the European clustering algorithm are shown in Figure 12b, and the different colored stereoscopic boxes are the detection results of pedestrians. The image data after data fusion are shown in Figure 12c. The red dots represent the pedestrian detection results based on YOLO V4, and the green dots represent the pedestrian detection results based on the Euclidean clustering algorithm. The results of face detection based on the YOLO V4 network trained by Celeba are shown in Figure 12d.

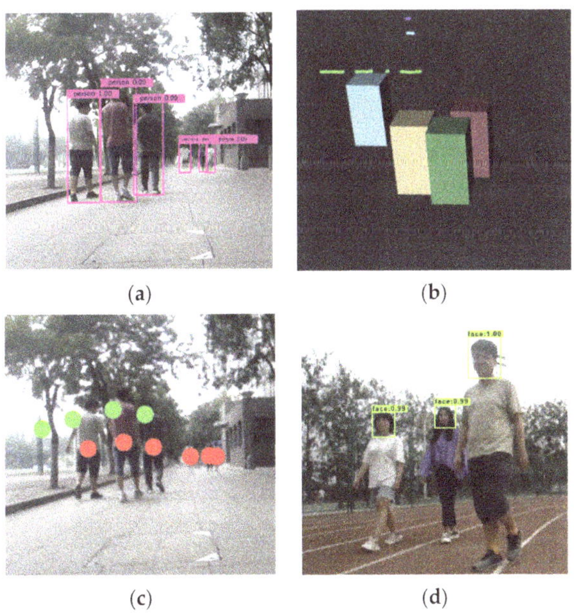

Figure 12. The results of pedestrian detection and face detection. (**a**) 2D detction box. (**b**) 3D detection box. (**c**) Fusion results. (**d**) Face detection results.

As shown in Figure 12, occluded pedestrians can be detected by the trained YOLO V4 network better. Long-distance pedestrians can be detected by the clustering algorithm, and the location information of pedestrians can be obtained. Combined with the multi-sensor features, the detection range and feasibility of the fusion algorithm are higher than those of a single sensor. As shown in Figure 12d, the face targets can be detected by a trained YOLO V4 network accurately.

The FP rate index is used to verify the accuracy of the pedestrian detection algorithm under different pedestrian occlusion rates and different scenarios. Considering the moving velocity of the automated patrol vehicle and the sensing range of the sensor, we fixed the distance between the automated patrol vehicle and the pedestrian target at 10 m. The academic building and outer space are used as the respective experimental scenarios to verify the effects of different lighting and environmental complexity on the detection of pedestrians. In order to verify the detection accuracy after data fusion, the occlusion of pedestrians is defined into four levels [36], including no occlusion for pedestrians, partial occlusion for pedestrians, severe occlusion for pedestrians and complete occlusion for pedestrians. The fitting curves of the false positive rates under different scenarios are shown in Figure 13.

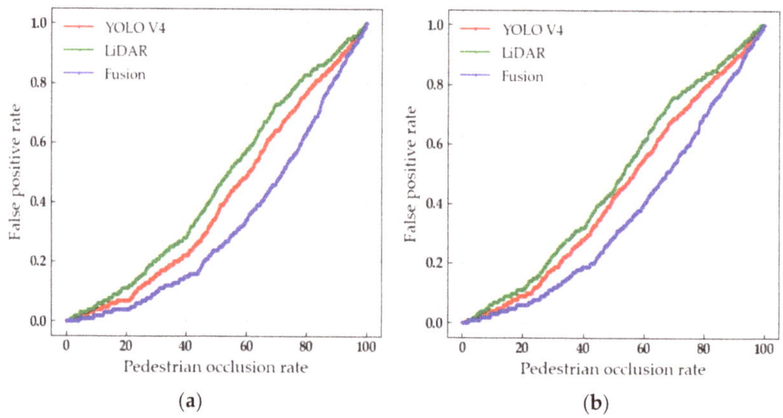

Figure 13. Fitting curves of false positive rate under different scenarios. (**a**) Outdoor environment. (**b**) Academic building.

The lower the FP rate of pedestrian targets, the better the detection effect of the fusion algorithm. With the increase in the pedestrian occlusion rate, the FP rate of pedestrian targets will gradually increase. The FP rate of the fusion algorithm is lower than 5% when the pedestrian occlusion rate is below 20% in the academic building and outdoor environment, as shown in Figure 13. Therefore, the fusion algorithm has higher accuracy and stronger robustness compared to the single-sensor detection algorithm. The accuracy of the pedestrian detection method based on the fusion algorithm satisfies the requirements of the crowd density estimation algorithm.

Moreover, we evaluate the effectiveness of crowd density estimation in experimental scenarios through a three-part experiment, which includes: (1) crowd clustering threshold selection and crowd detection implementation; (2) crowd density distribution verification; (3) the accuracy of crowd density estimation verification.

The clustering effect of the crowd target is directly related to the clustering threshold. In order to obtain the higher clustering results, a suitable threshold is very important. Therefore, we conduct experiments on the selection of clustering thresholds. The result of the crowd clustering detection with the clustering threshold set to 0.5 is shown in Figure 14. In Figure 14a, the red dots represent the pedestrian detection results based on YOLO V4, and the green dots represent the pedestrian detection results based on the Euclidean clustering algorithm. And in Figure 14b, different colored stereoscopic boxes represent the detection results of crowd targets.

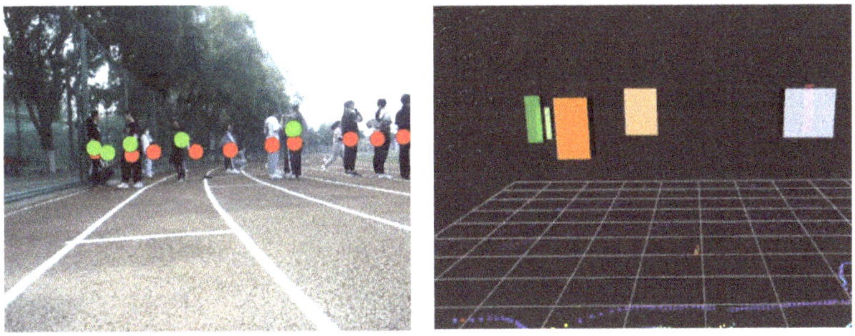

(a) (b)

Figure 14. Detection results of crowd target when the clustering threshold is set to 0.5. (**a**) The crowd aggregation state. (**b**) The detection results of crowd target.

The larger the clustering threshold is, the more blurred the boundary of the crowd target is, as shown in Figure 14. Additionally, the algorithm is more inclined to cluster the scattered points into the detection result, which does not satisfy our requirements. If a small clustering threshold is selected, the number of clustered crowd objects will be very large and will lead to a smaller number of pedestrians contained in a single cluster. The pedestrian targets that cause body temperature to be detected will be reduced, resulting in an increased probability of infection risk. Therefore, we take the clustering threshold as 0.5 m according to the comprehensive judgment of the correlation between the threshold and the number of clusters. The detection results of the crowd target clustering satisfy our requirements under this clustering threshold.

The number of pedestrians in the unit will be detected by the automated patrol vehicle. Additionally, combined with the area of unit, the crowd density in the unit will be estimated. We use the 3D thermodynamic chart to show the current distribution of the crowd density in the scenario, as shown in Figure 15. In order to show that our method has better performance, the initial density distribution matrix is used for comparison.

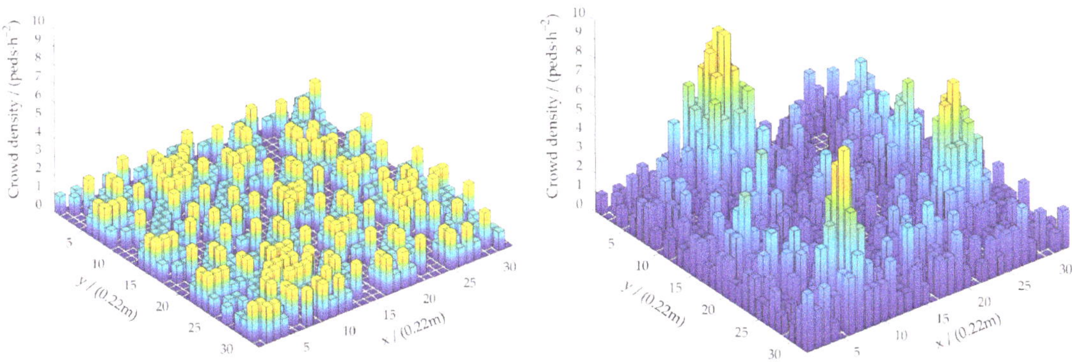

(a) (b)

Figure 15. Visualizations of crowd density distribution detection results. (**a**) Initial crowd density distribution M_{old}. (**b**) Final crowd density distribution M_{new}.

Crowd density is proportional to its thermal peak. Columns with low density values are represented in blue, and the color of the columns transfers to yellow as the density increases. The initial density distribution matrix M_{old} is shown in Figure 15a, and the final density distribution matrix M_{new} calculated by our algorithm is shown in Figure 15b. In M_{old}, the aggregation area is not prominent, and the density distribution is uniform. As shown in Figure 15b, there are three peaks in the current scenario. Each peak area represents a group of crowd aggregation. The detection result is consistent with the result of the manual measurement of density distribution. The crowd density in the scenario is estimated based on this distribution. We compare the detection results with the real values measured manually to obtain the density error, as shown in Table 2.

Table 2. Analysis of the accuracy of crowd density estimation.

Scenario Number n_s	The Values of Crowd Density Estimation ρ_{crowd}/(peds·h^{-2})	The Error of Crowd Density Estimation $E_{\rho_{crowd}}$/(%)
1	0.9	1.11
2	2.8	2.52
3	5.3	3.58
4	8.4	6.54
5	10.5	5.91

Five crowd groups were preset in the experimental scenario and recorded as five groups of scenarios. The larger the scenario number, the more pedestrians in a single scenario. As the number of pedestrians increases, the value of crowd density increases gradually, as shown in Table 2. The density error is the largest when the scenario number is 4 and the smallest when the scenario number is 1. Since three or more pedestrians are completely occluded in the crowd represented by scenario number 4, the error is the largest. The experimental results have shown that the error of the proposed algorithm in detecting crowd density is 6.54% at maximum and 1.11% at minimum. The detection accuracy satisfies the requirements of crowd density estimation.

The feasibility of the body temperature detection algorithm is verified through two-part experiments, which include: (1) real-time body temperature detection verification; (2) body temperature detection accuracy verification. The body temperature of pedestrians in real time is detected by the automated patrol vehicle. The temperature detection results are displayed visually, as shown in Figure 16.

(a) (b) (c)

Figure 16. Results of temperature detection. (**a**) Original image. (**b**) Face detection. (**c**) Temperature detection.

During the experiment, the air temperature is 21.4°. As shown in Figure 16, the real-time body temperature of the current pedestrian can be detected by the automated patrol vehicle, and it can be marked on the pedestrian's body accurately without abnormal values. Therefore, the proposed algorithm is feasible. Then, the accuracy of our algorithm in detecting pedestrian body temperature is verified through comparative experiments. Considering that the temperature of the same object detected by different devices at different times is not the same, the current results measured by thermometers are highly convincing. Therefore, we select the data measured by the thermometer as the accurate body temperature, and the detection error of the body temperature is calculated through comparing with the detection results, as shown in Figures 17–19.

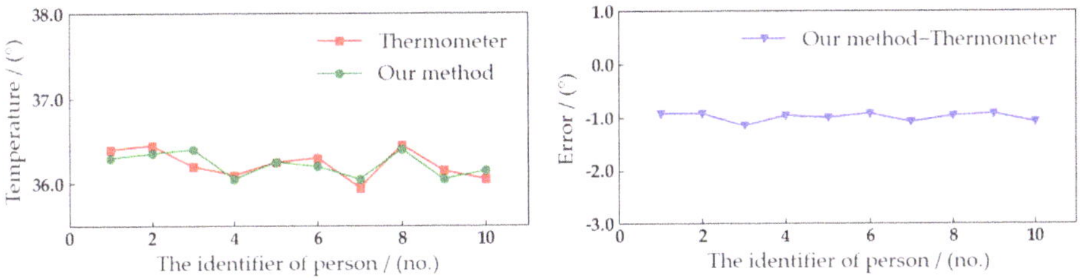

(**a**) (**b**)

Figure 17. Comparison of pedestrian body temperature detection results at a distance of 3 m. (**a**) Temperature detection results. (**b**) Detection error.

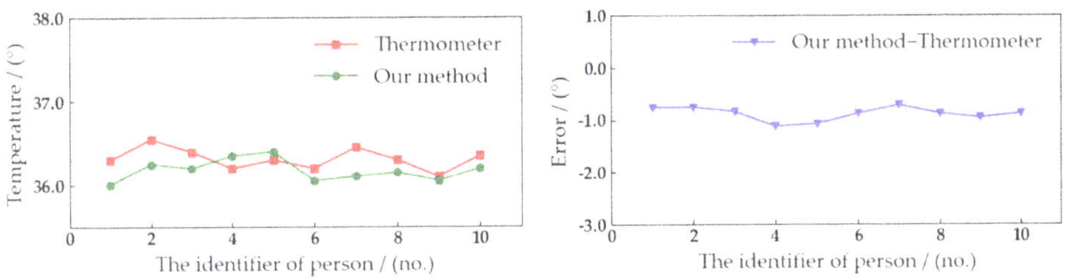

(**a**) (**b**)

Figure 18. Comparison of pedestrian body temperature detection results at a distance of 4 m. (**a**) Temperature detection results. (**b**) Detection error.

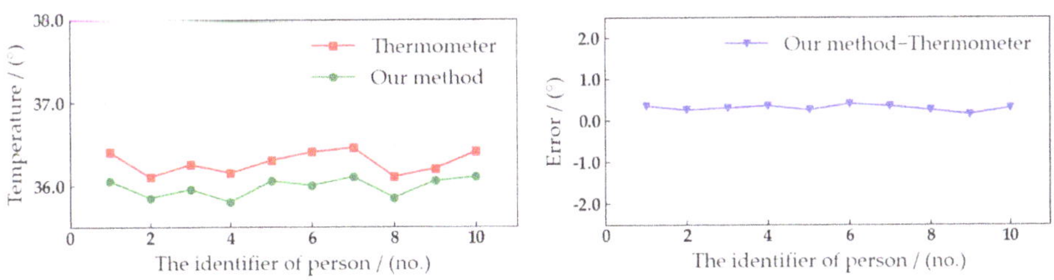

(**a**) (**b**)

Figure 19. Comparison of pedestrian body temperature detection results at a distance of 5 m. (**a**) Temperature detection results. (**b**) Detection error.

The detection distance is fixed at 3, 4 and 5 m. The interval is set to three seconds, and we select ten consecutive body temperature detection data to compare with the detection results of thermometer. Due to the position of the external bold body and the correction algorithm, the detection of pedestrian body temperature at 3 m is the most accurate, as shown in Figure 17. Comparing the detection results with the body temperature data measured by the thermometer, the absolute value of the error is less than 0.8° in ten records.

Considering the safety factors on campus, the automated patrol vehicle has to maintain a reasonable distance from pedestrians. Therefore, the distance between the automated patrol vehicle and the crowd target can be kept at about 3 m, and the body temperature of pedestrians can be detected stably and with high precision at this distance.

The feasibility and accuracy of the pedestrian data acquisition algorithm, the optimal threshold for crowd target clustering, the density estimation and the accuracy of body temperature detection are verified, in turn, after the discussion of the above experiments. The experimental results have shown that the method can work stably in the campus environment.

5. Conclusions

In this paper, in order to reduce the risk of cross-infection and improve the accuracy of sensing, we proposed a novel multi-sensor data fusion method, a multi-layer fusion method and a temperature detection method. Meanwhile, the above methods were applied in an automated patrol vehicle, which improves the flexibility of detection. The experimental results demonstrated that the pedestrian detection results are improved by more than 17% compared to the single-sensor algorithm; the density estimation results are more intuitive after applying the multi-layer fusion method. Compared with the thermometer, the error in body temperature detection is less than 0.8°. Due to the limitations of the equipment cost and the volume of the automated patrol vehicle, the detection effect of the proposed method on the lateral and rear targets of the vehicle body needs to be improved, which can be solved by increasing the number of sensors in the future.

Author Contributions: Conceptualization, P.W.; methodology, C.L.; software, C.L. and H.Y.; validation, Y.W. and H.Y.; formal analysis, Y.W.; investigation, P.W.; resources, P.W.; data curation, C.L.; writing—original draft preparation, C.L.; writing—review and editing, P.W.; visualization, C.L.; supervision, P.W.; project administration, P.W.; funding acquisition, P.W. All authors have read and agreed to the published version of the manuscript.

Funding: This research was funded by the Beijing Natural Science Foundation (grant number 4212034).

Institutional Review Board Statement: Not applicable.

Informed Consent Statement: Not applicable.

Data Availability Statement: All data and models used during the study appear in this article.

Acknowledgments: The authors would like to thank X. Liu and X. Wang for their technical assistance with the experiments and analysis.

Conflicts of Interest: The authors declare no conflict of interest.

References

1. Osama, Z.; Matthew, D. A navigation strategy for an autonomous patrol vehicle based on multi-fusion planning algorithms and multi-paradigm representation schemes. *Robot. Autonom. Syst.* **2017**, *96*, 133–142.
2. Wang, P.; Deng, H.; Zhang, J.; Wang, L.; Zhang, M.; Li, Y. Model predictive control for connected vehicle platoon under switching communication topology. *IEEE Trans. Int. Transp. Syst.* **2021**, *99*, 1–14. [CrossRef]
3. Ghosh, S.; Amon, P.; Hutter, A.; Kaup, A. Reliable pedestrian detection using a deep neural network trained on pedestrian counts. In Proceedings of the 2017 IEEE International Conference on Image Processing (ICIP), Beijing, China, 17–20 September 2017.
4. Wang, P.; Jiang, Y.; Xiao, L.; Zhao, Y.; Li, Y. A joint control model for connected vehicle platoon and arterial signal coordination. *J. Int. Transp. Syst.* **2020**, *24*, 81–92. [CrossRef]
5. Fang, W.; Chen, J.; Hu, R. Pedestrian attributes recognition in surveillance scenarios with hierarchical multi-task CNN models. *China Commun.* **2018**, *15*, 208–219.
6. Sigut, J.; Castro, M.; Arnay, R.; Sigut, M. OpenCV basics: A mobile application to support the teaching of computer vision concepts. *IEEE Trans. Educ.* **2020**, *63*, 328–335. [CrossRef]
7. Lecun, Y.; Boser, B.; Denker, J.S.; Henderson, D.; Howard, R.E.; Hubbard, W.; Jackel, L.D. Backpropagation applied to handwritten zip code recognition. *Neural Comput.* **1989**, *1*, 541–551. [CrossRef]
8. Ryu, j.; Kim, S. Feature map swap: Multispectral data fusion method for pedestrian detection. In Proceedings of the 2019 19th International Conference on Control, Automation and Systems (ICCAS), Jeju, Korea, 15–18 October 2019.

9. Luo, H. Pedestrian reidentification based on feature fusion and metric learning. *Int. J. Circuits Syst. Signal Proc.* **2022**, *16*, 105–114. [CrossRef]
10. Luo, J.; Zhang, C.; Wang, C. Indoor multi-floor 3d target tracking based on the multi-sensor fusion. *IEEE Access* **2020**, *8*,36836–36846. [CrossRef]
11. Mendez, J.; Molina, M.; Rodriguez, N.; Cuellar, M.P.; Morales, D.P. Camera-LiDAR multi-level sensor fusion for target detection at the network edge. *Sensors* **2021**, *21*, 3992. [CrossRef]
12. Han, J.; Liao, Y.; Zhang, J.; Wang, S.; Li, S. Target fusion detection of LiDAR and camera based on the improved YOLO algorithm. *Mathematics* **2018**, *6*, 213. [CrossRef]
13. Zhang, R.; Yang, Y.; Wang, W.; Zeng, L.; Chen, J.; McGrath, S. An algorithm for obstacle detection based on YOLO and light filed camera. In Proceedings of the 2018 12th International Conference on Sensing Technology (ICST), Limerick, Ireland, 4–6 December 2018.
14. Jiang, X.; Zhang, L.; Zhang, T.; Lv, P.; Zhou, B.; Pang, Y.; Xu, M.; Xu, C. Density-aware multi-task learning for crowd counting. *IEEE Trans. Multimedia* **2021**, *23*, 443–453. [CrossRef]
15. Liu, C.; Yao, Y.; Xu, C. Conventional and neural network-based water vapor density model for GNSS troposphere tomography. *GPS Sol.* **2022**, *26*, 4. [CrossRef]
16. Muhammed, V.A.; Santhosh, G.K. Deep learning framework for density estimation of crowd videos. In Proceedings of the 2018 8th International Symposium on Embedded Computing and System Design (ISED), Cochin, India, 13–15 December 2018.
17. Ziyadinov, V.V.; Kurochkin, P.S.; Tereshonok, M.V. Convolutional neural network training optimization for low point density image recognition. *J. Commun. Technol. Electr.* **2021**, *66*, 1363–1369. [CrossRef]
18. Ding, X.; He, F.; Lin, Z. Crowd density estimation using fusion of multi-layer features. *IEEE Trans. Int. Transp. Syst.* **2021**, *22*, 4776–4787. [CrossRef]
19. Huang, S.; Zhou, H.; Liu, Y.; Chen, R. High-resolution crowd density maps generation with multi-scale fusion conditional GAN. *IEEE Access* **2020**, *8*, 108072–108087. [CrossRef]
20. Sajid, U.; Ma, W.; Wang, G. Multi-resolution fusion and multi-scale input priors based crowd counting. In Proceedings of the 2020 25th International Conference on Pattern Recognition (ICPR), Milan, Italy, 10–15 January 2021.
21. Nguyen, M.N.; Tran, V.H.; Huynh, T.N. Depth embedded and dense dilated convolutional network for crowd density estimation. In Proceedings of the 2021 International Conference on System Science and Engineering (ICSSE), Ho Chi Minh City, Vietnam, 26–28 August 2021.
22. Wang, S.; Pu, Z.; Li, Q.; Guo, Y.; Li, M. Edge computing-enabled crowd density estimation based on lightweight convolutional neural network. In Proceedings of the 2021 IEEE International Smart Cities Conference (ISC2), Manchester, UK, 7–10 September 2021.
23. Zhang, Y.; Zhao, H.; Duan, Z.; Huang, L.; Deng, J.; Zhang, Q. Congested crowd counting via adaptive multi-scale context learning. *Sensors* **2021**, *21*, 3777. [CrossRef]
24. Wang, P.; Wang, Y.; Wang, X.; Liu, Y.; Zhang, J. An intelligent actuator of an indoor logistics system based on multi-sensor fusion. *Actuators* **2021**, *10*, 120. [CrossRef]
25. Rayanasukha, S.; Somboonkaew, A.; Sumriddetchkajorn, S.; Chaitavon, K.; Chanhorm, S.; Saekow, B.; Porntheeraphat, S. Self-compensation for the influence of working distance and ambient temperature on thermal imaging based temperature measurement. *IEEE Trans. Instrument. Meas.* **2021**, *70*, 1–6. [CrossRef]
26. Kim, S.; Kim, T. Radiometric temperature-based pedestrian detection for 24 hour surveillance. In Proceedings of the 2018 IEEE International Conference on Multimedia & Expo Workshops (ICMEW), San Diego, CA, USA, 23–27 July 2018.
27. Lee, P.J.; Bui, T.A.; Lo, C.C. Temperature model and human detection in thermal image. In Proceedings of the 2018 IEEE 7th Global Conference on Consumer Electronics (GCCE), Nara, Japan, 9–12 October 2018.
28. Mushahar, M.F.A.; Zaini, N. Human body temperature detection based on thermal imaging and screening using YOLO Person Detection. In Proceedings of the 2021 11th IEEE International Conference on Control System, Computing and Engineering (ICCSCE), Penang, Malaysia, 27–28 August. 2021.
29. Zhu, Z.; Huang, Y.; Chen, Z.; Huang, H. Non-contact infrared temperature detection and RFID technology access control design. *J. Phys. Conf. Ser.* **1982**, *1*, 012087. [CrossRef]
30. Bhogal, R.K.; Potharaju, S.; Kanagala, C.; Polla, S.; Jampani, R.V.; Yennem, V.B.R. Corona virus disinfectant tunnel using face mask detection and temperature monitoring. In Proceedings of the 2021 5th International Conference on Intelligent Computing and Control Systems (ICICCS), Madurai, India, 6–8 May 2021.
31. He, H.; Yan, Z.; Geng, Z.; Liu, X. Research on pedestrian tracking algorithm based on deep learning. In Proceedings of the 2021 International Conference on Computer Information Science and Artificial Intelligence (CISAI), Kunming, China, 17–19 September 2021.
32. Liu, S.; Yu, M.; Li, M.; Xu, Q. The research of virtual face based on deep convolutional generative adversarial networks using TensorFlow. *Phys. A Stat. Mech. Its Appl.* **2019**, *521*, 667–680. [CrossRef]
33. Zhang, Z. A flexible new technique for camera calibration. *IEEE Trans. Patt. Anal. Mach. Int.* **2000**, *22*, 1330–1334. [CrossRef]
34. Rubin, B.S.; Sathiesh, K. An efficient inception V2 based deep convolutional neural network for real-time hand action recognition. *IET Image Proc.* **2019**, *14*, 688–696.

35. Alqahtani, M.; Kumar, N.; Aljawder, D.; Abdulrahman, A.; Alnashaba, F.; Fayyad, M.A.; Alshaikh, F.; Alsahaf, F.; Saeed, S.; Almahroos, A.; et al. Randomized controlled trial of favipiravir, hydroxychloroquine, and standard care in patients with mild/moderate COVID-19 disease. *Sci. Rep.* **2022**, *12*, 4925. [CrossRef] [PubMed]
36. Dollár, P.; Appel, R.; Kienzle, W. Crosstalk cascades for framerate pedestrian detection. In Proceedings of the 12th European Conference on Computer Vision-Volume Part II, Heidelberg, Germany, 7 October 2012.

sensors MDPI

Article

MALS-Net: A Multi-Head Attention-Based LSTM Sequence-to-Sequence Network for Socio-Temporal Interaction Modelling and Trajectory Prediction

Fuad Hasan and Hailong Huang *

Department of Aeronautical and Aviation Engineering, The Hong Kong Polytechnic University, Hong Kong, China
* Correspondence: hailong.huang@polyu.edu.hk

Abstract: Predicting the trajectories of surrounding vehicles is an essential task in autonomous driving, especially in a highway setting, where minor deviations in motion can cause serious road accidents. The future trajectory prediction is often not only based on historical trajectories but also on a representation of the interaction between neighbouring vehicles. Current state-of-the-art methods have extensively utilized RNNs, CNNs and GNNs to model this interaction and predict future trajectories, relying on a very popular dataset known as NGSIM, which, however, has been criticized for being noisy and prone to overfitting issues. Moreover, transformers, which gained popularity from their benchmark performance in various NLP tasks, have hardly been explored in this problem, presumably due to the accumulative errors in their autoregressive decoding nature of time-series forecasting. Therefore, we propose MALS-Net, a Multi-Head Attention-based LSTM Sequence-to-Sequence model that makes use of the transformer's mechanism without suffering from accumulative errors by utilizing an attention-based LSTM encoder-decoder architecture. The proposed model was then evaluated in BLVD, a more practical dataset without the overfitting issue of NGSIM. Compared to other relevant approaches, our model exhibits state-of-the-art performance for both short and long-term prediction.

Keywords: vehicle trajectory prediction; autonomous driving; LSTM; transformer; multi-head attention

Citation: Hasan, F.; Huang, H. MALS-Net: A Multi-Head Attention-Based LSTM Sequence-to-Sequence Network for Socio-Temporal Interaction Modelling and Trajectory Prediction. *Sensors* **2023**, *23*, 530. https://doi.org/10.3390/s23010530

Academic Editor: Arturo de la Escalera Hueso

Received: 16 November 2022
Revised: 12 December 2022
Accepted: 19 December 2022
Published: 3 January 2023

1. Introduction

Autonomous driving has been gaining an increasing interest due to its guarantee of safer driving on the road. Recently, one of the core functions of autonomous driving that has been of particular interest in the literature is the future trajectory prediction of neighboring vehicles. In a congested highway driving scenario, where all cars are often driving at very high speeds, even a mildly reckless maneuver can consequentially result in a ripple effect causing a serious accident, which is why an autonomous car needs to forecast the future trajectories of its surrounding vehicles to assess the risk of its own future maneuvers.

However, predicting the future trajectories of surrounding vehicles has been extremely challenging in practice since the prediction is not only dependent on the historical trajectories of the vehicles but also the dynamic and complicated socio-temporal interdependence [1] of the dense web of vehicular traffic around the car, including other cars, buses, trucks, etc. For instance, in a dense highway setting, if one driver tries to change the lane, the neighboring lane driver might slow down to give way and the current lane driver will speed up to take the driver's place. Hence, to make an accurate prediction of the future trajectory, a proper model of the interaction between the participants needs to be utilized as one of the parameters for the prediction besides only raw historical trajectories of the surrounding vehicles.

To tackle this problem of complexity in modeling interaction, recent few years have seen the emergence of a plethora of deep learning-based data-driven models. To address the challenge of modeling the influence of neighboring vehicles on the target or ego vehicle, i.e. the social interaction, typical deep-learning-based operations that have been explored include various pooling techniques [2] with convolution [3] or graphs or both [4]. These have often been coupled with an RNN such as LSTM [5] or GRU [6] to exploit its inherent memory capability and extract temporal correlation. However, these models often struggle to model complex long-term temporal correlations. Transformers have been introduced to the literature with the promise of tackling the issue of long-term temporal correlation as well as parallelizing the decoding process. Inspired by its distinct attention mechanism, various attention-based techniques have been adopted in [1,7]. However, the multi-headed attention mechanism, originally proposed in the traditional transformer [8], has not extensively been explored in the highway trajectory prediction problem, mainly due to the problem of accumulative errors resulting from the autoregressive decoding procedure of transformers [9].

Moreover, most studies related to highway future trajectory prediction have mainly adopted the popular NGSIM dataset [10] to evaluate their performance. The NGSIM dataset was mainly collected to satisfy the need for data with explicit microscopic traffic flow and car-following information, for advancing traffic flow theory [11]. However, the NGSIM dataset was discovered to be extremely noisy in the literature by [12–14]. Various efforts have been made to smooth out the noise such as low pass filter, interpolation, etc. by [15]. Ref. [11] demonstrated that despite these efforts, the problem still persists and training on the NGSIM dataset thus has the potential possibility of resulting in a model with overfitted parameters, making it unfeasible to be deployed for practical application.

In this study, we thus propose a transformer-based LSTM sequence-to-sequence model to tackle the interaction-aware trajectory prediction problem. We exploit the main mechanism of the transformer, the multi-head attention (*MHA*), and implement it on an *MHA*-LSTM fused sequence-to-sequence architecture to tackle the autoregressive accumulative error problem of the transformer decoder. The encoder in our model is capable of encoding both the social and temporal interaction of the vehicles by implementing two Multi-Head Attention layers in the encoding process to put attention on both vehicle-vehicle and timestep-timestep graphs, followed by a traditional LSTM layer. This is then followed by another Multi-Head Attention-based decoder with an LSTM layer that is be able to decode by focusing on the socio-temporal interaction between the vehicles in successive decoding steps, which helps prevent the model from accumulating autoregressive decoding errors and also improve future trajectory prediction accuracy. We thus summarize the contribution and the academic as well as the practical novelty of our original work as follows.

1. In order to model the social dependence of past trajectories, we propose a Social Multi-Head Attention (*SMHA*) mechanism, and to model the temporal dependence, we use a successive Temporal Multi-Head Attention (*TMHA*) mechanism to focus attention on both social and temporal interaction and encode the input data.
2. A similar *MHA*-based LSTM decoding step is proposed to extract the predicted socio-temporal interaction in successive decoding steps, which improves the successive prediction accuracy and minimizes the accumulative errors of the transformer decoder.
3. The evaluation of the method has been performed on BLVD, a large-scale vehicle trajectory dataset that has less noise and is extracted from egocentric onboard-sensor data, to prevent overfitting. The experimental results demonstrate the superior performance of our model over state-of-the-art methods that have been implemented on both NGSIM and BLVD datasets.

The remainder of this article is constructed as follows. Section 2 provides an overview of the literature review on the general vehicle future trajectory prediction problem. Section 3 formulates the main trajectory prediction problem that we focus on and also provides an in-depth look on the model we propose. Section 4 mentions the implementation details and also reports our experimental results from both comparative and ablative studies. Finally, Section 5 concludes this paper and discusses future directions.

2. Related Work

The very first versions of the trajectory prediction literature focused on multiple physical model-based techniques. These essentially predict the future motion of a vehicle based on a form of kinematics, statistics or a fusion of both models. Some popular approaches include the Constant Velocity (CV) [16] and Constant Acceleration model (CA) [17], Kalman Filter-based motion model [18], Gaussian Mixture Model (GMM) [19] and Hidden Markov Model (HMM) [20]. This allowed [21] to utilize CV and CA to develop an intelligent driver model. Further works by [22] proposed a fusion of GMM with HMM to use parameters such as vehicle motion estimation, traffic patterns, and spatial interaction to predict the motion of the surrounding vehicles. The purpose of predicting the trajectories of surrounding or sometimes the ego vehicle has mainly been done to carry out risk assessments for safe automated driving, such as in [23,24], where Rapid Random Trees (RRT) and Linear Matrix Inequality (LMI) has often been implemented to analyze driving risk of lane change maneuvers or construct a human-machine shared control system [25]. However, it has been observed that physical models have a very short prediction horizon, beyond which the accuracy of the model becomes unfeasible [26]. Therefore, this has pushed the research toward the direction of data-driven methods.

The early works on data-driven models explored simpler learning-based models for interaction-aware trajectory prediction, such as support vector machines in [27,28] and Gaussian process regression [29]. However, these models need a handful of manually designed features to be completely constructed. As the incresing complexity of Spatio-temporal interdependence modelling was recognized, it became unfeasible to use handcrafted features and hyperparameters for these models. Thus deeper learning models, which can design and train their own features, proved to be a breakthrough.

At first, most deep learning models mainly focused on extracting temporal correlation via RNN architectures such as LSTM [30] and GRU [31], only focusing on the historical trajectory to make the future prediction [32,33]. Other deep-learning approaches adopted LSTMs to utilize their time-series modelling ability to make an energy-aware driving behavior analysis as well as predict the motion [34]. Due to the encoder-decoder sequence-to-sequence architecture of RNNs, they lacked the ability to model any sort of social or spatial interaction. More recent literature has thus mostly focused explicitly on coupling RNNs with some sort of social feature extraction architecture to model vehicle interaction. An approach adopted by [35] proposed a statistical method GMM fused with LSTMs to generate the leading-vehicle trajectory. Most popular such approaches, however, have mostly focused on other neural architectures such as CNNs and GNNs.

One of the widely used architectures to model social interaction has been the convolution and social pooling approach. One of the first social pooling approaches was proposed in [2] which used an LSTM along with a social pooling strategy (S-LSTM) to decode the prediction. In [36], a social GAN (SGAN) was used which utilizes both sequence-to-sequence architecture and a GAN to make the final prediction. A convolutional social pooling approach was proposed in [3] where the convolutional kernel-striding was applied to social pooling to improve on the LSTM social pooling approach in [2]. A CNN-LSTM approach was proposed in [37] which used an LSTM encoder-decoder architecture and performed a CNN operation on the hidden-layer tensors.

Graph-based architectures have also been frequently used for predicting future trajectories. Modelling the interaction-aware trajectory prediction of surrounding vehicles as graphs with nodes being the vehicles and the edges being the interaction between them has often been implemented in the literature. Such an approach was implemented in [6] which constructed a GNN to make the future trajectory prediction.

Attention-based networks, however, have been a recent breakthrough, because of their ability to rapidly extract important information from historical tracks. Ref. [38] applies a message-passing architecture to focus on pedestrian motion in order to make a future prediction. Another model which also focuses on pedestrian trajectory prediction is [39], wherein two attention layers are stacked to extract both spatial and temporal attention. Graphs have also been utilized in the attention mechanism. The reference [40] utilizes social graph attention to model both social and temporal interaction based on relative positions. However, the multi-head attention mechanism of the transformer does not appear to have been used as extensively in vehicle trajectory prediction. Its utilization mainly spans pedestrian trajectory prediction.

3. Methodology

3.1. Problem Formulation

In this paper, the future vehicle trajectory prediction problem is formulated as a non-linear regression task where the inputs to the model are the past trajectories of the observed neighbouring vehicles, which can be represented by

$$X = \{p_1, p_2, \ldots, p_T\}, \tag{1}$$

where

$$p_t = \left\{(x_t^0, y_t^0), (x_t^1, y_t^1), \ldots, (x_t^N, y_t^N)\right\}. \tag{2}$$

$p_t^i = (x_t^i, y_t^i)$ are the coordinates x and y of vehicle i at timestep t, where $t \in [1, T]$ and $i \in [1, N]$, T stands for the total length of the observed trajectory and N is the total number of observed neighbouring vehicles within 30 meters of the target vehicle, i.e. $\left|p_t^{target} - p_t^i\right| \leq 30$.

Based on the past trajectories, the objective of the proposed model is to learn a non-linear regression function that predicts the coordinates of all observed vehicles in the prediction horizon F, such that the predicted trajectories of the neighbouring vehicles are represented as $\hat{Y}$, as follows, which approximates the ground truth (GT) trajectory, $Y = \{q_{T+1}, q_{T+2}, \ldots, q_{T+F}\}$

$$\hat{Y} = \{\hat{q}_{T+1}, \hat{q}_{T+2}, \ldots, \hat{q}_{T+F}\}, \tag{3}$$

where

$$\hat{q}_t = \left\{(\hat{x}_t^0, \hat{y}_t^0), (\hat{x}_t^1, \hat{y}_t^1), \ldots, (\hat{x}_t^N, \hat{y}_t^N)\right\}. \tag{4}$$

The frame of reference of the dataset used in this paper is by default ego-centric as the trajectories were extracted from the onboard sensors. The longitudinal y-axis indicates the direction forward to the ego vehicle and the lateral x-axis direction indicates the axis perpendicular to it. The right-hand side is considered positive according to the dataset [41]. As a result, the model is independent of the curvature of the road which conveniently allows it to be used in the highway as long as an object-detection and a lane estimation algorithm is built on the target vehicle [3].

3.2. Network Architecture

The high-level architecture of the model we propose is shown in Figure 1. As discussed previously, it is structured as a sequence-to-sequence architecture capable of extracting socio-temporal correlation from past observed trajectories and then generating future trajectories based on that. Raw input $X = [p_1, p_2, \ldots, p_T]$ is first preprocessed into a suitable tensor. The encoder module then encodes the socio-temporal attention data into the input via two distinct multi-head-attention layers and then also encodes temporal memory via the LSTM layer. The decoder then decodes the encoded information into future predictions $\hat{Y} = [\hat{q}_{T+1}, \hat{q}_{T+2}, \ldots, \hat{q}_{T+F}]$. The hidden features are passed onto successive decoding steps via further *MHA* layers to improve further future predictions which also prevents the traditional transformer decoding accumulation error by enriching the hidden features.

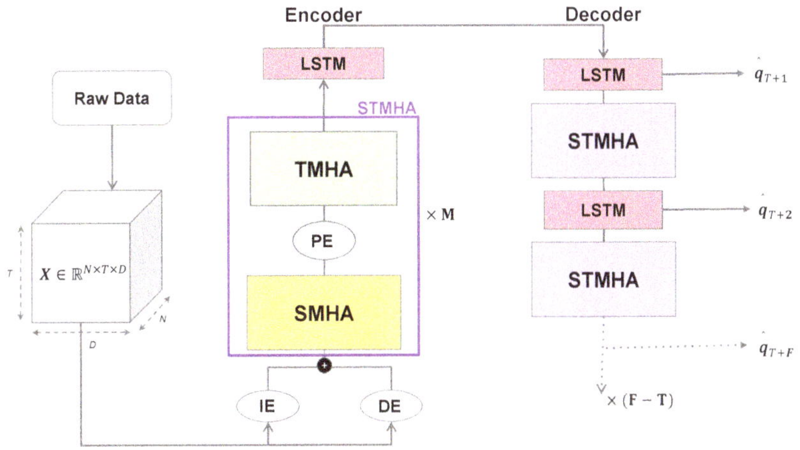

Figure 1. Proposed MALS-Net Architecture.

3.3. Input Representation

To input the trajectory data into our proposed model, the coordinates are first scaled to a range of (-1, 1) to aid the model to reach faster convergence [42]. The normalized coordinates are then preprocessed into a tensor $X \in \mathbb{R}^{N \times T \times D}$ where D is the feature space, in the case of our input having a value of 2 representing the (x, y) coordinates of the vehicles.

This tensor is then run through two different embedding layers simultaneously. The first embedding layer *PE* (see Figure 1 above), is a multi-layer perceptron (MLP) which, as shown in Equation (5), maps raw input features to a higher dimension D^{IE}. The second embedding layer *DE*, in Equation (6), maps the relative distance among each vehicle $p_t^i - p_t^j$, for $i, j \in [1, N]$ and $t \in [1, T]$, to a higher dimension D^{DE}. The two feature representations are then concatenated to produce the final input into the model with a feature space size $D_{model} = D^{IE} + D^{DE}$. The final input thus consists of both the embedded position and relative distance between vehicles as features before being run through the model.

$$IE_t^i = \text{MLP}\left(p_t^i, W_{IE}\right) \tag{5}$$

$$DE_t^i = \text{MLP}\left(p_t^i - p_t^j, W_{DE}\right) \tag{6}$$

3.4. Social Multi-Head Attention

The Social Multi-Head Attention (*SMHA*) layer, as shown in Figure 2, is proposed to extract the rich inter-vehicular social interaction information from the input. The features of the embedded input $X_{emb} \in \mathbb{R}^{N \times T \times D_{model}}$, through the last *IE* and *DE* layers, now contain rich embedded higher dimensional information about both the trajectories and relative distance of each vehicle. However, there is a multitude of features and extracting a viable correlation requires putting greater weight on features that actually contribute to a unit change in results. This work is done by an attention layer. This first multi-head attention layer enables the model to understand what features to put the most attention to. The embedded input is first transposed to $(X_{emb})^T \in \mathbb{R}^{T \times N \times D_{model}}$ so that we can extract attention weights from the last two dimensions $\mathbb{R}^{N \times D}$. The features of the input are then split into n_{heads} transformer heads as $\mathbb{R}^{n_{heads} \times T \times N \times \frac{D_{model}}{n_{heads}}}$ and fed into the Multi-Head Attention (*MHA*) layer. The *MHA* first computes the query (Q), key (K), and value (V) matrices at timestep t, as shown in Equation (7), via three MLPs which conserves the size of the feature space of the input as $\frac{D_{model}}{n_{heads}}$.

$$Q_t = \text{MLP}\left(\left\{x_t^i\right\}_{i=1}^{N}, w_q\right), \quad K_t = \text{MLP}\left(\left\{x_t^i\right\}_{i=1}^{N}, w_k\right), \quad V_t = \text{MLP}\left(\left\{x_t^i\right\}_{i=1}^{N}, w_v\right) \quad (7)$$

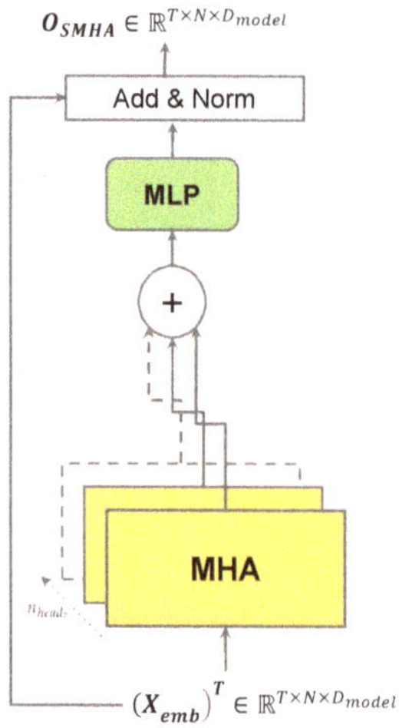

Figure 2. The *SMHA* block.

The weights w_q, w_k and w_v are initialized as samples from the Xavier Normal Distribution [43], $\mathcal{N}(0, \sigma^2)$ where

$$\sigma = k \times \sqrt{\frac{2}{l_{in} + l_{out}}}. \quad (8)$$

The value of k is the gain designed to be 1 and l_{in} and $lout$ are the input and output feature dimensions which in this case are both $\frac{D_{model}}{n_{heads}}$.

The inter-vehicular interaction is then represented by an undirected graph $G_t = \{V_t, E_t\}$ where the nodes V_t represent the observed vehicles $V_t = \{v \mid i = 1, 2, \ldots, N\}$ and the edges represent the interaction as binary values between vehicles i and j, $E_t[i][j] = 1$ if $p_t^i - p_t^j \leq d_{near}$, otherwise the value is zero. The interaction graph is used to mask social attention.

The masked attention of head h at timestep t is then computed as

$$\text{Attention}_h(Q_t, K_t, V_t) = \frac{\text{Softmax}\left(\left[(q_t^i)^T k_t^j\right]_{mask}\right)}{\sqrt{d_k}} \left[v_t^i\right]_{mask}, \quad (9)$$

where $mask = \{j \mid G_t[i,j] = 1, j \in [1, N]\}$. The masking based on the interaction graph G_t allows the attention scores to be calculated only when the vehicles are near the target, defined by the threshold d_{near}. In practicality, this is done to resemble a real driving scenario. A driver's decision is only based on cars that are observable and nearby. If it cannot observe a car and/or it is not nearby, it is not practical to calculate an interaction score between them as they do not share an interaction in that scenario. Therefore, we have to design the value of threshold d_{near} to resemble a region around the driver where the driving decision would be affected. The masked Multi-Head Attention on the social correlation can then be computed as

$$MHA(Q_t, K_t, V_t) = \text{Concat}(Attention_1, \ldots, Attention_{n_{heads}}). \quad (10)$$

The output from the Multi-Head Attention layer of size $O_{MHA} \in \mathbb{R}^{N \times T \times D_{model}}$ is then fed through an intra-layer MLP which helps boost training speed. This is then followed by an Add & Norm layer. The Add layer is simply a residual connection of the transposed input $(X_{emb})^T \in \mathbb{R}^{T \times N \times D_{model}}$ which adds the current output with the input. This connection greatly helps models such as transformers reach convergence faster by resolving the vanishing gradient problem so it is also extensively used in our model [44]. The output of the Social Multi-Head Attention Layer is thus computed as follows,

$$O_{SMHA} = \text{Norm}(\text{MLP}(O_{MHA}) + (X_{emb})^T). \tag{11}$$

The output O_{SMHA} has the same dimensions as the transposed Input $(X_{emb})^T \in T \times N \times D_{model}$.

3.5. Positional Encoding

To date, the *SMHA* layer has extracted the social correlation via the weights and biases of the layer. Thus, we now have to extract the temporal correlation from O_{SMHA}. However, in order to model temporal dependencies, the order is very important. Even though the multi-head attention mechanism is powerful enough to process a long sequence of data very quickly due to its attention graph-based architecture, in doing so, it loses its sense of the order of each point in a sequence. The order was not particularly useful in the *SMHA* step as it is redundant information for social interaction between vehicles, but it is, however, imperative for the output $O_{SMHA} \in \mathbb{R}^{T \times N \times D_{model}}$ to preserve its sense of order before being pushed through the *TMHA* step, otherwise it will produce huge errors in the decoding step. Therefore, we propose a Positional Encoding layer (*PE*) to inject the sequential order into the data via sinusoidal positional encoding. First we transpose O_{SMHA} back to $O_{SMHA} \in \mathbb{R}^{N \times T \times D_{model}}$ and then initialize the positional encoding matrix $P \in \mathbb{R}^{T \times D_{model}}$. Each scalar $P_t^d \mid t \in [0, T], d \in [0, D_{model}]$ of the Positional Encoding Matrix is a sinusoidal function as follows

$$\begin{cases} PE^i(t, 2d) = \sin\left(\dfrac{t}{10{,}000^{\frac{2d}{D_{model}}}}\right), \\ PE^i(t, 2d) = \cos\left(\dfrac{t}{10{,}000^{\frac{2d}{D_{model}}}}\right). \end{cases} \tag{12}$$

The above operation is then applied for $t \in [1, T]$ and then the positional encoding matrix $P \in \mathbb{R}^{T \times D_{model}}$ is added to each vehicle $i \subset [1, N]$, to obtain $O_{PE} \in \mathbb{R}^{N \times T \times D_{model}}$, which is the sequential order infused input to the *TMHA* layer.

3.6. Temporal Multi-Head Attention

In order to model and extract the temporal correlation of each vehicle from the input trajectory data, we propose another multi-head attention layer (*TMHA*), as illustrated in Figure 3, similar to the *SMHA*. The feature space of O_{PE} is first split into n_{heads} transformer heads as $\mathbb{R}^{n_{heads} \times T \times N \times \frac{D_{model}}{n_{heads}}}$ and fed into a Multi-Head Attention Layer (*MHA*). The *MHA* first feeds the input into three MLPs respectively in order to compute the query (Q), key (k), and value (V) matrices for vehicle i, as shown in Equation (13), which preserves the feature space of the input as $\frac{D_{model}}{n_{heads}}$.

$$Q_i = \text{MLP}\left(\left\{o_t^i\right\}_{t=1}^{T}, w_q\right), \quad K_t = \text{MLP}\left(\left\{o_t^i\right\}_{t=1}^{T}, w_k\right), \quad V_t = \text{MLP}\left(\left\{o_t^i\right\}_{t=1}^{T}, w_v\right) \tag{13}$$

The weights w_q, w_k and w_v are again initialized as samples from the Xavier Normal Distribution [43] again, $\mathcal{N}(0, \sigma^2)$, as discussed in Equation (8) above.

The masked attention of head h for the i-th vehicle is then computed as:

$$\text{Attention}_h(Q_i, K_i, V_i) = \frac{\text{Softmax}\left(\left[\left(q_t^i\right)^T k_u^i\right]_{u=1}^{T-1}\right)}{\sqrt{d_k}} \left[v_t^i\right]_{u=1}^{T-1}, \tag{14}$$

where $([(q_t^i)^T k_u^i]_{u=1}^{T-1})$ demonstrates the time-masking of the timesteps. Essentially, the time mask prevents the current steps from accessing features from the relative future. For example, if the current timestep is s then the features from timestep s to T will be masked as zero. This prevents the model from overfitting and making exclusive correlations on the training data.

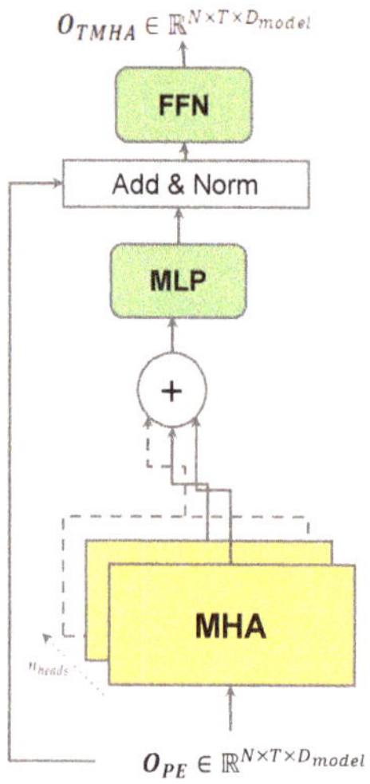

Figure 3. The *TMHA* block.

The Multi-Head Attention to the temporal dependency can then be calculated as

$$MHA(Q_i, K_i, V_i) = \text{Concat}(Attention_1, \ldots, Attention_{n_{heads}}). \tag{15}$$

The output from the Multi-Head Attention layer of size $O_{MHA} \in \mathbb{R}^{N\times T\times D_{model}}$ is then fed through an intra-layer MLP, followed by an Add & Norm layer, which provides the residual connection of the output from the last layer $O_{PE} \in \mathbb{R}^{N\times T\times D_{model}}$. After the residual connection and normalization, the output is fed through another multi-layer perceptron network (*FFN*). *FFN* is made up of two feed-forward networks that successively map the feature vectors to a higher dimension and then back to the original dimension such that output from MLP1 is $\mathbb{R}^{N\times T\times D_{FFN}}$ and MLP2 maps it back to $\mathbb{R}^{N\times T\times D_{model}}$. This is done to add more model parameters so that the temporal attention vectors can be fed through further layers such as an RNN. The output of the Temporal Multi-Head Attention Layer is thus computed as follows,

$$O_{TMHA} = FFN(\text{Norm}(\text{MLP}(O_{MHA}) + O_{PE}), W_{FFN}), \tag{16}$$

where

$$FFN = \text{MLP2}(\text{MLP1}(FFN_{in}, w_{MLP1}), w_{MLP2}). \tag{17}$$

Thus the output from the *TMHA* layer now has both social and temporal correlation encoded into it. We refer to the *SMHA*, *PE* and *TMHA* layers compounded together as *STMHA*. We then stack M number of *STMHA* layers successively to extract more complex socio-temporal dependence from the past trajectory information.

3.7. LSTM Encoder

We propose a traditional LSTM encoder after the *STMHA* layer stack to extract the hidden memory information from the output tensor $O_{STMHA} \in \mathbb{R}^{N\times T\times D_{model}}$. This is done to facilitate the hidden tensor passing to the LSTM decoder module. We propose this sequence-to-sequence architecture as a replacement for the traditional transformer architecture with an MHA-based transformer decoder. As mentioned before, due to the autoregressive nature of the transformer decoder, it carries the problem of accumulated errors. We thus use a sequence-to-sequence LSTM decoder to solve this problem. At first, the LSTM encoder layer recurrently computes the hidden-state tensor $H_t \in \mathbb{R}^{L \times N \times D_{hid}}$ at time t for $t \in [1, T]$ with L encoder layers and D_{hid} hidden dimensions as

$$o_t^i, h_t^i = \text{LSTM}(h_{t-1}^i, o_{t-1}^i, w_{t-1}^i). \tag{18}$$

After recurrently updating the hidden state for T timesteps, the $H_T \in \mathbb{R}^{L \times N \times D_{hid}}$ at the final timestep T is then passed on to the decoder module.

3.8. STMHA-*LSTM Decoder*

To decode the hidden-state tensor $H_T \in \mathbb{R}^{L \times N \times D_{hid}}$ and at the same time extract the rich socio-temporal dependence information encoded into the hidden tensor from the decoded information, we propose an $STMHA$-LSTM decoder, as demonstrated in Figure 4. This module is made up of an LSTM decoder architecture with a built-in multi-head attention mechanism to focus on the predicted behavior to improve further predictions at successive timesteps. The first layer of this architecture is another ordinary LSTM decoder which takes an input and a hidden tensor and decodes the output and updates the hidden tensor. The hidden input that is initialized for the decoder is the hidden tensor from the encoder, H_T, that was recurrently updated for timesteps $[1, T]$. As input, the decoder takes in the raw input data at timestep T, $X_T \in \mathbb{R}^{N \times 1 \times 2}$ with the last dimension denoting the coordinates of the position of the observed vehicles. It then decodes $H_{T+1} \in \mathbb{R}^{L \times N \times D_{hid}}$ for the next timestep $T + 1$, as follows,

$$\hat{q}_{T+1}^i, H_{T+1}^i = \text{LSTM}(H_T^i, o_T^i, w_T^i). \tag{19}$$

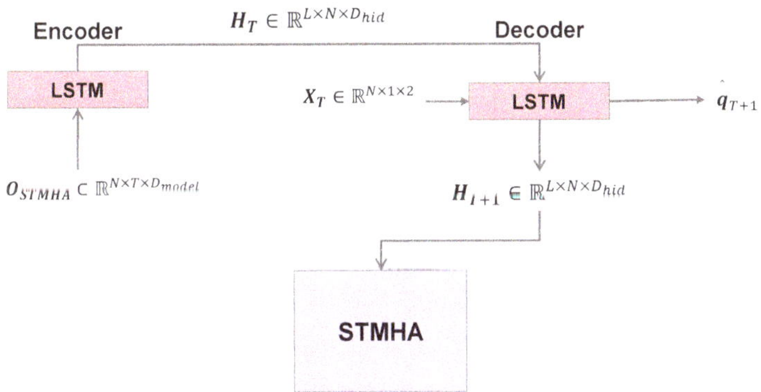

Figure 4. The RNN encoder-decoder block.

The hidden tensor at timestep $T + 1$ is then fed into an $STMHA$ layer to extract and update the socio-temporal dependence at the next decoding step. This was repeated for every following decoding step. This is done to further improve the accuracy of successive decoding steps, as well as decode the future trajectories from the hidden information of the transformer encoder with lower auto-regressive accumulated errors. Teacher-forcing is also applied to improve convergence. Each decoding step either takes the predicted output as inputs such as $\hat{q}_{T+1}$ or X_{T+1} depending on the teacher forcing ratio. This enables the model to decode the highly accurate future coordinates of each vehicle at every timestep.

4. Experiments

In this section, we will report our chosen parameters, hardware, and results from the experiments performed on the publicly available dataset BLVD in order to evaluate the performance of the predictions by our proposed MALS-Net architecture. We will also conduct comparative experiments with other state-of-the-art models as well as ablative comparisons with our chosen architecture.

4.1. Dataset

The proposed model was trained, validated, and tested on the BLVD dataset. It consists of a total of 654 high-resolution videos resulting in 120,000 points of data. It was extracted from Changshu city, Jiangsu province. Each frame consists of the ID, 3D coordinates, vehicle direction, and interactive behavior information of all observed vehicles by the ego. Due to the data being collected by onboard sensors, there is less filter noise in the data compared to NGSIM, with more realistic sensor noise, therefore making it more practical. To divide the dataset into training, validation, and testing sets, we follow [41]. The dataset contains various scenario categories of the ego vehicle. We only choose the scenario involving highways with both a high and low density of participants. Other than that, the dataset is also split between day and night. We concatenate and shuffle all these sub-datasets before feeding them into our model.

4.2. Implementation Details

We extensively used a desktop running Windows 11 with 3.8 GHz AMD Ryzen 7 CPU, 32 GB RAM, and an NVIDIA 3070Ti Graphics Card to build our model. To train our model, we utilized the parallel High-Performance Computing Service which is a Hong Kong Polytechnic University resource. The high-performance computing nodes are made up of several industry-grade GPUs, the exact model of which is unknown to the authors.

4.3. Hyperparameter Settings

For the observable region, we set it to be a 30 m radial region, based on the assumption that a human driver would not be able to see beyond a 30 m region around them. For the interaction graph G_t, we set the threshold d_{near} to be 15 m. We set the number of $STMHA$ layers $M = 2$ and n_{heads} of all the MHA modules to be 4 and the value of D_{model} in our model to be 128. The LSTM blocks all have a D_{hid} of 60. and a number of layers, $L = 2$. For the trajectory, we used the past timesteps T to be 3 s and future timestamps F to be 5 s. For model training, we used the Adam [45] optimizer with $\eta = 0.001$, $\beta = 0.999$. The learning rate used is 0.0001 and the batch size of 32. The teacher-forcing ratio used is 0.5.

4.4. Evaluation Metrics

In line with the existing literature [1–6,36,37,39,40] we adopted the root mean squared error (RMSE) between the prediction and the ground truth for ease of comparison of the model's performance with other state-of-the-art methods on the NGSIM dataset as well as for analyzing its performance with ground truth. RMSE at prediction time $t', t \in [T+1, \ldots, T+F]$ can be calculated as follows,

$$\text{RMSE}_t = \sqrt{\frac{1}{L}\sum_{l=1}^{L}(\hat{Y}_t^l - Y_t^l)^2}, \tag{20}$$

where $\hat{Y}$ is the predicted positions and Y is the Ground Truth Position of the l-th testing sample at timestep t' and L is the total length of the test set.

4.5. Ablative Analysis

To defend the effectiveness of our architecture, we perform a variety of different ablative experiments in this section. At first to verify the performance of our encoder module in the first encoding social correlation, then the temporal correlation and then memory information into the input, we carry out three distinct experiments. These three experiments are described as follows.

1. **MALSwoTA:** This variant of the model excludes the *TMHA* layer which extracts the temporal correlation. It thus also excludes the *PE* module and the output from the *SMHA* is directly fed into the LSTM encoder.
2. **MALSwoSA:** This variant of the model excludes the *SMHA* layer which extracts the social correlation. The *DE* module is also excluded from this variant as that contributed to the social correlation information included in the encoded tensor. The output from the *IE* layer is directly fed into the *PE* and then the *TMHA* layer with the dimension D_{model} after the input embedding (*IE*).
3. **MALSwoLE:** This variant of the model excludes the LSTM encoder. The hidden tensors for the LSTM decoder are prepared via some specific operations including an additional MLP that maps the D_{model} out of the *TMHA* layer to a size of D_{hid}.

The results in Table 1 compare the above three models with our proposed architecture MALS. The improvement score is based on the mean of the difference in RMSE over the 5 s of the prediction horizon. First, it can be observed that adding the *SMHA* layer improved the model performance by 39.6%. This shows that the significance of the *SMHA* layer in extracting the social context of the traffic scenario via the graph is crucial to the prediction and is thus a vital addition to our model. Additionally, the *TMHA* layer stands as even more important, with an improvement of almost 50% by adding the layer. This confirms that, in addition to the social context, more importantly, the prediction is mainly based on the historical trajectories and self-attention to the historical trajectories which our proposed model is proficient in extracting via the *TMHA* layer. Encoding the memory information also serves as important according to our ablative results. The LSTM encoder, encoding the hidden memory information improves the model performance by 23.5%. We can also infer that part of this improvement also comes from improving the accumulative errors caused by the transformer-based *STMHA* encoder if no LSTM encoder is used to subsequently extract the hidden memory information. Figure 5 illustrates the RMSE values of the three experiments and their corresponding improvement.

Table 1. Ablative Analysis of the Encoder Module.

Model	RMSE-1s	RMSE-2s	RMSE-3s	RMSE-4s	RMSE-5s	Average Improvement
MALSwoTA	0.89	1.78	3.05	4.66	6.48	47.7%
MALSwoSA	0.69	1.53	2.76	4.26	5.87	39.6%
MALSwoLE	0.58	1.26	2.11	3.17	4.48	23.5%
MALS-Net	0.48	1.01	1.60	2.31	3.36	

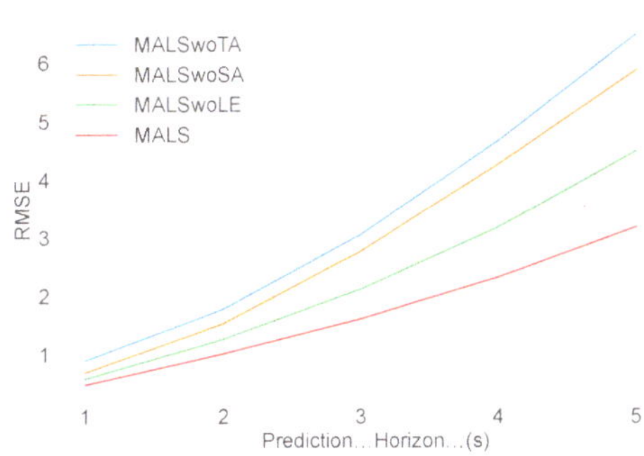

Figure 5. Comparison of RMSE values from the Ablative Analysis on the Encoder

Secondly, to assess the effectiveness of our proposed decoder architecture, we conduct two distinct experiments as follows.

1. **MALSwoLED:** This variant of the model is a version of the original transformer architecture. It excludes both the LSTM encoder and decoder. The encoded input from the *STMHA* is directly fed into another *STMHA*-based decoder similar to the transformer decoder and the right-shifted outputs are then fed into the decoder to make successive predictions.
2. **MALSwoLDA:** This variant excludes the *STMHA* block between successive decoding steps. The hidden information from the LSTM is passed onto the next decoding step without extracting further socio-temporal context from it in making successive timestep predictions.

At first, we can observe the effects of the LSTM encoder-decoder architecture. Compared to an ordinary transformer-based encoder-decoder architecture without any RNN, we can see the errors grow almost exponentially in successive decoding steps. This is the seeming effect of the accumulative errors that originates from the autoregressive nature of the transformer decoding. In Table 2, it is clear that adding the LSTM encoding-decoding to allow hidden information passing to make predictions is superior to the transformer-based encoding-decoding, especially in future timesteps. Adding an LSTM-based encoder-decoder thus improves both the model RMSE as well as autoregressive accumulating errors. Secondly, it is also evident how the *STMHA* layer for the LSTM decoder is also crucial in mitigating successive decoding errors. Adding this layer also showcases the increasing improvement of RMSE in successive decoding steps. This establishes that extracting the socio-temporal interdependence of the traffic in successive decoding steps is very important in further predicting trajectories further into the future. Figure 6 illustrates the RMSE values of these two experiments and their corresponding improvement.

Table 2. Ablative Analysis of the Decoder Module.

Model	RMSE-1s	RMSE-2s	RMSE-3s	RMSE-4s	RMSE-5s	Average Improvement
MALSwoLED	0.64	1.66	2.58	4.71	5.53	39.1%
MALSwoLDA	0.57	1.21	2.58	3.15	4.70	25.9%
MALS-Net	0.48	1.01	1.60	2.31	3.36	

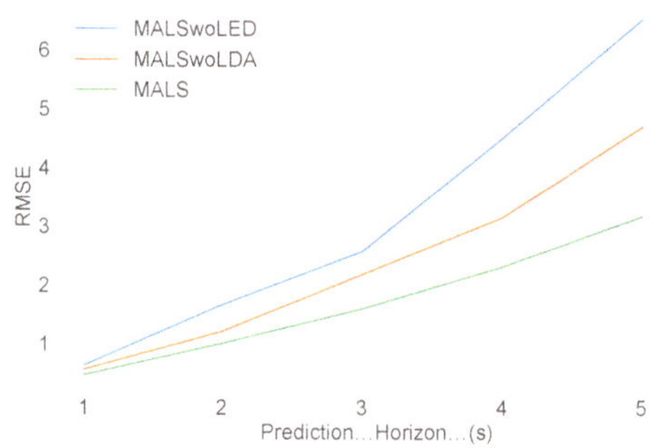

Figure 6. Comparison of RMSE values from the Ablative Analysis on the Decoder.

Our model also proposed a threshold to distinguish nearby vehicles from observed vehicles, which we call d_{near}. This threshold also limits the social influence range between vehicles, so properly designing this parameter is crucial to our proposed model. We thus also conducted further experiments to design the value of this threshold. We chose the value from a pool of four values $10, 20, 30, 40$. As shown in Table 3, excessively small values of d_{near} result in poor performance seemingly due to the fact that it ignores the realistic interaction between vehicles beyond the threshold. It is also observed that values larger than 30 do not produce a notable improvement, so we chose the value of 30 to represent d_{near}.

Table 3. Ablative Analysis of the Nearby Distance Threshold.

d_{near} (m) Mean	RMSE-1s	RMSE-2s	RMSE-3s	RMSE-4s	RMSE-5s
10	0.51	1.42	2.38	3.67	4.70
20	0.46	1.25	1.98	2.86	3.91
30	0.43	1.21	1.96	2.86	3.88
40	0.44	1.20	1.95	2.86	3.86

4.6. Comparative Analysis

We use the following models in the literature to compare our model.

- **CV** [16]: This method assumes a constant velocity and applies a Kalman Filter to predict future trajectories.
- **V-LSTM** [3]: This method uses a simple LSTM-based encoder-decoder model to make predictions.
- **S-LSTM** [3]: This method uses a social pooling technique to sum the neighboring vehicle features via an LSTM to predict trajectories.
- **CS-LSTM** [3]: This method models the traffic in grids and utilizes the convolution operation to extract social interaction and predict future trajectories.
- **DSCAN** [46]: This method uses a constraint network and models attention between vehicles to extract the weights to make future predictions.
- **SGAN** [36]: This method uses an adversarial network architecture that utilizes an encoder-decoder structure as well as a discriminator to make trajectory predictions.
- **HMNet** [47]: This model utilizes a hierarchical context-free LSTM encoder-decoder to forecast the trajectories.

We demonstrate the performance of our model compared to the above models in Table 4. The capability of our model to use the strength of the transformer network and model socio-temporal interaction without dealing with autoregressive errors is demonstrated in the table. Our model seemingly outperforms all other baselines, with specifically significant performance in the fourth and fifth-second prediction horizons, establishing the strength of our model in long-term predictions.

Table 4. Comparative Analysis.

Model Average Improvement	RMSE-1s	RMSE-2s	RMSE-3s	RMSE-4s	RMSE-5s
CV	0.73	1.78	3.13	4.78	6.68
V-LSTM	0.68	1.65	2.91	4.46	6.27
S-LSTM	0.65	1.31	2.16	3.25	4.55
CS-LSTM	0.61	1.27	2.09	3.10	4.37
DSCAN	0.58	1.26	2.03	2.98	4.13
SGAN	0.57	1.32	2.22	3.26	4.40
HMNet	0.50	1.13	1.89	2.85	4.04
MALS-Net	0.48	1.01	1.60	2.31	3.36

4.7. Prediction Visualization

RMSE is generally an effective indicator of performance but visualization is often needed to analyze the strengths and weaknesses of a model. Thus, in this section, we provide a visualization of some test cases to better understand our model's performance in depth, by comparing predicted trajectories with ground truth. Figure 7 demonstrates three distinct scenes each with different levels of our model performance, with the trajectories marked in different colors. The first scene demonstrates a typical congested highway driving scenario. The ego here is surrounded by five other vehicles, all contributing to the interaction-modeling of the ego. With enough social interaction context, our model seems to perform very well with minimal difference in the ground truth and predicted trajectories.

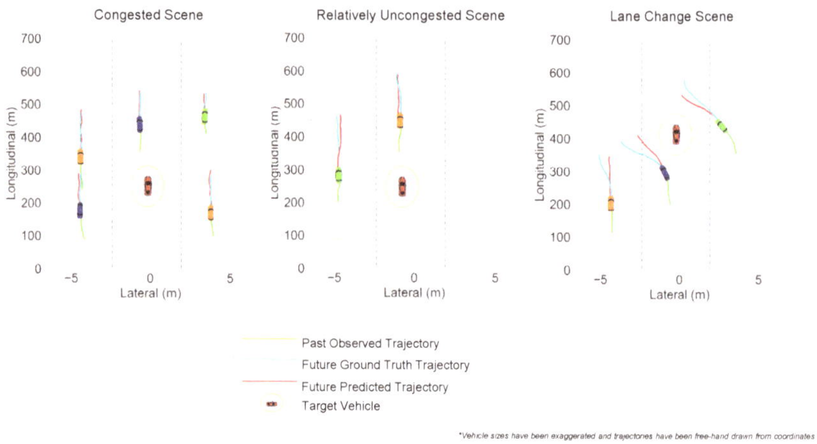

Figure 7. Prediction Visualization of three different scenes.

The second scene represents a fairly uncongested scene compared to the first. Our proposed model also seems to perform relatively well in this case with a negligible deviation of the predicted trajectory from the ground truth. However, the performance, in contrast to the first scene, is relatively lower. We suppose it is due to the relatively much lower interaction context. Because there are not enough cars, our proposed socio-temporal interaction modeling does not produce perfectly predicted trajectories with a negligible yet noticeable deviation between the predicted and the ground truth.

The third case illustrates a scenario where there are multiple lane change cases. The performance of the model, in this case, is relatively poor due to the fact that the exact trajectory is ambiguous even though the model predicts a possible lane change. It also failed to capture the deviation of the path of the yellow vehicle due to the blue one taking its lane in front. We also believe there are not enough lane change cases in the highway dataset of BLVD which possibly also contributes to the poorer performance. We think creating a behavior prediction branch in the model to predict behaviors and then feeding the behaviors back into the model to improve the interaction prediction can improve the model's performance on lane change scenarios, due to the fact that predicting the exact time when an intention-change will occur can improve the lane change path prediction. We also think that pre-training the model on some datasets such as HighD [48] with more lane change scenarios may mitigate some issues of extreme cases.

5. Conclusions

In this article, we proposed a transformer-based LSTM encoder-decoder network to model the socio-temporal interaction and predict the future trajectories of surrounding vehicles. We used the multi-head mechanism of transformers to efficiently extract the social and temporal interaction individually and have encoded it into the input as hidden information via the LSTM encoder in the encoder module. We have then used the decoder module to decode the hidden information, using another multi-head attention layer on the decoder to improve the successive decoding accuracy. We trained and evaluated our model on a practical, ego-centered, large-scale dataset that is derived from onboard sensor data. We conducted extensive studies, including both ablative and comparative studies. Our experiments verified that our model outperforms all other previous models. Ablative studies confirmed that our model solves the accumulative error caused by the transformer's autoregressive decoding behavior. The visualization results showed our model's strength in difficult congested scenarios, as well as its limitation in lane-change path predictions. The proposed model can be implemented in a practical driving scenario to predict the future trajectories of surrounding vehicles based on their historical tracks, provided that there is an object detection system such as YOLO is in place. In the future, we plan to enhance the model's performance on exact lane-change tracks by better modeling the driver-intention and feeding it into the interaction procedure. As another potential future path, this model can also be extended to involve more traffic participants such as cyclists and pedestrians and predict their behavior as well. Currently, this model can only be utilized in a freeway driving scenario. Adopting this model for urban driving, incorporating more complex information such as lane types, spatial HD maps, and traffic lights may be another future direction worth pursuing.

Author Contributions: Conceptualization, H.H.; methodology, F.H.; investigation, F.H.; resources, H.H.; writing—original draft preparation, F.H.; writing—review and editing, H.H.; visualization, F.H.; supervision, H.H.; project administration, H.H. All authors have read and agreed to the published version of the manuscript.

Funding: This research was funded by The Hong Kong Polytechnic University College of Undergraduate Researchers & Innovators (PolyU CURI)'s Undergraduate Research & Innovation Scheme (URIS). The Project Fund number is 9-AAE-44. Any details can be found in the following link: https://www.polyu.edu.hk/gs/ug-research/curi/about-curi/ (accessed on 15 November 2022).

Institutional Review Board Statement: Not applicable.

Informed Consent Statement: Not applicable.

Data Availability Statement: This study was evaluated on the publicly available dataset BLVD https://drive.google.com/file/d/1fbZsIHM8Yq8TE2avDzT8xuWIBfrcFgQC/view?usp=sharing, accessed on 15 August 2022.

Conflicts of Interest: The authors declare no conflict of interest.

Abbreviations

The following abbreviations are used in this manuscript:

CNN	Convolutional Neural Network
RNN	Recurrent Neuran Network
GNN	Graph Neural Network
NLP	Natural Language Processing
LSTM	Long Short-Term Memory
GRU	Gated Recurrent Unit

References

1. Messaoud, K.; Yahiaoui, I.; Verroust-Blondet A.; Nashashibi, F. Attention based Vehicle Trajectory Prediction. *IEEE Trans. Intell. Vehicles* **2021**, *6*, 175–185. [CrossRef]
2. Alahi, A.; Goel, K.; Ramanathan, V.; Robicquet, A.; Fei-Fei, L.; Savarese, S. Social LSTM: Human Trajectory Prediction in Crowded Spaces. In Proceedings of the IEEE Conference on Computer Vision and Pattern Recognition (CVPR), San Francisco, CA, USA, 18–20 June 2016; pp. 961–971.
3. Deo N.; Trivedi, M.M. Convolutional Social Pooling for Vehicle Trajectory Prediction. In Proceedings of the IEEE/CVF Conference on Computer Vision and Pattern Recognition Workshops (CVPRW), Salt Lake City, UT, USA, 18–22 June 2018; pp. 1468–1476.
4. Wang, R.; Li, M.; P. Zhang P.; Wen, F. Graph Partition Convolution Neural Network for Pedestrian Trajectory Prediction. In Proceedings of the 2021 IEEE 33rd International Conference on Tools with Artificial Intelligence (ICTAI), Washington, DC, USA, 1–3 November 2021; pp. 8994–9003.
5. Wang, Y.; Zhao, S.; Zhang, R.; Cheng, X.; Yang, L. Multi-Vehicle Collaborative Learning for Trajectory Prediction with Spatio-Temporal Tensor Fusion. *IEEE Trans. Intell. Transp. Syst.* **2020**, *23*, 236–248. [CrossRef]
6. Li, X.; Ying X.; Chuah, M. C. GRIP++: Enhanced Graph-based Interaction-Aware Trajectory Prediction for Autonomous Driving. *arXiv* **2019**, arXiv:1907.07792.
7. Zhou, H.; Ren, D.; Xia, H.; Fan, M.; Yang X.; Huang, H. AST-GNN: An Attention-based Spatio-Temporal Graph Neural Network for Interaction-Aware Pedestrian Trajectory Prediction. *Neurocomputing* **2021**, *445*, 298–308. [CrossRef]

8. Vaswani A.; Shazeer, N.; Parmar, N.; Uszkoreit, J.; Gomez, A. N.; Kaiser, L.; Polosukhin, I. Attention Is All You Need. IEEE *Adv. Neural Inf. Process. Syst.* **2017**, 5998–6008.
9. Chen, K.; Chen, G.; Xu, D.; Zhang, L.; Huang Y.; Knoll, A. NAST: Non-autoregressive Spatial-Temporal Transformer for Time Series Forecasting. *arXiv* **2021**, arXiv:2102.05624.
10. Next Generation Simulation (NGSIM). Available online: http://ops.fhwa.dot.gov/trafficanalysistools/ngsim.html (accessed on 20 September 2022).
11. Coifman B.; Li, L. A Critical Evaluation of the Next Generation Simulation (NGSIM) Vehicle Trajectory Dataset. *Transp. Res. Part B Methodol.* **2017**, *105*, 362–377. [CrossRef]
12. Thiemann, C.; Treiber, M.; Kesting A. Estimating Acceleration and Lane-Changing Dynamics from Next Generation Simulation Trajectory Data. *Trans. Res. Rec.* **2008**, *2088*, 90–101. [CrossRef]
13. Hamdar, S.; Mahmassani H. Driver Car-Following Behavior: From Discrete Event Process to Continuous Set of Episodes. In Proceedings of the 87th Annual Meeting of the Transportation Research Board, Washington, DC, USA, 13–17 January 2008.
14. Duret, A.; Buisson, C.; Chiabaut N. Estimating Individual Speed-Spacing Relationship and Assessing Ability of Newell's Car-following Model to Reproduce Trajectories. *Trans. Res. Rec.* **2008**, *2088*, 188–197. [CrossRef]
15. Montanino, M.; Punzo V. Trajectory Data Reconstruction and Simulation-based Validation against Macroscopic Traffic Patterns. *Trans. Res. Part B* **2015**, *80*, 82–106. [CrossRef]
16. Fei, R.; Li, S.; Hei, X.; Xu, Q.; Zhao, J.; Guo, Y. A Motion Simulation Model for Road Network based Crowdsourced Map Datum. *J. Intell. Fuzzy Syst.* **2020**, *38*, 391–407. [CrossRef]
17. Houenou, A.; Bonnifait, P.; Cherfaoui, V.; Yao, W. Vehicle Trajectory Prediction based on Motion Model and Maneuver Recognition. In Proceedings of the 2013 *IEEE*/RSJ International Conference on Intelligent Robots and Systems, Tokyo, Japan, 3–7 November 2013; pp. 4363–4369.
18. Elnagar, P. Prediction of Moving Objects in Dynamic Environments using Kalman Filters. In Proceedings pf the 2001 *IEEE* International Symposium on Computational Intelligence in Robotics and Automation (Cat. No.01EX515), Banff, AB, Canada, 29 July–1 August 2001; pp. 414–419.
19. Qiao, S.J.; Jin, K.; Han, N.; Tang, C.J.; Gesangduoji, G. Trajectory Prediction Algorithm based on Gaussian Mixture Model. *J. Softw.* **2015**, *26*, 1048–1063.
20. Qiao, S.; Shen, D.; Wang, X.; Han, N.; Zhu, W. A Self-Adaptive Parameter Selection Trajectory Prediction Approach via Hidden Markov Models. IEEE *Trans. Intell. Transp. Syst.* **2014**, *16*, 284–296. [CrossRef]
21. Treiber, M.; Hennecke, A.; Helbing, D.; Congested Traffic States in Empirical Observations and Microscopic Simulations. *Phys. Rev. E* **2000**, *62*, 1805. [CrossRef] [PubMed]
22. Deo, N.; Rangesh, A.; Trivedi, M.M. How Would Surround Vehicles Move? a unified framework for maneuver classification and motion prediction. IEEE *Trans. Intell. Veh.* **2018**, *3*, 129–140. [CrossRef]
23. Huang C.; Huang, H.; Zhang, J.; Hang, P.; Hu, Z.; Lv, C. Human-Machine Cooperative Trajectory Planning and Tracking for Safe Automated Driving. IEEE *Trans. Intell. Transp. Syst.* **2022**, *23*, 12050–12063. [CrossRef]
24. Huang C.; Hang, P.; Hu, A.; Lv, C. Collision-Probability-Aware Human-Machine Cooperative Planning for Safe Automated Driving IEEE *Trans. Veh. Technol.* **2021**, *70*, 9752–9763. [CrossRef]
25. Zhang, Y.; Hang, P.; Huang, C.; Lv, C. Human-Like Interactive Behavior Generation for Autonomous Vehicles: A Bayesian Game-Theoretic Approach with Turing Test. *Adv. Intell. Syst.* **2022**, *4*, 2100211. [CrossRef]
26. Gomes, I.; Wolf, D. A Review on Intention-aware and Interaction-aware Trajectory Prediction for Autonomous Vehicles. *TechRxiv* **2022**, *14*. Available online: https://www.techrxiv.org/articles/preprint/A_Review_on_Intention-aware_and_Interaction-aware_Trajectory_Prediction_for_Autonomous_Vehicles/19337447/1 (accessed on 15 Novenber 2022). [CrossRef]
27. Tomar, R.S.; Verma S.; Tomar, G.S. SVM Based Trajectory Predictions of Lane Changing Vehicles. In Proceedings of the 2011 International Conference on Computational Intelligence and Communication Networks, Gwalior, India, 7–9 October 2011; pp. 716–721.
28. Chen, X.; Yang, J.; Ye Q.; Liang, J. Recursive Projection Twin Support Vector Machine via Within-class Variance Minimization. *Pattern Recognit.* **2011**, *44*, 2643–2655. [CrossRef]
29. Goli, S.A.; Far B.H.; Fapojuwo, A.O. Vehicle Trajectory Prediction with Gaussian Process Regression in Connected Vehicle Environment. In Proceedings of the 2018 *IEEE* Intelligent Vehicles Symposium (IV), hangshu, China, 26–30 June 2018; pp. 550–555.
30. Hochreiter, S.; Schmidhuber, J. Long Short-Term Memory. *Neural Comput.* **1997**, *9*, 1735–1780. [CrossRef]
31. Chung, J.; Gulcehre, C.; Cho, K.; Bengio, Y. Empirical Evaluation of Gated Recurrent Neural Networks on Sequence Modeling. *arXiv* **2016**, arXiv:1412.3555.
32. Althoff, M.; Mergel, A. Comparison of Markov Chain Abstraction and Monte Carlo Simulation for the Safety Assessment of Autonomous Cars. IEEE *Trans. Intell. Transp. Syst.* **2011**, *12*, 1237–1247. [CrossRef]
33. Hillenbrand, J.; Spieker, A.M.; Kroschel, K. A Multilevel Collision Mitigation Approach—Its Situation Assessment, Decision Making, and Performance Tradeoffs. IEEE *Trans. Intell. Transp. Syst.* **2006**, *7*, 528–540. [CrossRef]
34. Xing, Y.; Lv, C.; Mo, X.; Hu, Z.; Huang, C.; Hang, P.; Toward Safe and Smart Mobility: Energy-Aware Deep Learning for Driving Behavior Analysis and Prediction of Connected Vehicles IEEE *Trans. Intell. Transp. Syst.* **2021**, *22*, 4267–4280. [CrossRef]
35. Xing, Y.; Huang, C.; Lv, C.; Liu, Y.; Wang, H.; Cao, D. *A Personalized Deep Learning Approach for Trajectory Prediction of Connected Vehicles*; SAE International: Warrendale, PA, USA, 2020. [CrossRef]

36. Gupta, A.; Johnson, J.; Fei-Fei, L.; Savarese, S.; Alahi, A. Social GAN: Socially Acceptable Trajectories with Generative Adversarial Networks. In Proceedings of the *IEEE* Conference on Computer Vision and Pattern Recognition, Salt Lake City, UT, USA, 18–23 June 2018; pp. 2255–2264
37. Mo, X.; Xing Y.; Lv, C. Interaction-Aware Trajectory Prediction of Connected Vehicles using CNN-LSTM Networks. In Proceedings of the *IECON* 2020 The 46th Annual Conference of the *IEEE* Industrial Electronics Society, Singapore, 18–21 October 2020; pp. 5057–5062.
38. Zhang, P.; Ouyang, W.; Zhang, P.; Xue, J.; Zheng, N. SR-LSTM: State Refinement for LSTM towards Pedestrian Trajectory Prediction. *arXiv* **2019**, arXiv:1903.02793.
39. Yu, C.; Ma, X.; Ren, J.; Zhao, H.; Yi, S. Spatio-Temporal Graph Transformer Networks for Pedestrian Trajectory Prediction. In *European Conference on Computer Vision*; Springer: Cham, Switzerland, 2020; Volume 12357, pp. 507–523.
40. Pang, Y.; Zhao, X.; Hu, J.; Yan, H.; Liu, Y. Bayesian Spatio-Temporal Graph Transformer Network (B-Star) for Multi-Aircraft Trajectory Prediction. Available online: https://ssrn.com/abstract=3981312 (accessed on 30 September 2022).
41. Xue, J.; Fang, J.; Li, T.; Zhang, B.; Zhang, P.; Ye, Z.; Dou, J. Blvd: Building a Large-Scale 5D Semantics Benchmark for Autonomous Driving. *arXiv* **2019**, arXiv:1903.06405.
42. Santurkar, S.; Tsipras, D.; Ilyas, A.; Madry, A. How Does Batch Normalization Help Optimization? *arXiv* **2018**, arXiv:1805.11604v5.
43. Glorot, X.; Bengio, Y. Understanding the difficulty of training deep feedforward neural networks. *J. Mach. Learn. Res. Proc. Track* **2010**, *9*, 249–256.
44. Pascanu, R.; Mikolov, T.; Bengio, Y. On the difficulty of training Recurrent Neural Networks. *arXiv* **2012**, arXiv:1211.5063.
45. Kingma, D.P.; Ba, J. Adam: A Method for Stochastic Optimization. *arXiv* **2012**, arXiv:1412.6980
46. Yu, J.; Zhou, M.; Wang, X.; Pu, G.; Cheng, C.; Chen, B. A Dynamic and Static Context-Aware Attention Network for Trajectory Prediction. *ISPRS Int. J. Geo-Inf.* **2021**, *10*, 336. [CrossRef]
47. Xue, Q.; Li, S.; Li, X.; Zhao, J.; Zhang, W. Hierarchical motion encoder–decoder network for trajectory forecasting. *arXiv* **2021**, arXiv:2111.13324.
48. Krajewski, R.; Bock, J.; Kloeker, L.; Eckstein, L. The highD dataset: A drone dataset of naturalistic vehicle trajectories on German highways for validation of highly automated driving systems. In Proceedings of the 2018 21st International Conference on Intelligent Transportation Systems (ITSC), Maui, HI, USA, 4–7 November 2018; pp. 2118–2125.

sensors

MDPI

Article

Criticality Assessment Method for Automated Driving Systems by Introducing Fictive Vehicles and Variable Criticality Thresholds

Demin Nalic [1], Tomislav Mihalj [2,*], Faris Orucevic [2], Martin Schabauer [2], Cornelia Lex [2], Wolfgang Sinz [3] and Arno Eichberger [2]

1 ADAS Department, MQS Automotive AT OG, 8074 Raaba, Austria
2 Institute of Automotive Engineering, Graz University of Technology, 8010 Graz, Austria
3 ADAS Simulation and Data Analysis Department, MAGNA Steyr Fahrzeugtechnik AG Co. & KG, 8041 Graz, Austria
* Correspondence: tomislav.mihalj@tugraz.at

Abstract: The safety approval and assessment of automated driving systems (ADS) are becoming sophisticated and challenging tasks. Because the number of traffic scenarios is vast, it is essential to assess their criticality and extract the ones that present a safety risk. In this paper, we are proposing a novel method based on the time-to-react (TTR) measurement, which has advantages in considering avoidance possibilities. The method incorporates the concept of fictive vehicles and variable criticality thresholds (VCTs) to assess the overall scenario's criticality. By introducing variable thresholds, a criticality scale is defined and used for criticality calculation. Based on this scale, the presented method determines the criticality of the lanes adjacent to the ego vehicle. This is performed by placing fictive vehicles in the adjacent lanes, which represent copies of the ego. The effectiveness of the method is demonstrated in two highway scenarios, with and without trailing vehicles. Results show different criticality for the two scenarios. The overall criticality of the scenario with trailing vehicles is higher due to the decrease in avoidance possibilities for the ego vehicle.

Keywords: fictive vehicles; safety assessment; scenario criticality; automated driving

Citation: Nalic, D.; Mihalj, T.; Orucevic, F.; Schabauer, M.; Lex, C.; Sinz, W.; Eichberger, A. Criticality Assessment Method for Automated Driving Systems by Introducing Fictive Vehicles and Variable Criticality Thresholds. *Sensors* **2022**, *22*, 8780. https://doi.org/10.3390/s22228780

Academic Editors: Yafei Wang, Chao Huang, Peng Hang, Zhiqiang Zuo and Bo Leng

Received: 29 September 2022
Accepted: 9 November 2022
Published: 14 November 2022

Publisher's Note: MDPI stays neutral with regard to jurisdictional claims in published maps and institutional affiliations.

1. Introduction

The assessment of ADS is essential for the approval and testing procedures. Due to a high scenario space and the complexity of ADS that needs to be tested for safety approval, scenario-based methods are promising and widely used approaches in the field of research. Recent studies, for example, [1,2], are focused on simulation-based testing and the generation of scenarios to reduce the testing effort for ADS. The reduction is achieved by identifying relevant safety-critical scenarios (SCS) using a variety of safety metrics and scenario-generation methods. A comprehensive review of current studies related to scenario-based approaches is presented in [3]. A scenario in the context of this work represents a series of actions occurring over a period of time, as defined in [4]. Other similar definitions are introduced and can be found, for example, in [5,6]. After extracting scenarios from a synthesized or real dataset, the scenarios are analyzed and assessed on their criticality. There are different definitions of what the criticality of a certain situation means. According to [7], accident probability combined with severity defines the criticality of a certain situation. The probability which is used here takes into account the avoidance of a particular situation. This approach enables a better interpretation of a specific situation. To calculate accurate avoidance possibilities, environment parameters, which are composed of static and dynamic objects, should be considered for evaluating the criticality of the situation. The available probabilistic methods, such as those in [8–10], calculate possible evasive maneuvers or accident probabilities; based on these, they estimate whether a situation is evasive, critical, or fulfills safety criteria. This consideration is advantageous on the one hand because it takes into account a large number of possibilities and the probability or feasibility of these possibilities in the calculation. On the other hand, due to complex solutions to mathematical problems, real-time capabilities and unique solutions are only sometimes given. To overcome such issues, we introduce a novel assessment method that introduces fictive ego vehicles on the left and right lanes adjacent to the tested ego vehicle. These fictive ego vehicles represent the exact copy of the ego vehicle, i.e., the vehicle configuration and the automated driving functions are the same. In this constellation, the ego vehicle and the neighboring fictive ego vehicles are assessed individually in the respective lane according to the introduced criticality scale. Combining the criticalities of the ego and fictive vehicle, an overall criticality of the situation can be calculated. To calculate the criticality, we take the avoidability of a collision into account using the TTR metric. The TTR metric consists of the time-to-brake (TTB) and time-to-steer (TTS) measurements, which represent the times left to avoid a critical situation by braking or steering, respectively. For SCS assessment, usually, the fixed metric thresholds are defined. However, metrics often depend on the kinematic relations between the ego and surrounding objects, such as distances and velocities. Although time-based metrics capture the influence of both distance and velocities, thresholds remain changeable depending on the object states. To overcome this issue with TTR, we are introducing variable criticality thresholds (VCTs) that determine a criticality scale. Finally, combining the concept of fictive ego vehicles, the TTR, and the VCTs, we derived the novel assessment method presented in this work. In the first Section 2, we present a detailed analysis of safety metrics used in scenario-based approaches for the assessment of SCS and a detailed description of the usage of the TTR. Using the TTR the criticality definition and the assessment method are presented in Section 3. As the last step of the work, we demonstrate the effectiveness of the presented method based on a safety-relevant test example.

2. Safety Metrics

According to [11], safety metrics are classified into temporal-, distance- and deceleration-based safety metrics. In [12], additional categorizations are introduced: the statistics-based, potential-field-based, and unexpected-driving-behavior-based metrics. These reviews show that temporal safety metrics are often and widely used to assess scenarios. The leading representative that is often mentioned in the context of the criticality of scenarios is the TTC. The main advantage of the TTC lies in its simplicity and efficiency for rear-end scenarios. The main disadvantages of the TTC criteria are the insensitivity to lateral collision risks and the fact that it cannot reflect the potential risk when the relative speed between two vehicles is zero. This implies that a possible intervention of a driver in safety-critical situations is not taken into account by the TTC. Therefore, various adjustments and adaptations of the classic TTC time were made; for example, a modified time-to-collision (MTTC) measure was introduced in [13]. Compared with the conventional TTC, the MTTC considers relative acceleration and relative velocity to improve collision prediction. In [14], the worst-time-to-collision (WTTC) metric was introduced. The WTTC considers the influence of vehicle dynamics and avoidance possibilities of the ego vehicle with the worst-case assumption regarding vehicle behavior. Another approach based on the TTC is presented in [15], where a time-exposed TTC (TET) and the time-integrated TTC (TIT) are introduced. These two metrics are extensions of the TTC and take the whole trajectory of vehicles over space and time into account. In [16], the TTR was introduced for the development and evaluation of forward-collision-avoidance systems, which consider a TTS, a TTS, and a TTK. In general, these reaction times determine the remaining time for successful collision avoidance by braking, steering, or acceleration. A similar approach is described in [17]; here, in addition to the TTS, a requested acceleration is introduced and used for the assessment of certain scenarios for ADS. The main advantages of these approaches compared with conventional temporal metrics such as the TTC and extensions of the TTC [18,19] are in consideration of avoidance possibilities and kinematic behavior of vehicles. The later the driver reacts successfully to an impending collision, the more the system is forced to exploit the vehicle's physical limits. Using the TTR as a safety metric offers the possibility to interpret and define the criticality in a wide range of safety-relevant scenarios. The present work represents a continuation of previous studies of [16,17,20], where longitudinal and lateral kinematics are taken into account for a criticality assessment of ADS.

Time-to-react is defined as $\mathrm{TTR} = \max\{\mathrm{TTB}, \mathrm{TTS}\}$, where TTB and TTS are the time-to-brake and time-to-steer metrics, respectively. They represent time reserves to the last point to brake (LPTB) and last point to steer (LPTS) [21]. The points at which the ego has to perform braking or steering to avoid a collision with a vehicle in front, also called a target vehicle. For practical reasons, the times at which TTB and TTS are equal to 0 will be denoted as τ_b and τ_s, respectively. In their general and simplest forms, they are defined by (1) and (2). The remaining distances between the ego and the target vehicle at τ_b and τ_s are defined as d_b and d_s by (3) and (4), respectively. The d_b is the distance that ego travels while reducing its velocity (v_ego) to the target's velocity v_t by applying the maximum braking $a_{\mathrm{min,x}}$ [21]. It is worth mentioning that $a_{\mathrm{min,x}}$ stands for the maximum longitudinal deceleration, wherein we respect the positive and negative signs of acceleration in the longitudinal direction. Here, d_s is calculated as equivalent to the evasion time t_ev (5) needed to change a lane with maximum lateral acceleration, $a_{\mathrm{max,y}}$ [17,21]. Unlike the longitudinal direction, in the lateral direction, we do not consider negative signs; therefore, $a_{\mathrm{max,y}}$ denotes the maximum acceleration, regardless of direction. These temporal and spatial relations are depicted in Figure 1, while all the parameters used to calculate the metrics and their thresholds are summarized in Table 1. Figure 1 illustrates that τ_s occurs at a later point in time than τ_b, which indicates the steering as the last emergency intervention to avoid a collision; however, this is not always the case. Figure 2 shows the dependency of the remaining distances d_b and d_s on the relative velocity between the ego vehicle and target vehicle. Here, braking leaves more time to avoid a collision at low velocities, but at higher relative velocities, steering would be more appropriate.

$$\tau_\mathrm{b} = \frac{d_0 - d_\mathrm{b}}{v_\mathrm{ego} - v_\mathrm{t}} \tag{1}$$

$$\tau_\mathrm{s} = \frac{d_0 - d_\mathrm{s}}{v_\mathrm{ego} - v_\mathrm{t}} \tag{2}$$

$$d_\mathrm{b} = -\frac{v_\mathrm{rel}^2}{2\,a_{\mathrm{min,x}}} \tag{3}$$

$$d_\mathrm{s} = \sqrt{\frac{2\,d_\mathrm{y}}{a_{\mathrm{max,y}}}}\,v_\mathrm{rel} \tag{4}$$

$$t_\mathrm{ev} = \sqrt{\frac{2\,d_\mathrm{y}}{a_{\mathrm{max,y}}}} \tag{5}$$

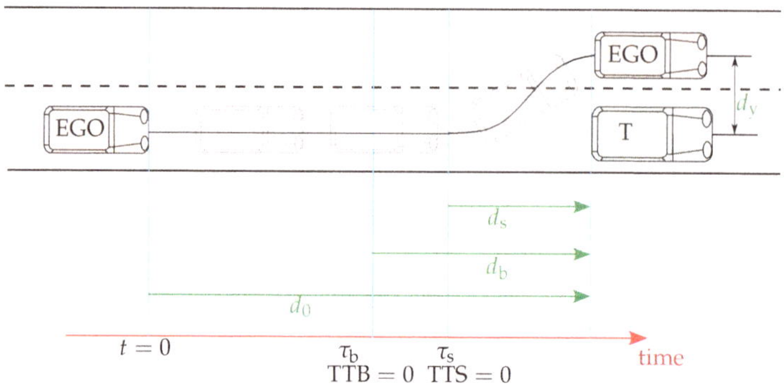

Figure 1. Visualization of steering and braking possibilities.

Table 1. Parameters used for safety metrics and thresholds.

State parameters	Description
v_{ego}	Velocity of the ego vehicle
v_t	Velocity of the target vehicle
v_{rel}	Relative velocity between the ego vehicle and the target vehicle in front
$a_{ego,x}$	Current longitudinal acceleration of the ego vehicle
$a_{min,x}$	Maximum longitudinal deceleration
$a_{max,y}$	Maximum lateral acceleration
$a_{req,x}$	Required longitudinal deceleration of the ego vehicle
$a_{req,y}$	Required lateral deceleration of the ego vehicle
Time parameters	
τ_b	Time to reach LPTB (TTB = 0)
τ_s	Time to reach LPTS (TTS = 0)
$\tau_{b,th}$	Threshold for the τ_b
$\tau_{s,th}$	Threshold for the τ_s
t_ρ	Response time of the ego vehicle
t_{ev}	Evasion time needed to change a lane
Distance parameters	
d_0	Initial distance between the ego and target vehicle at the point t=0
d_b	Distance to target vehicle at τ_b
d_s	Distance to target vehicle at τ_s
d_y	Lateral evasion distance
$d_{b,min}$	Minimum safety distance for braking maneuver
$d_{s,min}$	Minimum safety distance for steering maneuver
Remaining parameters	
g	Gravitational constant
μ	Road friction coefficient

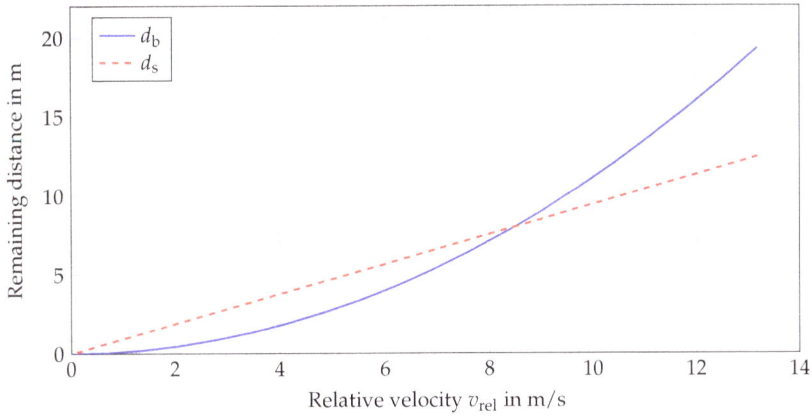

Figure 2. Remaining distance of an ego vehicle to a target vehicle for avoiding a rear-end collision in longitudinal traffic for a pure braking maneuver and a pure steering maneuver.

The previously defined relations in (1)–(4) consider only the constant velocities of the ego and target vehicles representing only a special case and resulting in a reduced precision in the presence of variable velocities. To increase the prediction precision while keeping it feasible regarding its complexity, a straight road and constant velocity v_t is assumed, while the ego vehicle keeps constant longitudinal acceleration $a_{ego,x}$ up to the point where evasive braking or evasive steering actions must be executed. Furthermore, considering ideal conditions, the possible maximum braking and lateral acceleration are $a_{min,x} = -\mu g$ and $a_{max,y} = \mu g$, respectively, where μ represents the maximum available tire road friction. Considering such steady states, the remaining distances d_b and d_s are given by (6) and (7):

$$d_b = -\frac{(v_{rel} + a_{ego,x}\,\tau_b)^2}{2\,a_{min,x}} \tag{6}$$

$$d_s = \sqrt{\frac{2\,d_y}{a_{max,y}}}\,(v_{rel} + a_{ego,x}\,\tau_s), \tag{7}$$

while τ_b and τ_s are determined by solving the following two quadratic equations.

$$\frac{1}{2}\left(a_{ego,x} - \frac{a_{ego,x}^2}{a_{min,x}}\right)\tau_b^2 + \left(v_{rel} - \frac{v_{rel}\,a_{ego,x}}{a_{min,x}}\right)\tau_b - d_0 - \frac{v_{rel}^2}{2\,a_{min,x}} = 0 \tag{8}$$

$$\frac{1}{2}\,a_{ego,x}\,\tau_s^2 + \left(v_{rel} + a_{ego,x}\sqrt{2\,\frac{d_y}{a_{max,y}}}\right)\tau_s - d_0 + v_{rel}\sqrt{2\,\frac{d_y}{a_{max,y}}} = 0 \tag{9}$$

The metrics described by (6)–(9) are used below to establish the thresholds and assess criticality.

3. Assessment Method

3.1. Concept of Fictive Vehicles

Figure 3 shows the novel concept of fictive vehicles on vehicle configurations taken from PEGASUS [22,23], where nine vehicles from the immediate surrounding of the ego vehicle are considered as safety relevant for scenario development or analysis. Furthermore, the vehicles EGO_f^r and EGO_f^l represent fictive duplicates of the ego vehicle and are needed for the assessment method described below. However, there is one condition under which a fictive vehicle can be placed in adjacent lanes—the spaces at the neighboring lanes next to the ego must be unobstructed. For better interpretability, the environment around the ego vehicle is divided into trailing traffic, leading traffic, and the part adjacent to the ego vehicle, which consists of fictive vehicles.

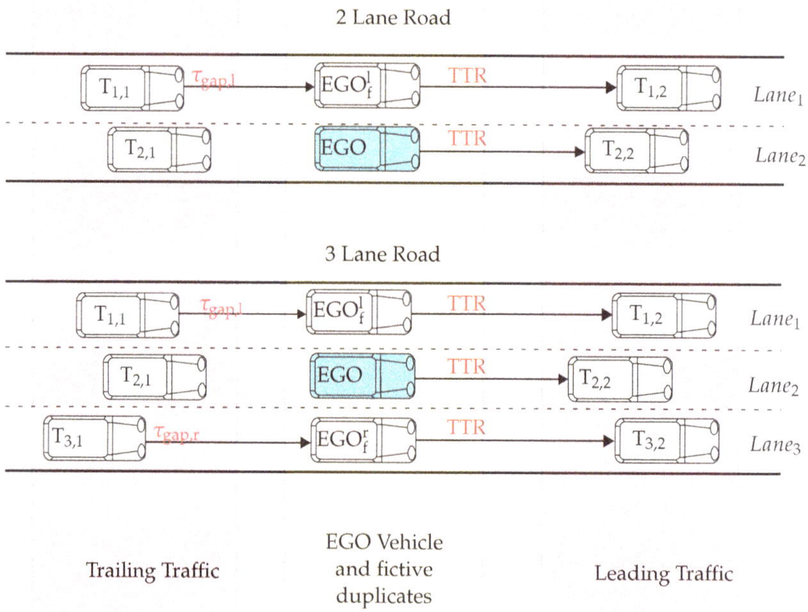

Figure 3. Graphical representation of the two-lane and three-lane scenarios.

3.2. *Definition of the Criticality*

The criticality definition consists of two parts. The first part deals with the definition of variable criticality threshold (VCT) based on the metrics introduced in Section 2. The second part deals with the quantification of the criticality based on the VCT.

3.2.1. Definition of Variable Criticality Thresholds

To determine the VCT, we build upon the required accelerations $a_{\mathrm{req,x}}$ and $a_{\mathrm{req,y}}$ for braking and steering, respectively, as well as on minimum safety distances for braking ($d_{\mathrm{b,min}}$) and steering ($d_{\mathrm{s,min}}$). Distances are calculated based on the mathematical model of responsibility-sensitive safety (RSS) [24], that accounts for the worst-case scenario, including the response time of the ego and the maximum theoretical braking to the standstill of a target vehicle. However, we introduced slight modifications for calculating $d_{\mathrm{b,min}}$, which are shown in (10). Unlike the RSS, we do not consider the maximum longitudinal acceleration of the ego during the response time; therefore, we use the current ego's acceleration instead. Another adjustment is regarding the ego's deceleration after the response time, where we use $a_{\mathrm{req,x}}$, which denotes braking but is not limited only to small values.

$$d_{\mathrm{b,min}} = v_{\mathrm{ego}}\, t_\rho + \frac{1}{2}\, a_{\mathrm{ego,x}}\, t_\rho^2 - \frac{(v_{\mathrm{ego}} + t_\rho\, a_{\mathrm{max,acc}})^2}{2\, a_{\mathrm{req,x}}} + \frac{v_{\mathrm{t}}^2}{2\, a_{\mathrm{min,x}}} \tag{10}$$

The safety distance for steering, $d_{s,min}$, is determined in a similar manner (11). The only difference is that the braking distance of the ego is substituted with the required distance to perform a lane change.

$$d_{s,min} = v_{ego}\, t_\rho + \frac{1}{2}\, a_{ego,x}\, t_\rho^2 + \sqrt{2\, \frac{d_y}{a_{req,y}}}\, (v_{ego} + a_{ego,x}\, t_\rho) + \frac{v_t^2}{2\, a_{min,x}} \tag{11}$$

Substituting the initial distance d_0 in (8) and (9) with the minimum distances $d_{b,min}$ and $d_{s,min}$, the threshold values for TTB (12) and TTS (13) can be determined. In this way, the safety distance is transferred into the time domain and steered by $a_{req,x}$ and $a_{req,y}$. With $a_{req,x}$ and $a_{req,y}$, it is possible to interpret the thresholds and link them to objective or subjective safety and comfort investigations. Now, at a certain τ_b, when $d_0 = d_{b,min}$, we can claim—if there is sudden maximum braking of the target vehicle—that the ego will need to brake with $a_{req,x}$ to avoid a collision; the same applies for steering.

$$\frac{1}{2}\left(a_{ego,x} - \frac{a_{ego,x}^2}{a_{min,x}}\right)\tau_{b,th}^2 + \left(v_{rel} - \frac{v_{rel}\, a_{ego,x}}{a_{min,x}}\right)\tau_{b,th} - d_{b,min} - \frac{v_{rel}^2}{2\, a_{min,x}} = 0 \tag{12}$$

$$\frac{1}{2}\, a_{ego,x}\, \tau_{s,th}^2 + \left(v_{rel} + a_{ego,x}\sqrt{2\, \frac{d_y}{a_{max,y}}}\right)\tau_{s,th} - d_{s,min} + v_{rel}\sqrt{2\, \frac{d_y}{a_{max,y}}} = 0 \tag{13}$$

In the present work, different safety thresholds were defined according to the different criticality values determined by the required accelerations $a_{req,x}$ and $a_{req,y}$, which were obtained from the literature and are provided in Table 2. Driving comfort was taken as the criterion to determine the lowest deceleration threshold level. Several studies, such as [25,26], have correlated the subjective evaluation of driving comfort to objective values, and one of the most significant is acceleration. Taking the relevant literature into account, $-2\,\mathrm{m/s^2}$ was considered as the lowest threshold for longitudinal acceleration. The other extreme is the highest threshold, which corresponds to emergency braking and depends on the current friction coefficient. In this study, we defined the emergency level by $-\mu\, g$. Inbetween these limits and regarding the criticality levels defined below, we defined two intermediate levels based on the partial braking of adaptive cruise control (ACC) [27]. Therefore, the lower intermediate level is $-3\,\mathrm{m/s^2}$ and the higher intermediate level is $-5\,\mathrm{m/s^2}$. Furthermore, the lateral direction ($a_{req,y}$) is determined based on the lane change duration. For a comfortable lane change, we have chosen 6 s, which is an acceptable value based on the driving simulator research in [28]. The intermediate values are an educated guess based on the analysis of the naturalistic driving data derived in [29]. According to [29], the lane change duration varies from 0.7 s to 16.1 s, with a mean of 3.81 s and a standard deviation of 2.03 s. For the first intermediate step, we have chosen 4 s because it fits with the studies conducted in [28,29]. The second intermediate step is 2 s, which is based on the standard deviation presented in [29]. The emergency maneuver is set as previously defined with $a_{max,y} = 7\,\mathrm{m/s^2}$, which also fits the order size of the lower values in [29]. It should be highlighted that the values for lateral acceleration in Table 2 are explicitly derived from subjective studies on lane change and calculated using a simplified equation for the evasion time t_{ev} (5), where the lateral evasion distance, d_y, is the lane width 3.5 m. Therefore, different lateral accelerations can be expected if other studies or advanced equations for t_{ev} are considered.

Table 2. Acceleration levels for lateral and longitudinal movement.

Acceleration Levels	$a_{\text{req},x}$	$a_{\text{req},y}$
Comfort	$-2\,\text{m/s}^2$	$0.2\,\text{m/s}^2$
Intermediate 1	$-3\,\text{m/s}^2$	$0.5\,\text{m/s}^2$
Intermediate 2	$-5\,\text{m/s}^2$	$1.9\,\text{m/s}^2$
Emergency	$-\mu g$	$7\,\text{m/s}^2$

3.2.2. Criticality Scale

Based on the VCT, the criticality is quantified with four criticality values, c_i. This procedure is illustrated in Figure 4, where the index $i \in \{1,2,3,4\}$ represents the values for a certain criticality level. To calculate those values, the relations from (8), (9), (12) and (13), and the TTRs are needed. Using the longitudinal and lateral acceleration levels from Table 2 as the required deceleration for the safety distance $d_{\text{b,min}}$ and $d_{\text{s,min}}$ in (10) and (11), the distance relations are calculated for the longitudinal and lateral direction in (14).

Longitudinal direction

$$\begin{aligned} d^{\text{L}}_{\text{b,min}} &= d_{\text{b,min}}(a_{\text{req},x} = -2\,\text{m/s}^2) \\ d^{\text{int,1}}_{\text{b,min}} &= d_{\text{b,min}}(a_{\text{req},x} = -3\,\text{m/s}^2) \\ d^{\text{int,2}}_{\text{b,min}} &= d_{\text{b,min}}(a_{\text{req},x} = -5\,\text{m/s}^2) \\ d^{\text{H}}_{\text{b,min}} &= d_{\text{b,min}}(a_{\text{req},x} = -\mu g) \end{aligned}$$

Lateral direction

$$\begin{aligned} d^{\text{L}}_{\text{s,min}} &= d_{\text{s,min}}(a_{\text{req},y} = 0.2\,\text{m/s}^2) \\ d^{\text{int,1}}_{\text{s,min}} &= d_{\text{s,min}}(a_{\text{req},y} = 0.5\,\text{m/s}^2) \\ d^{\text{int,2}}_{\text{s,min}} &= d_{\text{s,min}}(a_{\text{req},y} = 1.9\,\text{m/s}^2) \\ d^{\text{H}}_{\text{s,min}} &= d_{\text{s,min}}(a_{\text{req},y} = \mu g) \end{aligned} \quad (14)$$

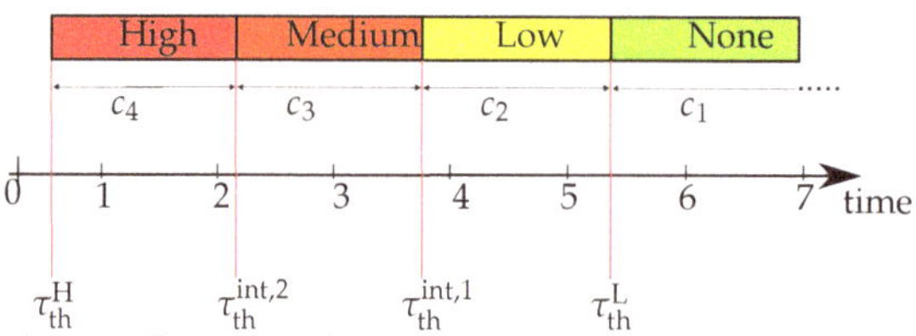

Figure 4. Illustration of the criticality levels.

The minimum distances $d^j_{\text{b,min}}$ and $d^j_{\text{s,min}}$ with $j \in \{\text{L, int,1 , int,2 ,H}\}$ from (14) represent the minimum distance between the ego and target vehicle at which evasive maneuver must be performed with corresponding $a_{\text{req},x}$ or $a_{\text{req},y}$ to avoid a collision. Applying these distances in the threshold relation (12) and (13), the VCTs are defined. The VCT levels are calculated according to (15).

TTB $\geq$ TTS

$$\begin{aligned}
\tau_{\text{th}}^{\text{L}} &= \tau_{\text{b,th}}(d_{\text{b,min}} = d_{\text{b,min}}^{\text{L}}) \\
\tau_{\text{th}}^{\text{int,1}} &= \tau_{\text{b,th}}(d_{\text{b,min}} = d_{\text{b,min}}^{\text{int,1}}) \\
\tau_{\text{th}}^{\text{int,2}} &= \tau_{\text{b,th}}(d_{\text{b,min}} = d_{\text{b,min}}^{\text{int,2}}) \\
\tau_{\text{th}}^{\text{H}} &= \tau_{\text{b,th}}(d_{\text{b,min}} = d_{\text{b,min}}^{\text{H}})
\end{aligned}$$

TTB $<$ TTS

$$\begin{aligned}
\tau_{\text{th}}^{\text{L}} &= \tau_{\text{s,th}}(d_{\text{s,min}} = d_{\text{s,min}}^{\text{L}}) \\
\tau_{\text{th}}^{\text{int,1}} &= \tau_{\text{s,th}}(d_{\text{s,min}} = d_{\text{s,min}}^{\text{int,1}}) \\
\tau_{\text{th}}^{\text{int,2}} &= \tau_{\text{s,th}}(d_{\text{s,min}} = d_{\text{s,min}}^{\text{int,2}}) \\
\tau_{\text{th}}^{\text{H}} &= \tau_{\text{s,th}}(d_{\text{s,min}} = d_{\text{s,min}}^{\text{H}})
\end{aligned} \tag{15}$$

These threshold values represent four intervals for the TTR. Depending on the range in which the TTR time is, a criticality value according to (16) is determined. The less time available for an evasive maneuver, the more critical the situation is. It is important to notice that threshold intervals are changeable over time and depend on the vehicle states, such as v_{ego}, a_{ego} and v_{t}.

$$c_i = \begin{cases} c_1 = 1, & \text{TTR} \geq \tau_{\text{th}}^{\text{L}} \\ c_2 = 2, & \tau_{\text{th}}^{\text{L}} > \text{TTR} \geq \tau_{\text{th}}^{\text{int,1}} \\ c_3 = 3, & \tau_{\text{th}}^{\text{int,1}} > \text{TTR} \geq \tau_{\text{th}}^{\text{int,2}} \\ c_4 = 4, & \tau_{\text{th}}^{\text{int,2}} > \text{TTR} \geq \tau_{\text{th}}^{\text{H}} \end{cases} \tag{16}$$

3.3. Criticality Assessment Method

To assess the overall criticality of the scenario we are combining the criticality definition from (16) with the concept of the fictive vehicles depicted in Figure 3. Considering the three-lane configuration from Figure 3, it is possible to place two fictive vehicles as the condition of unoccupied adjacent lanes is satisfied. For each of those fictive vehicles, as well as for ego, the criticality values according to (16) are determined. However, for the fictive vehicle, only braking as an evasive maneuver is considered which leads to TTR=TTB, while for the ego, TTR = max{TTB, TTS}. Now, the overall criticality c_{s} is determined by (17). This equation takes mean criticality values calculated between ego (c_{ego}) and each fictive vehicle ($c_{\text{ego,f}}^{\text{r}}$ and $c_{\text{ego,f}}^{\text{l}}$). The mean values are rounded to the next higher value, and the lower is chosen as the overall criticality. In the case of equal mean values for the left and right lanes, the one with a higher TTR is taken.

$$c_{\text{s}} = \min\left\{ \left\lceil \frac{c_{\text{ego}} + c_{\text{ego,f}}^{\text{r}}}{2} \right\rceil, \left\lceil \frac{c_{\text{ego}} + c_{\text{ego,f}}^{\text{l}}}{2} \right\rceil \right\} \tag{17}$$

Additionally, if a trailing vehicle exists, the time gaps $\tau_{\text{gap,l}}$ and/or $\tau_{\text{gap,r}}$ are calculated and considered for the criticality assessment. The time gap $\tau_{\text{gap,l}}$ represents a time gap between the trailing vehicle at the left side and $\tau_{\text{gap,r}}$ at the right side. The time gap itself represents the time interval between the fictive ego vehicles $\text{EGO}_{\text{f}}^{\text{l}}$ and $\text{EGO}_{\text{f}}^{\text{r}}$ and the trailing vehicles T_{l} and T_{r}. The calculation of the time gaps for the left and right sight are presented in

$$\tau_{\text{gap,l}} = \frac{(s_{\text{T}_{\text{l}}} - s_{\text{ego}_{\text{f}}^{\text{l}}})}{v_{\text{T}_{\text{l}}}}, \tag{18}$$

$$\tau_{\text{gap,r}} = \frac{(s_{\text{T}_\text{r}} - s_{\text{ego}_\text{f}^\text{r}})}{v_{\text{T}_\text{r}}} \tag{19}$$

with s_{T_l} and s_{T_r} being the trailing vehicle positions; v_{T_l} and v_{T_r} being the trailing vehicle velocities; $s_{\text{ego}_\text{f}^\text{l}}$ and $s_{\text{ego}_\text{f}^\text{r}}$ being the fictive ego vehicle positions. Here, $\tau^{\text{th}}_{\text{gap,r}}$ and $\tau^{\text{th}}_{\text{gap,l}}$ represent the threshold values for the time gap. If the time gape values are lower than these thresholds, the possibility of steering will not be considered for the overall assessment. In this case, TTR is set to be TTB.

4. Simulation Results

For demonstrating the introduced assessment method, two exemplary test scenarios shown in Figure 5 are taken into account. In the first scenario, the ego vehicle is approaching the target vehicle $\text{T}_{2,2}$ in lane 2, while vehicles $\text{T}_{1,2}$ and $\text{T}_{3,2}$ are placed adjacent to $\text{T}_{2,2}$. In the second scenario, the trailing $\text{T}_{1,1}$ is added. Threshold for the time gap ($\tau^{\text{th}}_{\text{gap,l}}$) between $\text{T}_{1,1}$ and $\text{EGO}^\text{l}_\text{f}$ is set to 3 s. This means that, if $\tau^{\text{th}}_{\text{gap,l}} \leq 3$, then the ego cannot consider steering in lane 1. Scenarios are created and simulated within the simulation software CarMaker from IPG Automotive, and the initial positions and velocities of vehicles are given in Table 3.

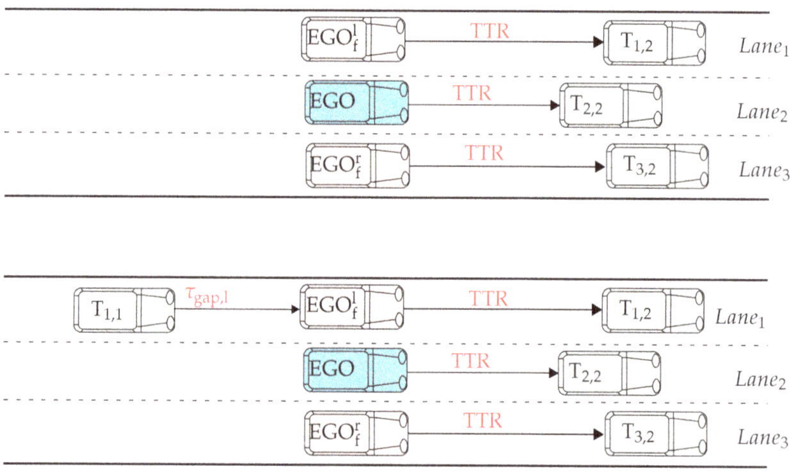

Figure 5. Graphical representation of test scenarios.

Table 3. Initial states and parameters for the test scenarios.

Scenario Description				
	Start position on the lane in m	Start velocity in km/h	Speed reduction time in s	Desired speed reduction in km/h
Scenario 1	Trailing vehicle is not considered			
EGO	50	130	None	None
$\text{T}_{1,2}$	250	105	15	90
$\text{T}_{2,2}$	250	100	15	50
$\text{T}_{2,3}$	250	105	15	70
Scenario 2	Trailing vehicle is considered.			
Trailing vehicle	0	130	15	100

Figure 6 shows the traveled distance and velocities of vehicles; while the ego vehicle keeps its speed constant, the target vehicles reduce their speed after 10 s to the defined velocities. Finally, scenarios will lead to the collision of the ego with $T_{2,2}$, as the ego vehicle does not brake and has no assistance functions. Collision is not mitigated on purpose, in order to demonstrate how the criticality of the scenario changes as the ego approaches $T_{2,2}$.

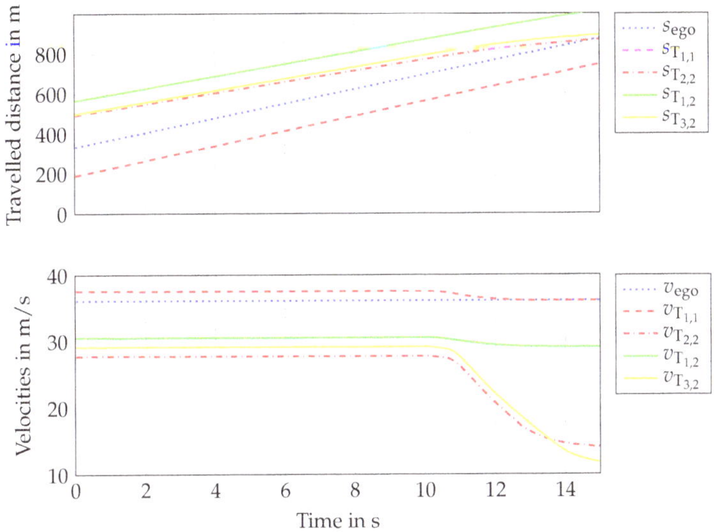

Figure 6. Distances traveled and velocities in the tested two scenarios.

Velocity reductions are different for each lane which cause different criticality values, and this behavior is shown in Figure 7. It can be seen that the criticality of the ego lane has the longest criticality phase with a criticality value of 4 (high criticality). This is because of the aggressive speed-reduction maneuver of the front vehicle, $T_{2,2}$. The right lane has a slightly lower criticality, and the left one has the lowest criticality over time. Because the left lane has a criticality of 2 (low criticality) until 7.72 s, it represents a potentially avoidable situation for the ego vehicle in this time interval. Therefore, the overall criticality c_s for the ego vehicle is one level lower between 5.92 s and 7.62 s, as shown in Figure 7. This is only valid up to 7.62 s, where the ego vehicle comes closer to the vehicle $T_{2,2}$ and the potential evasive maneuver space becomes smaller. The last point where the ego vehicle could make evasive maneuver is at the point where the TTR time intersects the τ_{th}^{H} at 11.6 s. After that, a collision cannot be avoided anymore.

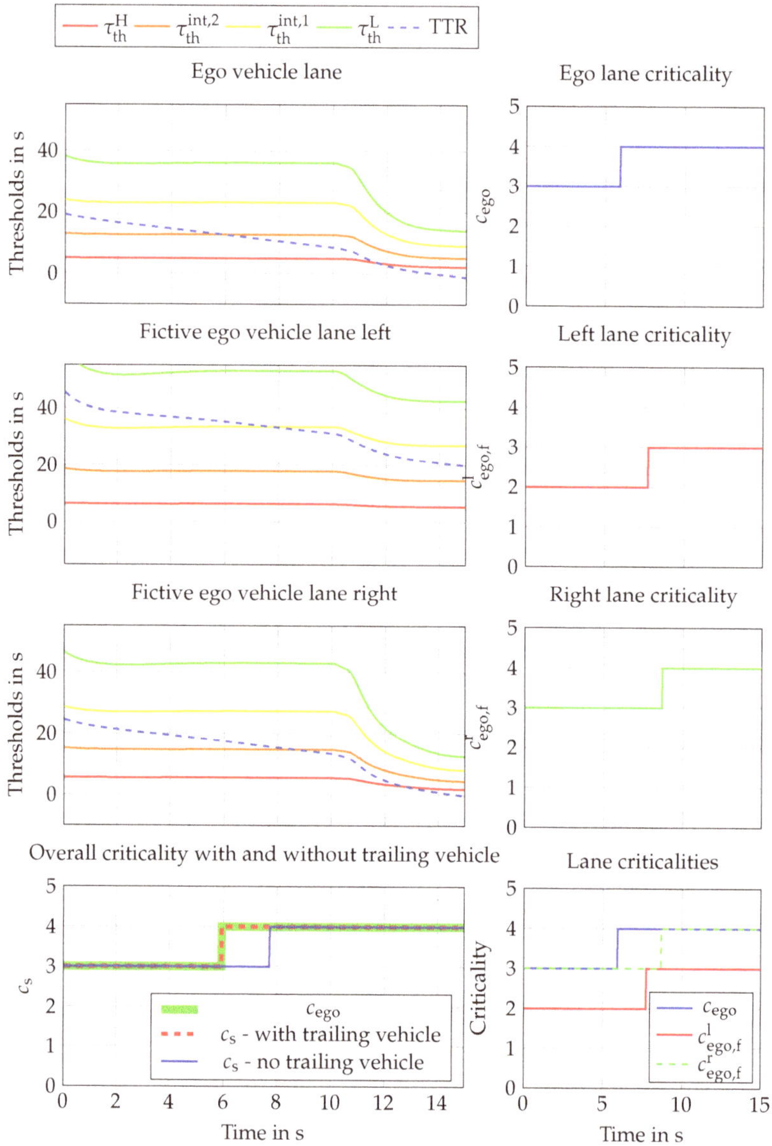

Figure 7. Criticality assessment results for the two introduced scenarios.

The whole scenario behaves slightly differently when a trailing vehicle is considered. The small time gap between the $T_{1,1}$ and EGO_f^l eliminates the possibility of ego to steer in the left lane so the total criticality rises to the maximum criticality for the time interval 5–8 s, see Figure 8.

In summary, these two examples demonstrate the functionality of the criticality assessment method for the assumptions made to calculate the TTR and the chosen VCT. In addition, these particular examples were chosen to show the potential of the method by considering and analyzing the influences of the ego vehicles surrounding for the criticality calculation.

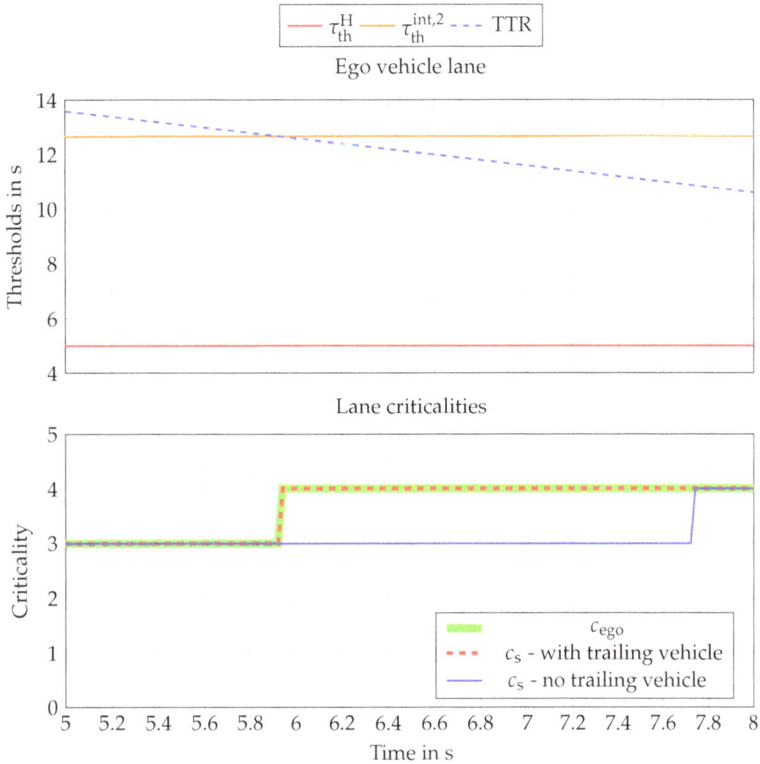

Figure 8. Criticality assessment results for the time frame between 5 s and 8 s.

5. Conclusions

In the current paper, we introduced a criticality assessment method for ADS. The novelty of the method lies in the introduction of fictive ego vehicles from the surrounding of the ego. In order to determine the overall criticality, the collision time reserve as TTR metric was introduced and evaluated using four criticality levels. The criticality levels are determined using variable threshold values, which are converted into the time domain from minimal safety distance metrics and acceleration values. Considered acceleration values present a linkage between objective investigation and threshold values, while the derived VCT fills the research gap by taking into account the kinematic relations between vehicles, and thus variable nature of the criticality. To derive feasible and intelligible method, certain assumption has been made. The method is derived for straight highway roads, while formulation assumes constant velocity of a target vehicle.

Finally, the method is demonstrated on two highway scenarios that defer in the presence of a trailing vehicle. The results show higher criticality of scenarios in which trailing vehicles obstruct the adjacent lane of the ego. This narrowed down the alternative options for the ego and increased the overall criticality. The potential and advantage of the proposed criticality assessment method reside in simplified consideration of the surrounding ego vehicles and the evaluation of possible evasive actions; in addition to the criticality assessment, it could be used for decision-making applications. For future works, the proposed criticality metric and threshold calculation will be applied in a wide range of scenarios, while the mathematical formulation will be extend to cover considered assumptions and to make realistic lane change models for steering maneuvers.

Author Contributions: Conceptualization, D.N., T.M. and M.S.; methodology, D.N. and T.M.; software, D.N. and F.O.; validation, D.N., F.O., T.M. and M.S.; formal analysis, F.O., A.E., C.L. and W.S.; investigation, D.N. and T.M.; resources, D.N. and T.M.; data curation, D.N. and F.O.; writing—original draft preparation, D.N., T.M., F.O., M.S., C.L. and W.S.; writing—review and editing, D.N., T.M., F.O., M.S., C.L., A.E. and W.S.; visualization, D.N.; supervision, A.E.; project administration, A.E.; funding acquisition, A.E. and W.S. All authors have read and agreed to the published version of the manuscript.

Funding: The project was part of the Austrian Research Promotion Agency FFG Program "EFREtop": 885754.

Institutional Review Board Statement: Not applicable.

Informed Consent Statement: Not applicable.

Data Availability Statement: Not applicable.

Acknowledgments: Supported by TU Graz Open Access Publishing Fund.

Conflicts of Interest: The authors declare no conflict of interest.

Abbreviations

ADS	automated driving systems
SCS	safety-critical scenario
TTC	time to collision
TTR	time to react
TTB	time to brake
TTS	time to steer
TTK	time to kickdown
TTx	time to x
LPTB	last point to brake
LPTS	last point to steer
VCT	variable criticality threshold

References

1. Feng, S.; Feng, Y.; Sun, H.; Zhang, Y.; Liu, H.X. Testing Scenario Library Generation for Connected and Automated Vehicles: An Adaptive Framework. *IEEE Trans. Intell. Transp. Syst.* **2020**, 23, 1213–1222. [CrossRef]
2. Neurohr, C.; Westhofen, L.; Henning, T.; de Graaff, T.; Möhlmann, E.; Böde, E. Fundamental Considerations around Scenario-Based Testing for Automated Driving. In Proceedings of the 2020 IEEE Intelligent Vehicles Symposium (IV), Las Vegas, NV, USA, 19 October–13 November 2020; pp. 121–127. [CrossRef]
3. Nalic, D.; Mihalj, T.; Bäumler, M.; Lehmann, M.; Eichberger, A.; Bernsteiner, S. Scenario Based Testing of Automated Driving Systems: A Literature Survey. In Proceedings of the FISITA Web Congress 2020, Virtual Event, 24 November 2020.
4. Wen, M.; Park, J.; Cho, K. A scenario generation pipeline for autonomous vehicle simulators. *Hum.-Centric Comput. Inf. Sci.* **2020**, *10*, 24. [CrossRef]
5. Andreotti, E.; Boyraz, P.; Selpi. Mathematical Definitions of Scene and Scenario for Analysis of Automated Driving Systems in Mixed-Traffic Simulations. *IEEE Trans. Intell. Veh.* **2020**, *6*, 366–375. [CrossRef]
6. Leitner, A. ENABLE-S3: Project Introduction. In *Validation and Verification of Automated Systems*; Springer: Cham, Switzerland, 2020; pp. 13–23.
7. Rodemerk, C.; Habenicht, S.; Weitzel, A.; Winner, H.; Schmitt, T. Development of a general criticality criterion for the risk estimation of driving situations and its application to a maneuver-based lane change assistance system. In Proceedings of the 2012 IEEE Intelligent Vehicles Symposium, Madrid, Spain, 3–7 June 2012; pp. 264–269. [CrossRef]
8. Stumper, D.; Knapp, A.; Pohl, M.; Dietmayer, K.C.J. Towards Characterization of Driving Situations via Episode-Generating Polynomials. In *Advanced Microsystems for Automotive Applications 2016*; Springer: Cham, Switzerland, 2016; pp. 165–173.
9. Riedmaier, S.; Ponn, T.; Ludwig, D.; Schick, B.; Diermeyer, F. Survey on Scenario-Based Safety Assessment of Automated Vehicles. *IEEE Access* **2020**, *8*, 87456–87477. [CrossRef]
10. Broadhurst, A.; Baker, S.; Kanade, T. Monte Carlo road safety reasoning. In Proceedings of the IEEE Proceedings. Intelligent Vehicles Symposium, Las Vegas, NV, USA, 6–8 June 2005; pp. 319–324.
11. Mahmud, S.S.; Ferreira, L.; Hoque, M.S.; Tavassoli, A. Application of proximal surrogate indicators for safety evaluation: A review of recent developments and research needs. *IATSS Res.* **2017**, *41*, 153–163. [CrossRef]

12. Li, Y.; Zheng, Y.; Morys, B.; Pan, S.; Wang, J.; Li, K. Threat Assessment Techniques in Intelligent Vehicles: A Comparative Survey. *IEEE Intell. Transp. Syst. Mag.* **2020**, *13*, 71–91. [CrossRef]
13. Ozbay, K.; Yang, H.; Bartin, B.; Mudigonda, S. Derivation and validation of new simulation-based surrogate safety measure. *Transp. Res. Rec.* **2008**, *2083*, 105–113. [CrossRef]
14. Wachenfeld, W.; Junietz, P.; Wenzel, R.; Winner, H. The worst-time-to-collision metric for situation identification. In Proceedings of the 2016 IEEE Intelligent Vehicles Symposium (IV), Gothenburg, Sweden, 19–22 June 2016; pp. 729–734. [CrossRef]
15. Minderhoud, M.M.; Bovy, P.H. Extended time-to-collision measures for road traffic safety assessment. *Accid. Anal. Prev.* **2001**, *33*, 89–97. [CrossRef]
16. Hillenbrand, J. Fahrerassistenz zur Kollisionsvermeidung. Ph.D. Thesis, Universität Karlsruhe (TH), Karlsruhe, Germany, 2007. [CrossRef]
17. Junietz, P.M. Microscopic and Macroscopic Risk Metrics for the Safety Validation of Automated Driving. Ph.D. Thesis, Technische Universität, Darmstadt, Germany, 2019.
18. Van Der Horst, R.; Hogema, J. Time-to-collision and collision avoidance systems. In Proceedings of the 6th ICTCT WorkshopSafety Evaluation of Traffic Systems: Traffic Conflicts and Other Measures, Salzburg, Austria, 27–29 October 1993; pp. 109–121.
19. Chen, R.; Sherony, R.; Gabler, H.C. Comparison of Time to Collision and Enhanced Time to Collision at Brake Application during Normal Driving. In Proceedings of the SAE 2016 World Congress and Exhibition. SAE International, Detroit, MI, USA, 12–14 April 2016. [CrossRef]
20. Hillenbrand, J.; Kroschel, K.; Schmid, V. Situation assessment algorithm for a collision prevention assistant. In Proceedings of the IEEE Proceedings. Intelligent Vehicles Symposium 2005, Las Vegas, NV, USA, 6–8 June 2005; pp. 459–465. [CrossRef]
21. Eckert, A.; Hartmann, B.; Sevenich, M.; Rieth, P. Emergency steer & brake assist: A systematic approach for system integration of two complementary driver assistance systems. In Proceedings of the 22nd International Technical Conference on the Enhanced Safety of Vehicles (ESV), Washington, DC, USA, 13–16 June 2011; pp. 13–16.
22. Winner, H.; Lemmer, K.; Form, T.; Mazzega, J. Pegasus—First steps for the safe introduction of automated driving. In *Road Vehicle Automation 5*; Springer: Cham, Switzerland, 2019; pp. 185–195.
23. Weber, H.; Bock, J.; Klimke, J.; Roesener, C.; Hiller, J.; Krajewski, R.; Zlocki, A.; Eckstein, L. A framework for definition of logical scenarios for safety assurance of automated driving. *Traffic Inj. Prev.* **2019**, *20*, S65–S70. [CrossRef] [PubMed]
24. Shalev-Shwartz, S.; Shammah, S.; Shashua, A. On a formal model of safe and scalable self-driving cars. *arXiv* **2017**, arXiv:1708.06374.
25. Nguyen, T.; NguyenDinh, N.; Lechner, B.; Wong, Y.D. Insight into the lateral ride discomfort thresholds of young-adult bus passengers at multiple postures: Case of Singapore. *Case Stud. Transp. Policy* **2019**, *7*, 617–627. [CrossRef]
26. Hoberock, L.L. A survey of longitudinal acceleration comfort studies in ground transportation vehicles. *J. Dyn. Syst. Meas. Control* **1977**, *99*, 76–84. [CrossRef]
27. Maurer, M.; Eskandarian, A. Forward collision warning and avoidance. In *Handbook of Intelligent Vehicles*; Springer: London, UK, 2012; pp. 657–687.
28. Rogic, B.; Nalic, D.; Eichberger, A.; Bernsteiner, S. A novel approach to integrate human-in-the-loop testing in the development chain of automated driving: The example of automated lane change. *IFAC-PapersOnLine* **2020**, *53*, 10188–10195. [CrossRef]
29. Yang, M.; Wang, X.; Quddus, M. Examining lane change gap acceptance, duration and impact using naturalistic driving data. *Transp. Res. Part C Emerg. Technol.* **2019**, *104*, 317–331. [CrossRef]

sensors

MDPI

Article

Nonlinear Predictive Motion Control for Autonomous Mobile Robots Considering Active Fault-Tolerant Control and Regenerative Braking

Peng Hang, Baichuan Lou and Chen Lv *

School of Mechanical and Aerospace Engineering, Nanyang Technological University, Singapore 639798, Singapore; peng.hang@ntu.edu.sg (P.H.); baichuan.lou@ntu.edu.sg (B.L.)
* Correspondence: lyuchen@ntu.edu.sg

Abstract: To further advance the performance and safety of autonomous mobile robots (AMRs), an integrated chassis control framework is proposed. In the longitudinal motion control module, a velocity-tracking controller was designed with the integrated feedforward and feedback control algorithm. Besides, the nonlinear model predictive control (NMPC) method was applied to the four-wheel steering (4WS) path-tracking controller design. To deal with the failure of key actuators, an active fault-tolerant control (AFTC) algorithm was designed by reallocating the driving or braking torques of the remaining normal actuators, and the weighted least squares (WLS) method was used for torque reallocation. The simulation results show that AMRs can advance driving stability and braking safety in the braking failure condition with the utilization of AFTC and recapture the braking energy during decelerations.

Keywords: autonomous mobile robot; motion control; nonlinear model predictive control; active fault-tolerant control; regenerative braking

Citation: Hang, P.; Lou, B.; Lv, C. Nonlinear Predictive Motion Control for Autonomous Mobile Robots Considering Active Fault-Tolerant Control and Regenerative Braking. *Sensors* **2022**, *22*, 3939. https://doi.org/10.3390/s22103939

Academic Editor: Steven L. Waslander

Received: 4 May 2022
Accepted: 21 May 2022
Published: 23 May 2022

1. Introduction

Compared with the traditional automated guided vehicles (AGVs), autonomous mobile robots (AMRs) have higher flexibility and intelligence, representing a more sophisticated, flexible, and cost-effective technology, in favor of smart manufacturing, smart factory, and intelligent logistics [1]. AMRs are usually equipped with multiple actuators for steering, drive and brake. Therefore, AMR is an over-actuated system since each wheel can provide independent traction force [2]. It is a critical issue to realize the coordinated control between multiple actuators [3,4].

In recent years, different kinds of advance control methods have been applied to the motion control of robots, including optimal control [5], model predictive control (MPC) [6], Reinforcement Learning (RL)-based control approach [7], adaptive neural network [8], and neuroadaptive learning algorithms [9]. The chassis control of AMR usually consists of longitudinal motion control and lateral motion control [10]. Longitudinal motion control is associated with the drive and brake actuators, e.g., in-wheel motors (IWMs) and electro-mechanical brake (EMB) systems. In longitudinal motion control of AMRs, velocity-tracking control is in favor of the autonomous driving [11]. In [12], a parameter-varying controller was designed for velocity tracking, which showed high robustness. In [13], the MPC method was used for velocity-tracking controller design, which could recover the braking energy with brake torque allocation. In [14], an adaptive sliding mode control (ASMC) algorithm using Radial Basis Function (RBF) neural network was applied to the velocity-tracking controller design, which could deal with external disturbances. Besides, Antilock Braking System (ABS), Acceleration Slip Regulation (ASR), and traction control have also been widely studied in longitudinal motion control for AMRs [15–17]. In the lateral motion control of AMRs, path-tracking is the main task for autonomous driving [18]. In [19], a linear quadratic regulator (LQR) technique was used for the four-wheel steering (4WS) path-tracking controller design. However, it showed poor robustness in dealing with uncertainties and disturbances. To reduce the effect of uncertainties in vehicle parameters, a robust path-tracking controller was designed with a μ-synthesis approach [20]. The MPC approach has been widely used in the path-tracking control of AMRs [21]. In [22], an adaptive path-tracking strategy was proposed based on MPC and fuzzy rules, which could guarantee vehicle stability under high-speed and large-curvature conditions. In [23], a Tube-based MPC method was applied to the path-tracking controller design, which showed strong robustness to address uncertainties and disturbances. In [24], an iterative learning control (ILC) method was used for the path-tracking control of AMR, which could improve the path-tracking performance significantly.

To deal with the failure of actuators, a fault-tolerant control has been widely studied [25–27]. In [28], a synthesis method was applied to the reconfigurable fault-tolerant control system, which could deal with the failure of steering actuators. With the driving force allocation control method, the vehicle can reconstruct the distribution control strategy on-line under fault conditions, realizing active fault tolerance [29]. In [30], the linear-quadratic control method and the control Lyapunov function technique were used to design the hybrid fault-tolerant control algorithm for the four-wheel-driving vehicle, which can address the actuator failure in the path-tracking process. In [31], a robust fault-tolerant control scheme was designed for distributed actuated electric vehicles, which integrated cooperative game and terminal sliding mode control (SMC) into the framework of the feedback linearization method (FLM). In [32], a fault tolerant sliding mode predictive control (SMPC) strategy was proposed to address the actuator failure, in which SMC was used to improve the robustness of the MPC in the presence of modeling uncertainties and disturbances. In [33], a novel quantized SMC strategy based on switching mechanism was proposed to compensate for actuator failure effects. In [34], the minimax MPC in the delta-domain was deployed to achieve the tracking performance under the actuator fault, system uncertainties, and disturbance.

Most studies only consider the failure of one actuator, which cannot cover all failure conditions. In this research, all kinds of failure conditions of IWMs were studied. Besides, few studies consider the regenerative braking and actuator failure in the motion control process of AMR at the same time. The contributions of this research are summarized as follow: (1) To deal with the system nonlinearity and external disturbances, an integrated feedforward and feedback control algorithm was designed for longitudinal motion control of AMR; (2) To realize the collaborative steering of 4WS, the nonlinear model predictive control (NMPC) method was applied to the path-tracking controller design; (3) To address the braking failure of actuators, an active fault-tolerant control (AFTC) algorithm was designed for AMR by redistributing the braking torques of the rest normal actuators.

The rest of this paper is organized as follows. Section 2 gives the problem description and control framework for AMR. The modelling work for control algorithm design is described in Section 3. Section 4 presents the control algorithm design for AMR. Then, the simulation tests are described in Section 5. Finally, Section 6 provide some conclusions and suggests future work.

2. Problem Description and Control Framework

2.1. Control Problem Description for AMR

To realize autonomous driving, the motion control for AMR mainly consists of longitudinal motion control and lateral motion control. Lateral motion control is reflected by the path-tracking issue. Longitudinal motion control is related to the drive and brake control, which is a critical issue in this study.

IWMs are the key components for AMR. On one hand, in-wheel motors can be used to drive the AMR. On the other hand, regenerative braking can be realized with in-wheel motors, recovering the braking energy. AMR is usually equipped with four in-wheel motors for independent drive, and four EMB systems for independent braking. Due to so many actuators, the reliability of the system is decreased. Therefore, safety is a critical issue for AMR. In the braking process, if braking failure of actuators occurs, this reduces safety. To maximize regenerative braking energy, IWMs have higher braking priority than EMBs. EMBs are usually used to compensate the rest braking force. Therefore, we mainly discuss the braking failure of IWMs in this paper.

Figure 1 shows the braking failure conditions of IWMs divided into five types, i.e., failure of one IWM, failure of two IWMs on two sides, failure of two IWMs on the same side, failure of three IWMs, and failure of four IWMs. In this paper, the AFTC algorithm is proposed to deal with all kinds of braking failure of IWMs.

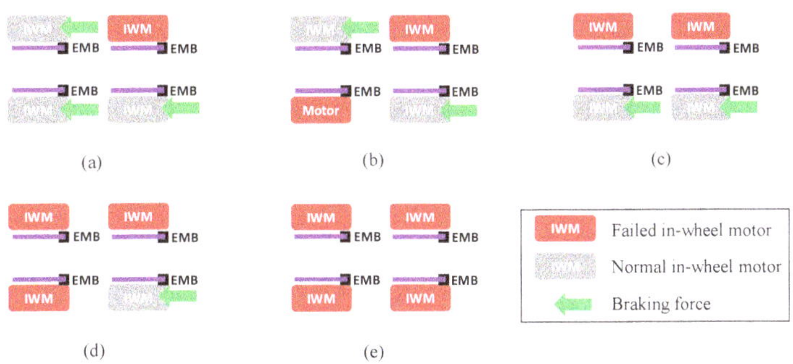

Figure 1. Braking failure of in-wheel motors: (**a**) Failure of one in-wheel motor; (**b**) failure of two in-wheel motors on two sides; (**c**) failure of two in-wheel motors on the same side; (**d**) failure of three in-wheel motors; (**e**) failure of four in-wheel motors.

2.2. Chassis Control Framework for AMR

The chassis control framework for AMR is illustrated in Figure 2, which mainly consists of longitudinal motion control and lateral motion control, i.e., the velocity-tracking control and the path-tracking control. In the path-tracking control module, NMPC is applied to the controller design. Based on the target path and the feedbacked vehicle state, the path-tracking controller outputs the front and rear wheel steering angels. In the velocity-tracking control module, an integrated feedforward and feedback controller is designed. To deal with the braking failure of IWMs, an AFTC module is designed after the velocity-tracking controller. With the torque redistribution of IWMs and EMBs, the AFTC algorithm is able to maximize the regenerative braking energy and guarantee safety at the same time.

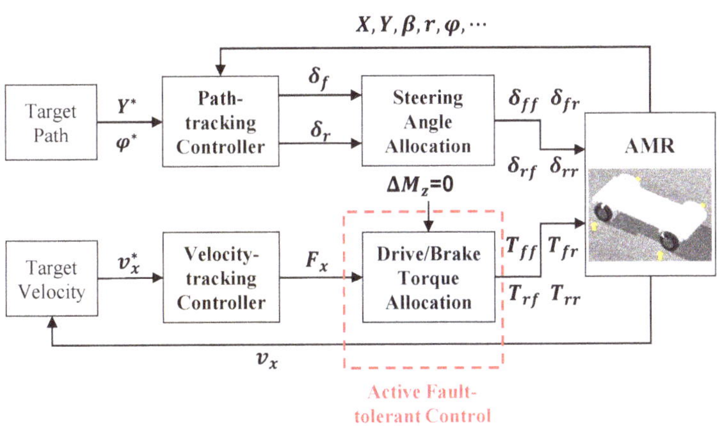

Figure 2. Chassis control framework for AMR.

3. Modelling

3.1. Vehicle Dynamic Model

Some assumptions are made in this paper. First, only seven degrees of freedom are considered for the vehicle dynamic model, i.e., longitudinal motion, lateral motion, yaw motion of the vehicle and the four wheels' motion. Pitch motion, roll motion, and vertical motion of AMR are ignored. Drive anti-skid control is not considered in the longitudinal motion control strategy. This paper mainly focuses on the velocity-tracking control and braking control. Additionally, the longitudinal acceleration of the wheel center is considered equal to the longitudinal acceleration of the AMR at CG.

The longitudinal dynamic model is derived as follows [35].

$$m(\dot{v}_x - v_y r) = F_x - F_w - F_f \tag{1}$$

$$F_x = F_{xfl}\cos\delta_{fl} + F_{xfr}\cos\delta_{fr} + F_{xrl}\cos\delta_{rl} + F_{xrr}\cos\delta_{rr} \tag{2}$$

$$F_w = C_D A \rho v_x^2/2 \tag{3}$$

$$F_f = f_r m g \tag{4}$$

where v_x and v_y denote the longitudinal and lateral velocities, r denotes the yaw rate at the center of gravity (CG), F_x denotes the total longitudinal tire force acting on the vehicle. F_w and F_f denote the wind resistance and the rolling resistance, respectively. m denotes the vehicle mass, δ_i ($i = fl, fr, rl, rr$) denotes the steering angle of each wheel (*fl* denotes the front left wheel, *fr* denotes the front right wheel, *rl* denotes the rear left wheel, and *rr* denotes the rear right wheel). F_{xi} ($i = fl, fr, rl, rr$) denotes the longitudinal force of each tire, C_D, A and ρ denote the air resistance coefficient, windward area and air density, respectively., and f_r and g denote the rolling resistance coefficient and the gravitational acceleration.

The lateral dynamic model is expressed by [36]

$$m(\dot{v}_y + v_x r) = F_y \tag{5}$$

$$F_y = F_{yfl}\cos\delta_{fl} + F_{yfr}\cos\delta_{fr} + F_{yrl}\cos\delta_{rl} + F_{yrr}\cos\delta_{rr} \tag{6}$$

where F_y denotes the total lateral tire force acting on the vehicle. F_{yi} ($i = fl, fr, rl, rr$) denotes the lateral force of each tire, which is expressed with the Dugoff tire model [37].

The yaw dynamic model is written according to [38]

$$I_z\dot{r} = M_z \tag{7}$$

$$M_z = (F_{yfl}\cos\delta_{fl} + F_{yfr}\cos\delta_{fr})l_f - (F_{yrl}\cos\delta_{rl} + F_{yrr}\cos\delta_{rr})l_r + \Delta M_z \tag{8}$$

where M_z denotes the total yaw moment acting on the vehicle, I_z denotes the yaw inertia moment, l_f denotes the distance from the front axle to CG, and l_r denotes the distance from the rear axle to CG. ΔM_z is the external yaw moment, which is created by the torque difference between left and right wheels.

$$\Delta M_z = [-F_{xfl}\cos\delta_{fl} + F_{xfr}\cos\delta_{fr} - F_{xrl}\cos\delta_{rl} + F_{xrr}\cos\delta_{rr}]\frac{B}{2} \tag{9}$$

where B denotes the vehicle track.

Additionally, the dynamic model of each wheel is derived by

$$I_w\dot{\omega}_i = T_i - F_{xi}R_w \tag{10}$$

where T_i denotes the wheel torque, $T_i = T_{di} - T_{bi}$, T_{di} and T_{bi} denote the drive and brake torques, respectively, ω_i and R_w denote the angular velocity of each wheel and the rolling radius of the tire, respectively, and I_w denotes the wheel moment of inertia.

3.2. Path-Tracking Model

As Figure 3 shows, the 4-wheel vehicle model is usually simplified to be a single-track model to simplify the controller design [39]. The steering angle transformation relationship between the two models follows the Ackerman steering geometry [40].

$$\begin{array}{ll} \tan\delta_{fl} = \frac{\tan\delta_f}{1-\frac{B}{2l}\left(\tan\delta_f - \tan\delta_r\right)}, & \tan\delta_{fr} = \frac{\tan\delta_f}{1+\frac{B}{2l}\left(\tan\delta_f - \tan\delta_r\right)} \\ \tan\delta_{rl} = \frac{\tan\delta_r}{1-\frac{B}{2l}\left(\tan\delta_f - \tan\delta_r\right)}, & \tan\delta_{rr} = \frac{\tan\delta_r}{1+\frac{B}{2l}\left(\tan\delta_f - \tan\delta_r\right)} \end{array} \tag{11}$$

where δ_f and δ_r denote the front and rear steering angles, and l denotes the distance from the front axle to the rear axle.

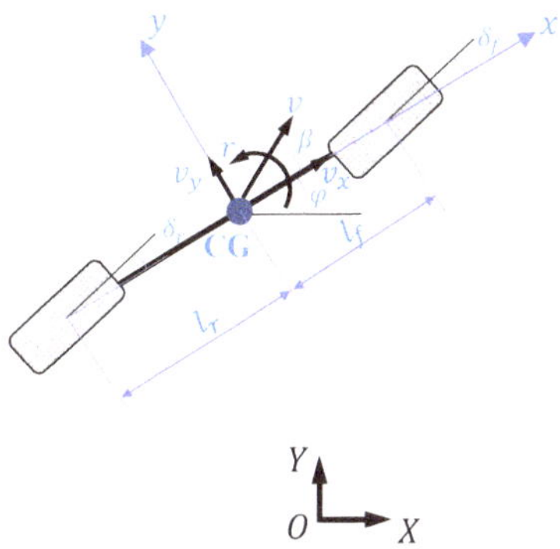

Figure 3. Single-track vehicle model.

The yaw angel φ and lateral position Y of AMR at CG are expressed as

$$\begin{cases} \dot{\varphi} = r \\ \dot{Y} = v_x \sin\varphi + v_y \cos\varphi \end{cases} \tag{12}$$

A combination of (5), (7), (11) and (12) yields the following path-tracking model for AMR.

$$\begin{array}{l} \dot{x}(t) - f(x(t), u(t)) \\ y(t) = g(x(t), u(t)) \end{array} \tag{13}$$

$$f(x(t), u(t)) = \begin{bmatrix} -v_x r + \frac{\sum F_y}{m} \\ \frac{\sum M_z}{I_z} \\ r \\ v_x \sin\varphi + v_y \cos\varphi \end{bmatrix} \tag{14}$$

$$g(x(t), u(t)) = \begin{bmatrix} 0 & 0 & 1 & 0 \\ 0 & 0 & 0 & 1 \end{bmatrix} x(t) \tag{15}$$

where the state vector $x = \left[v_y,\ r,\ \varphi,\ Y\right]^T$, the output vector $y = \left[\ \varphi,\ Y\right]^T$, and the control vector $u = \left[\delta_f, \delta_r\right]^T$.

4. Control Algorithm Design

4.1. Velocity-Tracking Control Algorithm

For velocity-tracking controller design, (1) is rewritten as follows.

$$m\dot{v}_x = F_x - F_w - F_f + F_c \tag{16}$$

where $F_c = mv_y r$.

After Taylor expansion of (2) regarding $\cos\delta_i$:

$$F_x = F_{xfl} + F_{xfr} + F_{xrl} + F_{xrr} + F_d \tag{17}$$

where higher-order terms are placed in F_d.

Then, the simplified longitudinal dynamic model can be expressed as

$$m\dot{v}_x = F_{xfl} + F_{xfr} + F_{xrl} + F_{xrr} - F_w - F_f + F_c + F_d \tag{18}$$

Based on the wheel dynamic model (10), it can be derived that

$$F_{xi} = \frac{T_i - I_{wi}\dot{v}_{xi}/R_w}{R_w} \quad (i = fl, fr, rl, rr) \tag{19}$$

Substitution of (19) into (18) yields

$$\left(m + \frac{\sum I_{wi}}{R_w^2}\right)\dot{v}_x = \frac{\sum T_i}{R_w} - F_w - F_f + F_c + F_d \tag{20}$$

The total torque of four wheels $\sum T_i$ is defined as the longitudinal control vector, which is made up of the feedforward and feedback controllers, i.e.,

$$\sum T_i = u_{ff} + u_{fb} \tag{21}$$

According to the model (20), the feedforward controller is derived as follows.

$$u_{ff} = \left(mR_w + \frac{\sum I_{wi}}{R_w}\right)\dot{v}_x^* + (F_w + F_f - F_c)R_w \tag{22}$$

where v_x^* denotes the target velocity. The feedforward controller is mainly used to compensate the control error caused by the nonlinearity of the system.

Substitution of (22) into (20) yields

$$\left(m + \frac{\sum I_{wi}}{R_w^2}\right)\dot{e}_{v_x} = \frac{u_{fb}}{R_w} + F_d \tag{23}$$

where e_{v_x} denotes the velocity tracking error, i.e., $e_{v_x} = v_x - v_x^*$.

The feedback controller is designed by PID. Furthermore, defining the state vector $x_l = \left[\int_0^t e_{v_x} d\tau,\ e_{v_x}, \dot{e}_{v_x}\right]^T$, control vector $u_l = [u_{fb}, \dot{u}_{fb}]^T$, disturbance vector $d_l = [F_d, \dot{F}_d]^T$, then, (23) can be written in the state-space form.

$$\begin{aligned} \dot{x}_l &= A_l x_l + B_l u_l + E_l d_l \\ y_l &= C_l x_l \end{aligned} \tag{24}$$

where $A_l = \begin{bmatrix} 0 & 1 & 0 \\ 0 & 0 & 0 \\ 0 & 0 & 0 \end{bmatrix}$, $B_l = \begin{bmatrix} 0 & 0 \\ B_s & 0 \\ 0 & B_s \end{bmatrix}$, $C_l = \begin{bmatrix} 0 & 1 & 0 \end{bmatrix}$, $E_l = \begin{bmatrix} 0 & 0 \\ E_s & 0 \\ 0 & E_s \end{bmatrix}$, $B_s = \frac{R_w}{mR_w^2 + \sum I_{wi}}$, and $E_s = \frac{R_w^2}{mR_w^2 + \sum I_{wi}}$.

To solve the feedback PID controller, the following performance index function is constructed:

$$J_{PID} = \int_0^\infty \left(y_l^T Q_l y_l + u_l^T R_l u_l\right) dt \tag{25}$$

where Q_l and R_l are weighting matrix, $Q_l = 10^3$, $R_l = I_{2\times 2}$.

Furthermore, the solution problem of the feedback PID controller can be transformed into the minimization of the performance index function, i.e.,

$$\min J_{PID}(K) \tag{26}$$

Finally, the linear-quadratic optimization approach is used to solve the feedback PID controller [41,42].

4.2. Path-Tracking Control Algorithm

For the path-tracking controller design, the path-tracking model (13) is expressed in the discreate state-space form as follows.

$$\begin{aligned} x(k+1) &= F(x(k), u(k)) \\ y(k) &= G(x(k), u(k)) \end{aligned} \tag{27}$$

where $F(x(k), u(k)) = x(k) + Tf(x(k), u(k))$, $G(x(k), u(k)) = g(x(k), u(k))$, T denotes the sampling time, and $T = 0.02s$.

Based on the discreate model (27), NMPC is applied to the path-tracking controller design. The prediction horizon and the control horizon are defined by N_p and N_c, $N_p \geq N_c$. $N_p = 10$, and $N_c = 5$. Then, the predictive outputs are derived as follows.

$$\begin{aligned} y(k+1) &= G(x(k+1), u(k+1)) \\ y(k+2) &= G(x(k+2), u(k+2)) \\ &\vdots \\ y(k+N_c) &= G(x(k+N_c), u(k+N_c)) \\ y(k+N_c+1) &= G(x(k+N_c+1), u(k+N_c)) \\ &\vdots \\ y(k+N_p) &= G(x(k+N_p), u(k+N_c)) \end{aligned} \tag{28}$$

Based on (28), this yields the output sequence as follows.

$$\boldsymbol{y}(k+1) = [y(k+1), y(k+2), \cdots, y(k+N_p)]^T \tag{29}$$

Besides, the reference output sequence is expressed by

$$\hat{\boldsymbol{y}}(k+1) = [\hat{y}(k+1), \hat{y}(k+2), \cdots, \hat{y}(k+N_p)]^T \tag{30}$$

where $\hat{y}(k+p) = [\varphi^*(k+p),\ Y^*(k+p)]^T$, $p = 1, \cdots, N_p$, $\varphi^*(k+p)$ and $Y^*(k+p)$ denote the reference values of yaw angle and lateral position.

Moreover, the control sequence is expressed as follows.

$$\boldsymbol{u}(k+1) = [u(k+1), u(k+2), \cdots, u(k+N_c)]^T \tag{31}$$

The proposed path tracking controller aims to minimize the tracking error $\|\boldsymbol{y}(k+1) - \hat{\boldsymbol{y}}(k+1)\|_2$ with the smallest control energy $\|\boldsymbol{u}(k+1)\|_2$. Furthermore, the following cost function is constructed.

$$\begin{aligned} J(k) = \sum_{i=1}^{N_p} [y(k+i|k) - \hat{y}(k+i|k)]^T Q [y(k+i|k) - \hat{y}(k+i|k)] \\ + \sum_{i=0}^{N_c-1} [u(k+i|k)]^T R [u(k+i|k)] \end{aligned} \tag{32}$$

where Q and R are diagonal weighting matrices, $Q = \text{diag}\{8 \times 10^3,\ 10^4\}$, $R = \text{diag}\{5 \times 10^5,\ 10^6\}$.

Finally, the NMPC path-tracking controller can be solved with the following optimization.

$$\begin{aligned}
&\min_{u(k)} \boldsymbol{J}(k) \\
&s.t. \\
&x(k+i|k) = F(x(k+i-1|k), u(k+i-1|k)) \\
&u_{\min} \leq u(k+i|k) \leq u_{\max}
\end{aligned} \tag{33}$$

4.3. Active Fault-Tolerant Control Algorithm

In this section, we only discuss the braking failure of IWMs. If IWMs have failure in the driving process, the AFTC mechanism is triggered immediately. After that, the AMR starts braking to guarantee safety. Therefore, we do not discuss the driving failure of IWMs independently.

Since the total torque of four wheels $\sum T_i$ has been worked out based on Section 4.1., it yields that

$$\sum T_i = T^{IWMs} + T^{EMBs} \tag{34}$$

where T^{IWMs} and T^{EMBs} denote the total torques of four IWMs and four EMBs, respectively, i.e., $T^{IWMs} = T_{fl}^{IWM} + T_{fr}^{IWM} + T_{rl}^{IWM} + T_{rr}^{IWM}$, $T^{EMBs} = T_{fl}^{EMB} + T_{fr}^{EMB} + T_{rl}^{EMB} + T_{rr}^{EMB}$.

Besides, the external yaw moment is generated by IWMs and EMBs, i.e.,

$$\Delta M_z = \Delta M_z^{IWMs} + \Delta M_z^{EMBs} \tag{35}$$

where ΔM_z^{IWMs} and ΔM_z^{EMBs} denote the external yaw moment generated by IWMs and EMBs, respectively.

To guarantee yaw stability, $\Delta M_z = 0$. The following work aims to distribute the torque for each IWM and EMB based on (34) and (35). Figure 4 shows the AFTC flowchart to deal with all kinds of braking failure of IWMs.

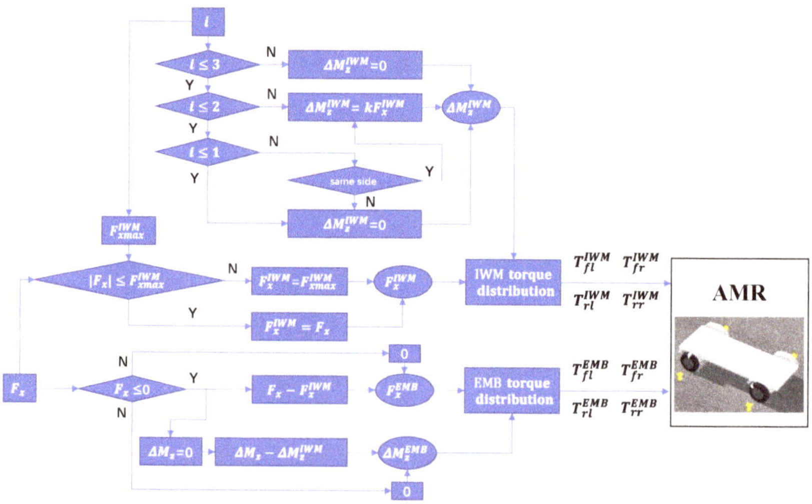

Figure 4. AFTC flowchart for all kinds of braking failure of IWMs.

To maximize the regenerative braking energy, IWMs has higher braking priority than EMBs. Therefore, the first step is to determine if $|F_x| \leq F_{xmax}^{IWM}$, F_{xmax}^{IWM} denotes the braking force boundaries of all normal IWMs, which is related to the failure number of IWM, i.e., i in Figure 4. If $|F_x| \leq F_{xmax}^{IWM}$, $F_x^{IWM} = F_x$, else $F_x^{IWM} = F_{xmax}^{IWM}$ and EMBs will compensate the rest braking force, i.e., $F_x^{EMB} = F_x - F_x^{IWM}$, where F_x^{IWM} and F_x^{EMB} denote the total braking force of four IWMs and four EMBs.

Since ΔM_z^{IWM} cannot be zero under some failure conditions, e.g., failure of two IWMs on the same side and failure of three IWMs, the generated $-\Delta M_z^{IWM}$ will be compensated by ΔM_z^{EMB}.

Once F_x^{IWM}, F_x^{EMB}, ΔM_z^{IWM} and ΔM_z^{EMB} are determined, the torque distribution algorithm will work to work out T_{fl}^{IWM}, T_{fr}^{IWM}, T_{rl}^{IWM}, T_{rr}^{IWM}, T_{fl}^{EMB}, T_{fr}^{EMB}, T_{rl}^{EMB}, T_{rr}^{EMB}. F_x^{IWM} and F_x^{EMB} can be derived from T^{IWM} and T^{EMB} based on (19).

For IWMs, the following torque distribution model is derived.

$$\Lambda^{IWM} = \eta^{IWM}\Theta^{IWM} \tag{36}$$

$$\eta^{IWM} = \begin{bmatrix} 1 & 1 & 1 & 1 \\ -\frac{B}{R_w} & \frac{B}{R_w} & -\frac{B}{R_w} & \frac{B}{R_w} \end{bmatrix}\lambda \tag{37}$$

$$\lambda = \begin{bmatrix} \lambda_{fl} & 0 & 0 & 0 \\ 0 & \lambda_{fr} & 0 & 0 \\ 0 & 0 & \lambda_{rl} & 0 \\ 0 & 0 & 0 & \lambda_{rr} \end{bmatrix} \tag{38}$$

$$\lambda_i = \begin{cases} 1, & \text{normal} \\ 0, & \text{failure of IWM}_i \end{cases} \quad (i = fl, fr, rl, rr) \tag{39}$$

where $\Lambda^{IWM} = \left[T^{IWM}, \Delta M_z^{IWM}\right]^T$ and $\Theta^{IWM} = \left[T_{fl}^{IWM}, T_{fr}^{IWM}, T_{rl}^{IWM}, T_{rr}^{IWM}\right]^T$.

Based on (36), the weighted least squares (WLS) method is used to distribute the torques of IWMs. The cost function for IWM torque distribution is constructed as follows.

$$\begin{aligned} &\Psi^{IWM} = \rho^{IWM}\left\|\omega_\Lambda^{IWM}\left(\eta^{IWM}\Theta^{IWM} - \Lambda^{IWM}\right)\right\|_2^2 + \left\|\omega_\Theta^{IWM}\left(\Theta^{IWM} - \Theta_d^{IWM}\right)\right\|_2^2 \\ &s.t.\ \Theta_{\min}^{IWM} \le \Theta^{IWM} \le \Theta_{\max}^{IWM} \end{aligned} \tag{40}$$

where ρ^{IWM} denotes the weighting coefficient, which is usually set very large to minimize the torque distribution error, $\rho^{IWM} = 10^6$. Θ_d^{IWM} denotes the desired control vector, $\Theta_d^{IWM} = [0,\ 0,\ 0,\ 0]^T$. $\Theta_{\min}^{IWM}$ and $\Theta_{\max}^{IWM}$ denote the minimum and maximum control boundaries of Θ^{IWM}, which is shown in Figure 5. ω_Λ^{IWM} and ω_Θ^{IWM} denote the weighting matrices. In this paper, T^{IWM} and ΔM_z^{IWM} have the same allocation weights, i.e., $\omega_\Lambda^{IWM} = \text{diag}\,[1,1]$, T_i^{IWM} $(i = fl,\ fr,\ rl,\ rr)$ and F_{zi} are positively correlated, where F_{zi} denotes the vertical load of each wheel. Thus, $\omega_\Theta^{IWM} = \text{diag}\left[\frac{1}{F_{zfl}}, \frac{1}{F_{zfr}}, \frac{1}{F_{zrl}}, \frac{1}{F_{zrr}}\right]$.

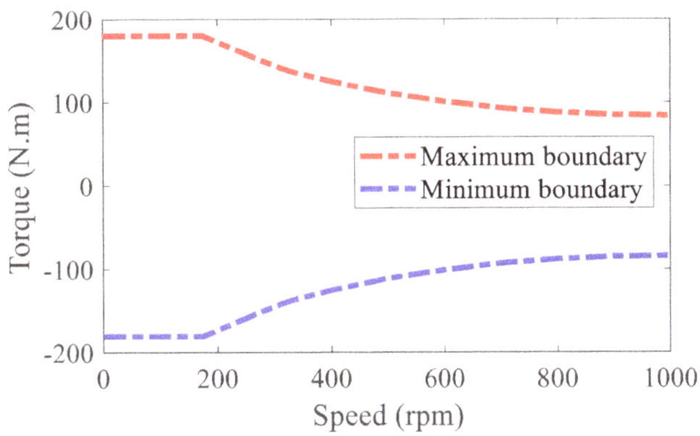

Figure 5. Control boundaries of IWM.

Furthermore, (40) is rewritten as

$$\Psi^{IWM} = \left\| \underbrace{\begin{bmatrix} \left(\rho^{IWM}\right)^{1/2} \omega_{\Lambda}^{IWM} \eta^{IWM} \\ \omega_{\Theta}^{IWM} \end{bmatrix}}_{A^{IWM}} \Theta^{IWM} - \underbrace{\begin{bmatrix} \left(\rho^{IWM}\right)^{1/2} \omega_{\Theta}^{IWM} \Lambda^{IWM} \\ \omega_{\Theta}^{IWM} \Theta_d^{IWM} \end{bmatrix}}_{B^{IWM}} \right\|_2^2 \quad (41)$$

Then, the WLS method for IWMs torque distribution is described as follows.

$$\begin{array}{l} \min\limits_{\Theta^{IWM}} \|A^{IWM}\Theta^{IWM} - B^{IWM}\|_2^2 \\ s.t.\ \Theta_{\min}^{IWM} \leq \Theta^{IWM} \leq \Theta_{\max}^{IWM} \end{array} \quad (42)$$

Based on (42), the torques for four IWMs, i.e., T_{fl}^{IWM}, T_{fr}^{IWM}, T_{rl}^{IWM}, T_{rr}^{IWM}, can be worked out.

For EMBs, the following torque distribution model is derived.

$$\Lambda^{EMB} = \eta^{EMB}\Theta^{EMB} \quad (43)$$

$$\eta^{EMB} = \begin{bmatrix} 1 & 1 & 1 & 1 \\ -\frac{B}{R_w} & \frac{B}{R_w} & -\frac{B}{R_w} & \frac{B}{R_w} \end{bmatrix} \quad (44)$$

where $\Lambda^{EMB} = \left[T^{EMB}, \Delta M_z^{EMB}\right]^T$ and $\Theta^{EMB} = \left[T_{fl}^{EMB}, T_{fr}^{EMB}, T_{rl}^{EMB}, T_{rr}^{EMB}\right]^T$.

Based on (43), the cost function for EMBs torque distribution is derived as follows.

$$\begin{array}{l} \Psi^{EMB} = \rho^{EMB}\|\omega_{\Lambda}^{EMB}\left(\eta^{EMB}\Theta^{EMB} - \Lambda^{EMB}\right)\|_2^2 + \|\omega_{\Theta}^{EMB}\left(\Theta^{EMB} - \Theta_d^{EMB}\right)\|_2^2 \\ s.t.\ \Theta_{\min}^{EMB} \leq \Theta^{EMB} \leq \Theta_{\max}^{EMB} \end{array} \quad (45)$$

where ρ^{EMB} denotes the weighting coefficient, which is usually set very large to minimize the torque distribution error, $\rho^{EMB} = 10^6$, Θ_d^{EMB} denotes the desired control vector, $\Theta_d^{EMB} = [0,\ 0,\ 0,\ 0]^T$, $\Theta_{\min}^{EMB}$ and $\Theta_{\max}^{EMB}$ denote the minimum and maximum control boundaries of Θ^{EMB}, $\Theta_{\min}^{EMB} = -200$, $\Theta_{\max}^{EMB} = 0$, ω_{Λ}^{EMB} and ω_{Θ}^{EMB} denote the weighting matrices, $\omega_{\Lambda}^{EMB} = \text{diag}\,[1, 1]$, $\omega_{\Theta}^{EMB} = \text{diag}\left[\frac{1}{F_{zfl}}, \frac{1}{F_{zfr}}, \frac{1}{F_{zrl}}, \frac{1}{F_{zrr}}\right]$.

Furthermore, (45) is rewritten as

$$\Psi^{EMB} = \left\| \underbrace{\begin{bmatrix} \left(\rho^{EMB}\right)^{1/2} \omega_{\Lambda}^{EMB} \eta^{EMB} \\ \omega_{\Theta}^{EMB} \end{bmatrix}}_{A^{EMB}} \Theta^{EMB} - \underbrace{\begin{bmatrix} \left(\rho^{EMB}\right)^{1/2} \omega_{\Theta}^{EMB} \Lambda^{EMB} \\ \omega_{\Theta}^{EMB} \Theta_d^{EMB} \end{bmatrix}}_{B^{EMB}} \right\|_2^2 \quad (46)$$

Then, the WLS method for EMB torque distribution is derived as follows.

$$\begin{array}{l} \min\limits_{\Theta^{EMB}} \|A^{EMB}\Theta^{EMB} - B^{EMB}\|_2^2 \\ s.t.\ \Theta_{\min}^{EMB} \leq \Theta^{EMB} \leq \Theta_{\max}^{EMB} \end{array} \quad (47)$$

Based on (47), the torques for four IWMs, i.e., T_{fl}^{EMB}, T_{fr}^{EMB}, T_{rl}^{EMB}, T_{rr}^{EMB}, can be worked out.

5. Simulation Results and Analysis

Three simulation cases were designed and carried out via the co-simulation platform based on Carsim and Simulink as shown in Figure 6. Figure 6a shows the Simulink algorithm structure in the co-simulation platform, including the path-tracking control algorithm, longitudinal velocity-tracking control algorithm and the AFTC algorithm. All the control algorithms were carried out in the Simulink software. The real AMR model was built in Carsim software. With the co-simulation of Carsim and Simulink, the effectiveness and feasibility of the proposed algorithm were verified. Figure 6b shows the simulation scenario in Carsim.

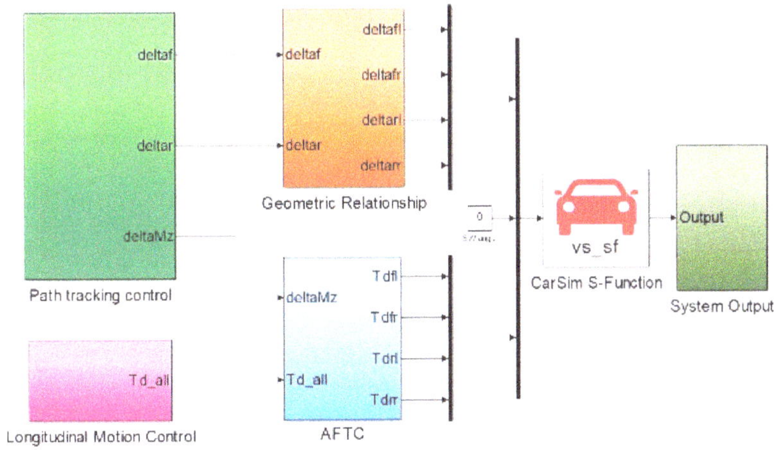

(a)

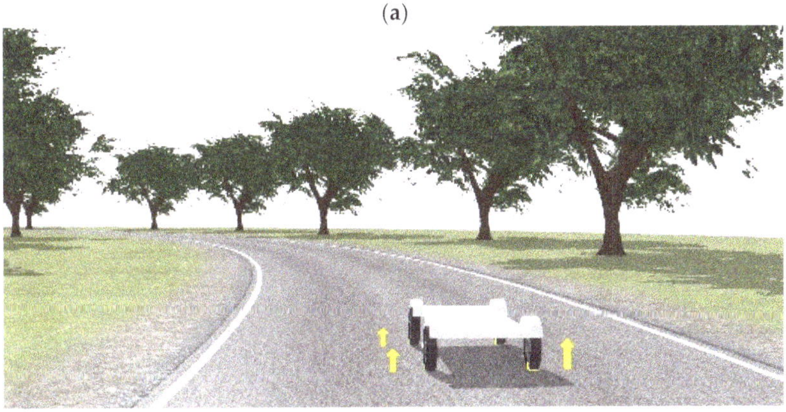

(b)

Figure 6. Co-simulation platform based on Carsim and Simulink: (**a**) Control algorithm; (**b**) simulation scenario.

5.1. Simulation Case 1

In this case, a straight-line braking condition was carried out. The AMR accelerated to 15 m/s and then started to brake after the 10th second. Three kinds of braking modes were compared in this case, i.e., regenerative braking (IWM), mechanical braking (EMB) and hybrid braking (IWM + EMB). The three kinds of braking modes were realized based on the same AMR with the parameters in Table 1 and the same simulation platform in Figure 6. The same velocity-tracking control algorithm and path-tracking control algorithm were utilized. In this case, braking failure was not considered.

Table 1. AMR parameters for simulation.

Parameters	Value	Parameters	Value
m (kg)	431	l_f (m)	0.829
C_D	0.28	l_r (m)	0.705
A (m^2)	0.97	l (m)	1.534
ρ (kg/m^3)	1.2258	B (m)	0.97
f_r	0.008	I_w (kg·m^2)	0.67
I_z (kg·m^2)	217	R_w (m)	0.298

The path lengths of AMR with different kinds of braking modes are illustrated in Figure 7. It was found that regenerative braking had the longest braking distance. The second was mechanical braking, and the shortest was hybrid braking. A detailed analysis is shown in Table 1. The braking distances for the three kinds of braking modes were 42.74 m, 33.15 m, 27.33 m, respectively, and the braking times for the three kinds of braking modes were 4.04 s, 3.09 s, 2.32 s, respectively. Figure 8 shows the velocities of AMR with different kinds of braking modes. Hybrid braking showed the largest deceleration among the three kinds of braking modes. It can be concluded that hybrid braking can shorten the braking distance and braking time remarkably, improving braking safety.

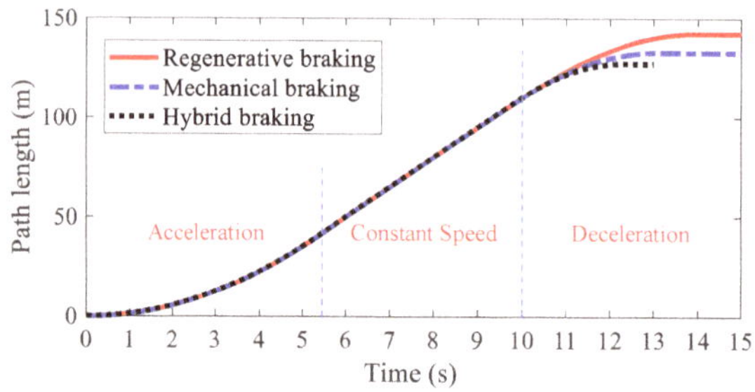

Figure 7. Path length of AMR in Case 1.

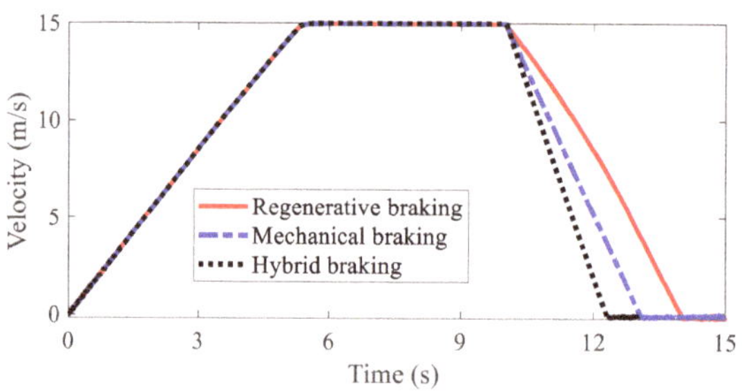

Figure 8. Velocity of AMR in Case 1.

Regenerative braking powers with different kinds of braking modes are depicted in Figure 9. Mechanical braking cannot recover braking energy. Regenerative braking has larger regenerative braking power than hybrid braking. As shown in Table 2 regenerative braking energies for regenerative braking, mechanical braking, and hybrid braking were 6.10×10^4 J, 0 J, and 3.29×10^4 J, respectively. Due to the application of EMB in hybrid braking, the hybrid braking mode had smaller regenerative braking energy than the regenerative braking mode.

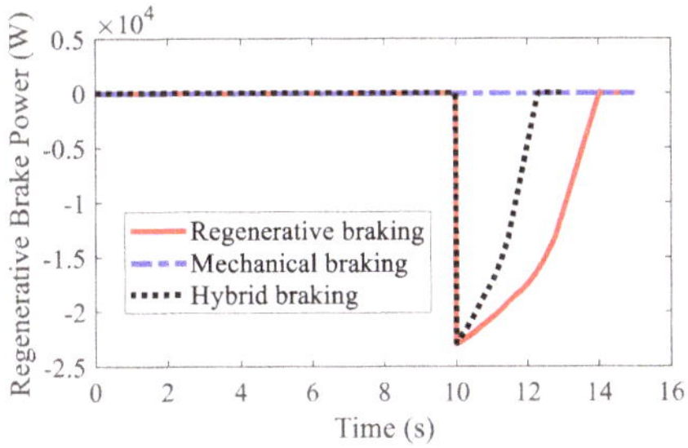

Figure 9. Regenerative braking power of AMR in Case 1.

Table 2. Comparative studies of three braking modes in Case 1.

	Regenerative Braking	Mechanical Braking	Hybrid Braking
Braking distance (m)	42.74	33.15	27.33
Braking time (s)	4.04	3.09	2.32
Regenerative energy (J)	6.10×10^4	0	3.29×10^4

The wheel torques of AMR for three kinds of braking modes are displayed in Figures 10–12, respectively. In the regenerative braking mode, only IWMs worked, in charge of both drive and control. In the mechanical braking mode, IWMs were only used for drive, and EMBs were used for braking. Therefore, the torques of IWMs changed to zero after 10th second. In the hybrid braking mode, both IWMs and EMBs were used for braking. EMBs could compensate the rest braking force for IWMs, shortening the braking time and braking distance.

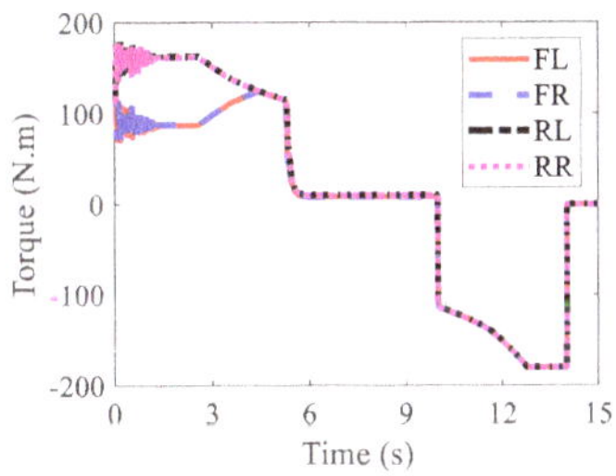

Figure 10. Wheel torques of AMR with regenerative braking in Case 1.

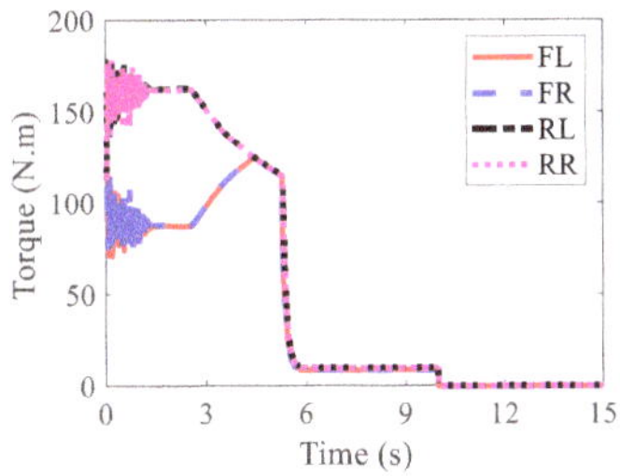

(**a**)

Figure 11. *Cont.*

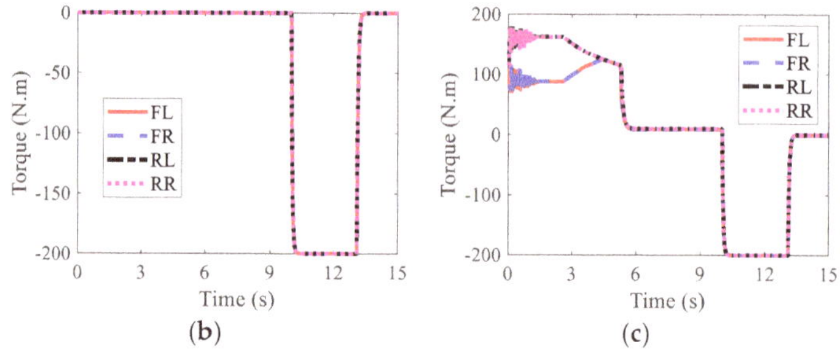

Figure 11. Wheel torques of AMR with mechanical braking in Case 1: (**a**) IWM; (**b**) EMB; (**c**) sum.

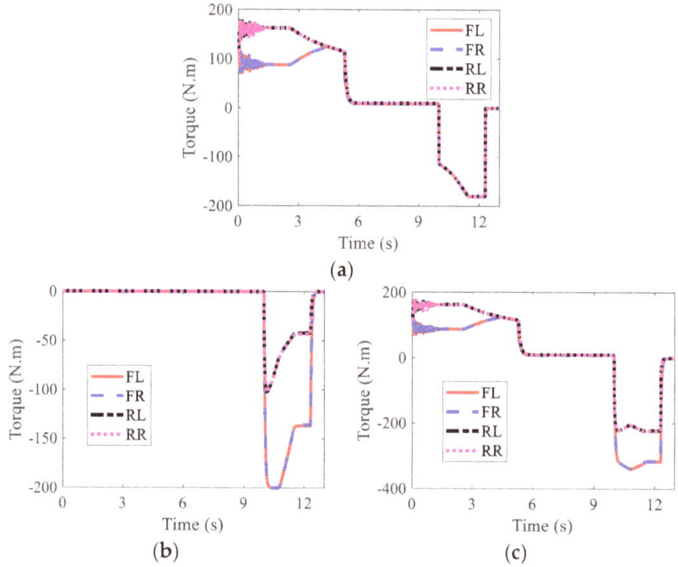

Figure 12. Wheel torques of AMR with hybrid braking in Case 1: (**a**) IWM; (**b**) EMB; (**c**) sum.

From the above simulation results, it can be seen that the regenerative braking mode was beneficial to braking energy recovery. However, it led to longer braking distance, which reduces braking safety. The mechanical braking mode could shorten the braking distance but not recover the braking energy. In general, the hybrid braking mode had the advantages of the above two kinds of braking modes, i.e., maximizing the regenerative braking efficiency and advancing the braking safety.

5.2. Simulation Case 2

This case aimed to validate the AFTC algorithm for the AMR on a curved road; the hybrid braking mode was used. The AMR accelerated to 20 m/s and then started to brake after the 12th second. However, failure of the FL IWM occurred at the 10th second and failure of the RL IWM at the 12th second.

Figure 13 shows the path-tracking results of AMR under three kinds of conditions, i.e., normal (no failure), failure (without AFTC), and AFTC. It can be seen from Figure 13b that without AFTC, the AMR departed from its target path after braking failure, showing a large lateral offset. With AFTC, the AMR could realize lane-keeping after the braking failure and brake safely until stopped, as in the normal condition. The steering angles of AMR are illustrated in Figure 14. After the braking failure of IWMs, the AMR showed very large steering angles to realize lane-keeping when without AFTC. However, with AFTC, the AMR could use torque redistribution to guarantee brake safety and lateral stability.

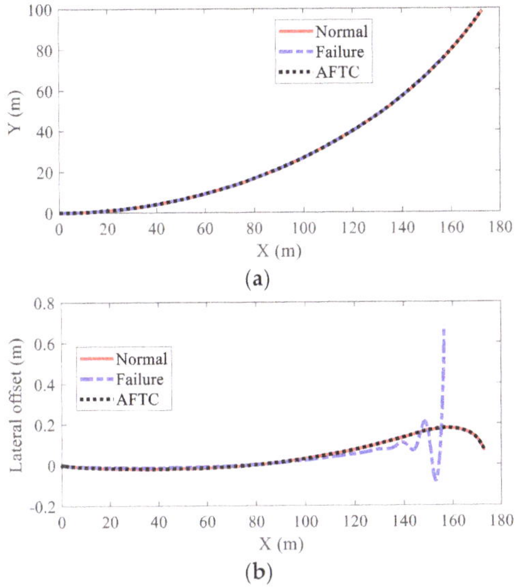

Figure 13. Path-tracking result of AMR in Case 2: (**a**) moving trajectories; (**b**) lateral offset.

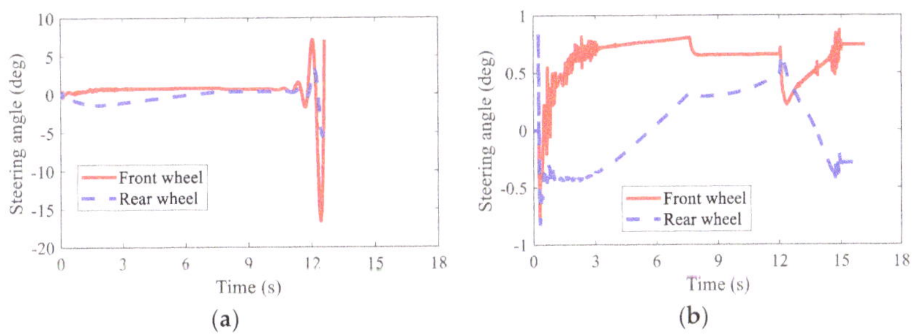

Figure 14. Steering angles of AMR in Case 2: (**a**) braking failure; (**b**) AFTC.

The velocities of AMR under three kinds of conditions are depicted in Figure 15. Due to the loss of stability, the simulation was stopped at the 12.6 s when without AFTC. The AMR could not finish the braking process after the braking failure of the IWMs. With AFTC, the AMR could realize safe braking as in the normal condition.

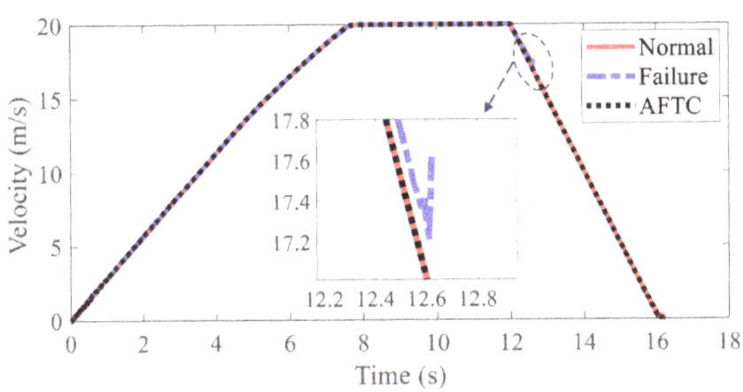

Figure 15. Velocity of AMR in Case 2.

The regenerative braking results are shown in Figure 16 and Table 3. In spite of the braking failure, the AFTC algorithm could help the AMR recover the braking energy up to 3.43×10^4 J. Due to the braking failure of FL and RL IWMs, the recovered braking energy was smaller than in the normal condition.

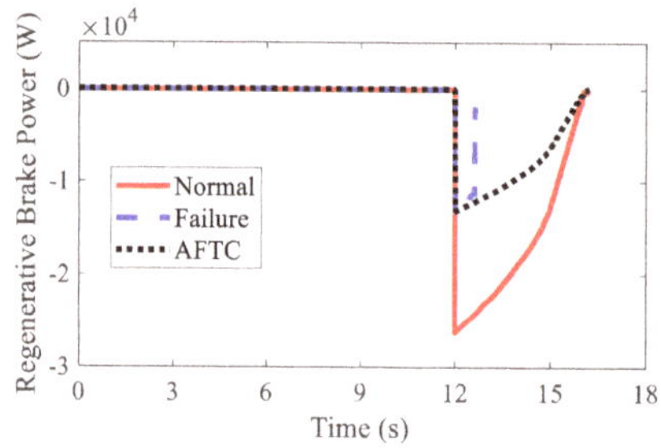

Figure 16. Regenerative braking power of AMR in Case 2.

Table 3. Regenerative braking energy of AMR in Case 2.

	Normal	Failure	AFTC
Regenerative energy (J)	6.85×10^4	7.67×10^3	3.43×10^4

The wheel torques of AMR with failure and with AFTC are illustrated in Figures 17 and 18, respectively. Due to the failure of FL and RL IWMs, the torques of the two IWMs changed to zero after the 10th second and the 12th second, respectively. Without AFTC, the AMR could not adjust its torque distribution to guarantee lateral stability. However, with AFTC, the EMBs redistributed the brake torque to compensate the braking force and overcome the external yaw moment caused by the braking failure of IWMs (Figure 18a,b).

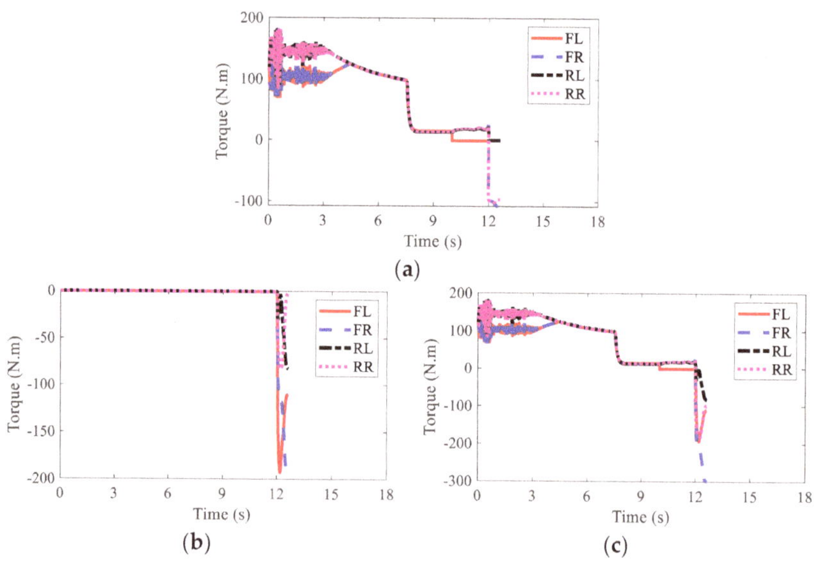

Figure 17. Wheel torques of AMR with failure in Case 2: (**a**) IWM; (**b**) EMB; (**c**) sum.

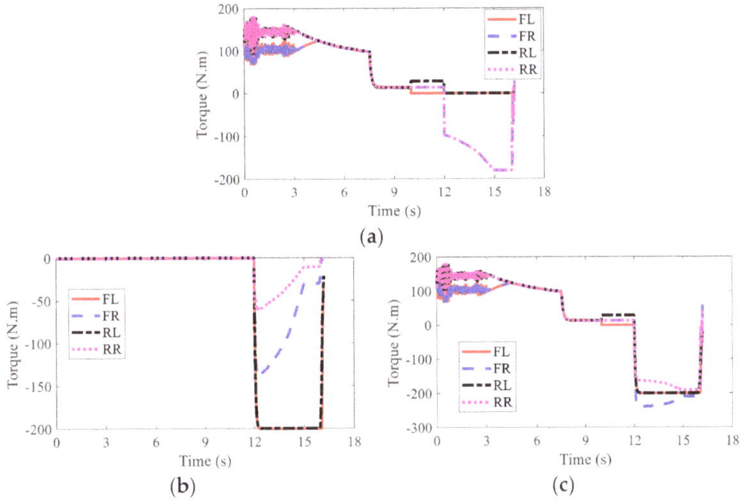

Figure 18. Wheel torques of AMR with AFTC in Case 2: (**a**) IWM; (**b**) EMB; (**c**) sum.

5.3. Simulation Case 3

In this case, the braking failure condition of three IWMs was studied, further validating the effectiveness of the AFTC algorithm. The AMR accelerated to 20 m/s and then started to brake after the 12th second. However, the FL IWM had a failure at the 10th second, and the RL and RR IWMs had a braking failure at the 12th second.

The path-tracking results of the AMR in this case are illustrated in Figure 19. This was similar to Case 2 in that without AFTC, the AMR departed from its original trajectory and lost stability after the braking failure of the IWMs. Moreover, the lateral offset was larger than that in Case 2. In spite of the increased failure numbers of IWMs, AFTC can help the AMR realize lane-keeping and safe braking. Figure 20 shows the steering angels of the AMR. It was found that the AMR had very large steering angles after the braking failure of IWMs, reaching the control boundaries. Despite this, the AMR could not guarantee stability and braking safety.

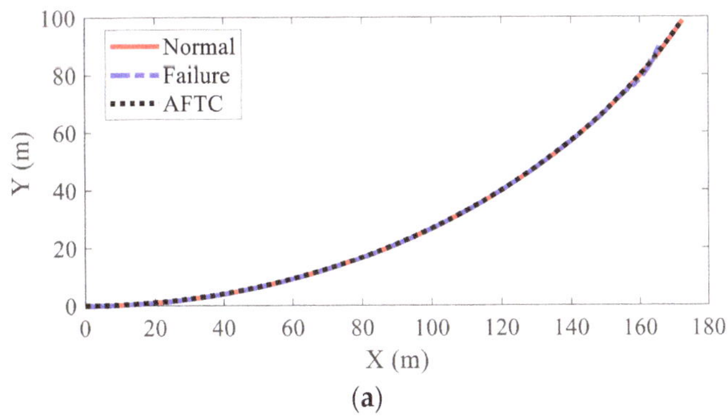

Figure 19. *Cont.*

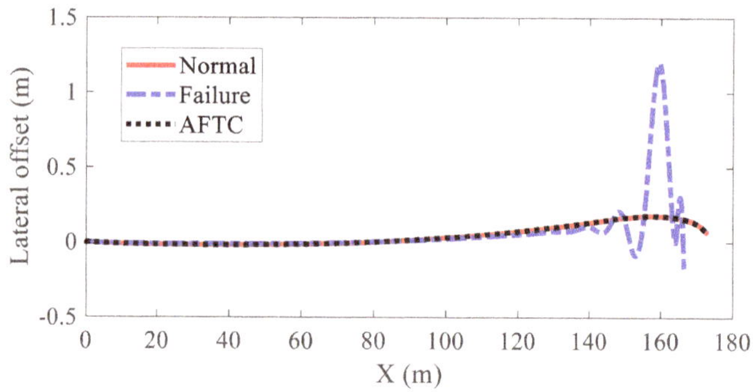

(**b**)

Figure 19. Path-tracking result of AMR in Case 3: (**a**) moving trajectories; (**b**) lateral offset.

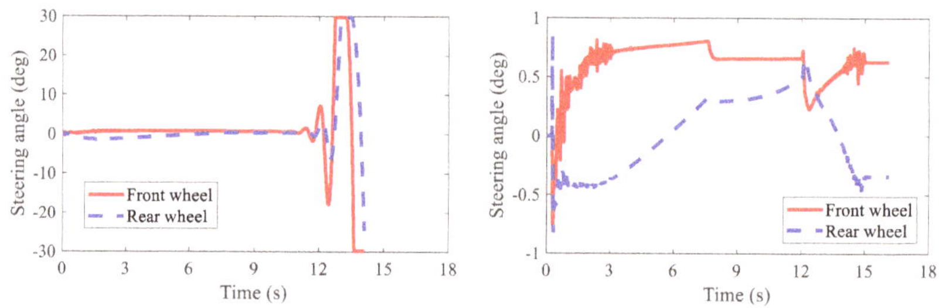

(**a**) (**b**)

Figure 20. Steering angles of AMR in Case 3: (**a**) braking failure; (**b**) AFTC.

Figure 21 shows the velocities of AMR under different conditions. Under the failure condition, the simulation was stopped at the 14.1 s due to the loss of stability of the AMR. However, the AFTC algorithm could help AMR address the braking failure of IWMs and finish the braking process safely.

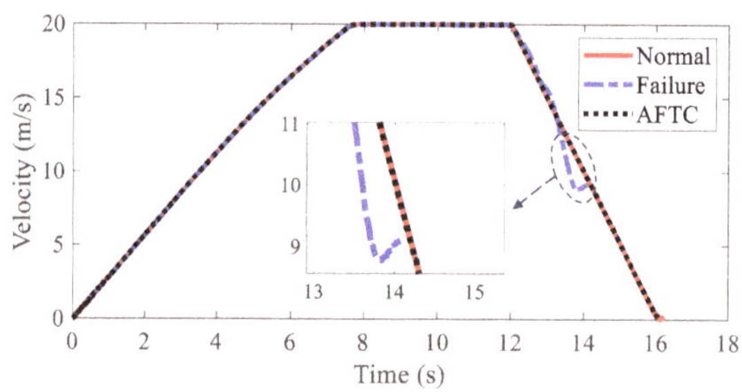

Figure 21. Velocity of AMR in Case 3.

The regenerative braking results of AMR are shown in Figure 22 and Table 4. In spite of the braking failure of three IWMs, the AFTC algorithm could help AMR recover braking energy up to -1.72×10^4 J using the normal IWM.

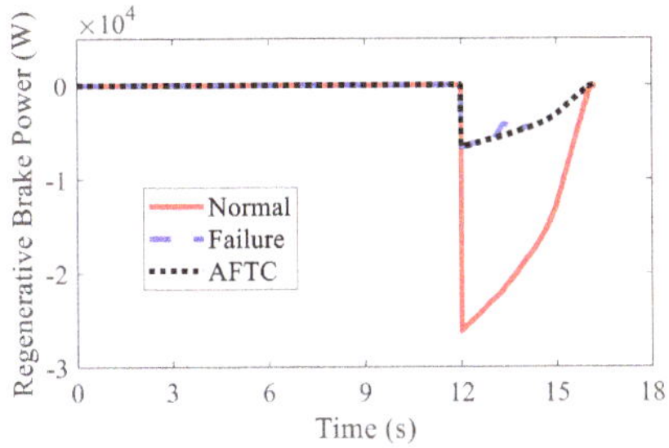

Figure 22. Regenerative braking power of AMR in Case 3.

Table 4. Regenerative braking energy of AMR in Case 3.

	Normal	Failure	AFTC
Regenerative energy (J)	6.84×10^4	1.14×10^4	1.72×10^4

The wheel torques of AMR under the failure condition and the AFTC condition are displayed in Figures 23 and 24, respectively. After the braking failure of three IWMs, the original torque distribution algorithm could not guarantee stability and braking safety. However, AFTC could help redistribute the torque of the normal IWM and four EMBs, recovering braking energy and guaranteeing braking safety and stability.

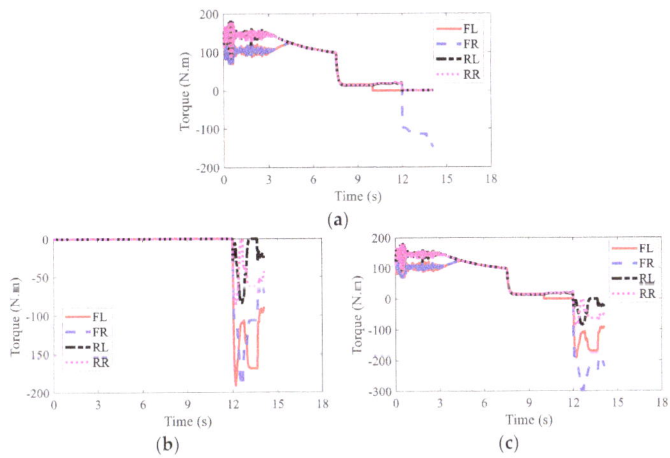

Figure 23. Wheel torques of AMR with failure in Case 3: (**a**) IWM; (**b**) EMB; (**c**) sum.

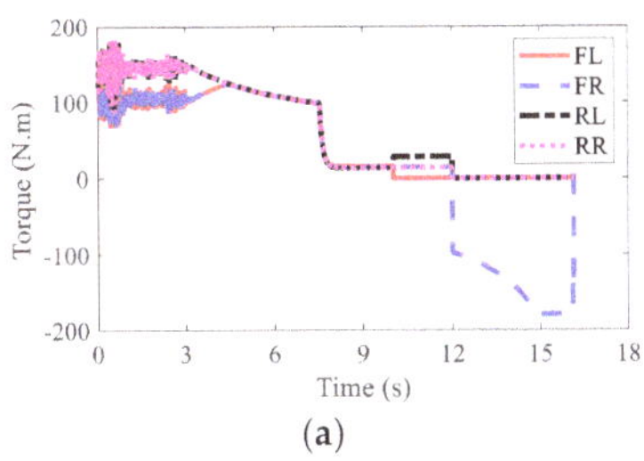

Figure 24. *Cont.*

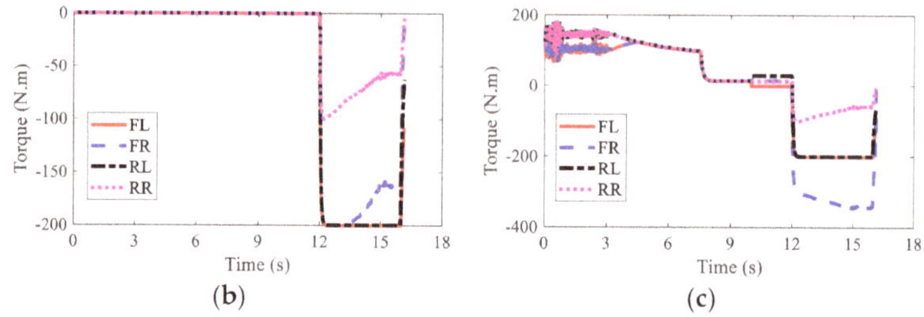

Figure 24. Wheel torques of AMR with AFTC in Case 3: (**a**) IWM; (**b**) EMB; (**c**) sum.

6. Conclusions

A chassis control framework was designed for an AMR. To address the braking failure of IWMs, an AFTC algorithm was studied by redistributing the braking torques of normal IWMs and four EMBs. Torque redistribution was carried out based on the WLS method. Three simulation cases were conducted to evaluate the feasibility and effectiveness of the proposed control algorithms. The simulation results indicate that the hybrid braking mode can help AMR recover the braking energy and advance braking safety. Moreover, the AFTC algorithm can deal with the braking failure of IWMs and realize braking energy recovery at the same time.

The hybrid conditions of IWM braking failure and EMB braking failure will be studied in future work.

Author Contributions: Writing—original draft preparation, P.H.; writing—review and editing, B.L.; supervision, C.L. All authors have read and agreed to the published version of the manuscript.

Funding: This research was supported by the Agency for Science, Technology and Research (A*STAR) under its IAF-ICP Programme ICP1900093 and the Schaeffler Hub for Advanced Research at NTU.

Institutional Review Board Statement: Not applicable.

Informed Consent Statement: Not applicable.

Data Availability Statement: The data presented in this study are available on request from the corresponding author. The data are not publicly available due to privacy reason.

Conflicts of Interest: The authors declare no conflict of interest.

References

1. Raharja, N.M.I.; Ma'arif, A.; Adiningrat, A.; Nurjanah, A.; Rijalusalam, D.U.; Sánchez-López, C. Empowerment of mosque community with ultraviolet light sterilisator robot. *J. Pengabdi. Dan Pemberdaya. Masy. Indones.* **2021**, *1*, 95–102.
2. Ni, J.; Hu, J.; Xiang, C. An AWID and AWIS X-by-wire UGV: Design and hierarchical chassis dynamics control. *IEEE Trans. Intell. Transp. Syst.* **2018**, *20*, 654–666. [CrossRef]
3. De Ryck, M.; Versteyhe, M.; Debrouwere, F. Automated guided vehicle systems, state-of-the-art control algorithms and techniques. *J. Manuf. Syst.* **2020**, *54*, 152–173. [CrossRef]
4. Le-Anh, T.; De Koster, M.B.M. A review of design and control of automated guided vehicle systems. *Eur. J. Oper. Res.* **2006**, *171*, 1–23. [CrossRef]
5. Chen, T.; Babanin, A.; Muhammad, A.; Chapron, B.; Chen, C. Modified evolved bat algorithm of fuzzy optimal control for complex nonlinear systems. *Rom. J. Inf. Sci. Technol.* **2020**, *23*, T28–T40.
6. Preitl, Z.; Precup, R.E.; Tar, J.K.; Takács, M. Use of multi-parametric quadratic programming in fuzzy control systems. *Acta Polytech. Hung.* **2006**, *3*, 29–43.
7. Zamfirache, I.A.; Precup, R.E.; Roman, R.C.; Petriu, E.M. Policy Iteration Reinforcement Learning-based control using a Grey Wolf Optimizer algorithm. *Inf. Sci.* **2022**, *585*, 162–175. [CrossRef]
8. Yang, G.; Yao, J.; Ullah, N. Neuroadaptive control of saturated nonlinear systems with disturbance compensation. *ISA Trans.* **2022**, *122*, 49–62. [CrossRef]
9. Yang, G.; Yao, J.; Dong, Z. Neuroadaptive learning algorithm for constrained nonlinear systems with disturbance rejection. *Int. J. Robust Nonlinear Control*, 2022; *in press*. [CrossRef]
10. Witczak, M.; Majdzik, P.; Stetter, R.; Lipiec, B. A fault-tolerant control strategy for multiple automated guided vehicles. *J. Manuf. Syst.* **2020**, *55*, 56–68. [CrossRef]

11. Zhang, S.; Zhuan, X. Research on Tracking Improvement for Electric Vehicle during a Car-following Process. In Proceedings of the 2020 Chinese Control and Decision Conference (CCDC), Hefei, China, 22–24 August 2020.
12. Li, M.; Xu, Y.; Lei, M.; Zhou, B. Velocity Tracking Control Based on Throttle-Pedal-Moving Data Mapping for the Autonomous Vehicle. *IEEE Access* **2019**, *7*, 176712–176718. [CrossRef]
13. Xu, W.; Chen, H.; Wang, J.; Zhao, H. Velocity optimization for braking energy management of in-wheel motor electric vehicles. *IEEE Access* **2019**, *7*, 66410–66422. [CrossRef]
14. Hang, P.; Chen, X.; Zhang, B.; Tang, T. Longitudinal velocity tracking control of a 4WID electric vehicle. *IFAC-Pap.* **2018**, *51*, 790–795. [CrossRef]
15. Ivanov, V.; Savitski, D.; Shyrokau, B. A survey of traction control and antilock braking systems of full electric vehicles with individually controlled electric motors. *IEEE Trans. Veh. Technol.* **2014**, *64*, 3878–3896. [CrossRef]
16. Chen, Q.; Kang, S.; Chen, H.; Liu, Y.; Bai, J. Acceleration slip regulation of distributed driving electric vehicle based on road identification. *IEEE Access* **2020**, *8*, 144585–144591. [CrossRef]
17. Guo, L.; Xu, H.; Zou, J.; Jie, H.; Zheng, G. Variable gain control-based acceleration slip regulation control algorithm for four-wheel independent drive electric vehicle. *Trans. Inst. Meas. Control* **2021**, *43*, 902–914. [CrossRef]
18. Peng, H.; Chen, X. Active Safety Control of X-by-Wire Electric Vehicles: A Survey. *SAE Int. J. Veh. Dyn. Stab. NVH* **2022**, *6*, 20. [CrossRef]
19. Mashadi, B.; Ahmadizadeh, P.; Majidi, M. Integrated controller design for path following in autonomous vehicles. *SAE Tech. Pap.* **2011**, 10. [CrossRef]
20. Mashadi, B.; Ahmadizadeh, P.; Majidi, M.; Mahmoodi-Kaleybar, M. Integrated robust controller for vehicle path following. *Multibody Syst. Dyn.* **2015**, *33*, 207–228. [CrossRef]
21. Hang, P.; Lv, C.; Huang, C.; Xing, Y.; Hu, Z. Cooperative Decision Making of Connected Automated Vehicles at Multi-Lane Merging Zone: A Coalitional Game Approach. *IEEE Trans. Intell. Transp. Syst.* **2022**, *23*, 3829–3841. [CrossRef]
22. Tian, Y.; Yao, Q.; Hang, P.; Wang, S. Adaptive Coordinated Path Tracking Control Strategy for Autonomous Vehicles with Direct Yaw Moment Control. *Chin. J. Mech. Eng.* **2022**, *35*, 1–15. [CrossRef]
23. Hang, P.; Xia, X.; Chen, G.; Chen, X. Active safety control of automated electric vehicles at driving limits: A tube-based MPC approach. *IEEE Trans. Transp. Electrif.* **2022**, *8*, 1338–1349. [CrossRef]
24. Zhao, Y.M.; Lin, Y.; Xi, F.; Guo, S. Calibration-based iterative learning control for path tracking of industrial robots. *IEEE Trans. Ind. Electron.* **2015**, *62*, 2921–2929. [CrossRef]
25. Huang, C.; Naghdy, F.; Du, H. Delta operator-based model predictive control with fault compensation for steer-by-wire systems. *IEEE Trans. Syst. Man Cybern. Syst.* **2018**, *50*, 2257–2272. [CrossRef]
26. Huang, C.; Naghdy, F.; Du, H. Delta operator-based fault estimation and fault-tolerant model predictive control for steer-by-wire systems. *IEEE Trans. Control Syst. Technol.* **2017**, *26*, 1810–1817. [CrossRef]
27. Huang, C.; Naghdy, F.; Du, H. Sliding mode predictive tracking control for uncertain steer-by-wire system. *Control Eng. Pract.* **2019**, *85*, 194–205. [CrossRef]
28. Wada, N.; Fujii, K.; Saeki, M. Reconfigurable fault-tolerant controller synthesis for a steer-by-wire vehicle using independently driven wheels. *Veh. Syst. Dyn.* **2013**, *51*, 1438–1465. [CrossRef]
29. Wang, C.; Heng, B.; Zhao, W. Yaw and lateral stability control for four-wheel-independent steering and four-wheel-independent driving electric vehicle. *Proc. Inst. Mech. Eng. Part D J. Automob. Eng.* **2020**, *234*, 409–422. [CrossRef]
30. Yang, H.; Cocquempot, V.; Jiang, B. Optimal fault-tolerant path-tracking control for 4WS4WD electric vehicles. *IEEE Trans. Intell. Transp. Syst.* **2009**, *11*, 237–243. [CrossRef]
31. Zhang, B.; Lu, S.; Wu, W.; Li, C.; Lu, J. Robust fault-tolerant control for four-wheel individually actuated electric vehicle considering driver steering characteristics. *J. Frankl. Inst.* **2021**, *358*, 5883–5908. [CrossRef]
32. Huang, C.; Naghdy, F.; Du, H. Fault tolerant sliding mode predictive control for uncertain steer-by-wire system. *IEEE Trans. Cybern.* **2019**, *49*, 261–272. [CrossRef] [PubMed]
33. Hao, L.Y.; Zhang, H.; Li, T.S.; Lin, B.; Chen, C.L.P. Fault tolerant control for dynamic positioning of unmanned marine vehicles based on TS fuzzy model with unknown membership functions. *IEEE Trans. Veh. Technol.* **2021**, *70*, 146–157. [CrossRef]
34. Huang, C.; Naghdy, F.; Du, H. Observer-based fault-tolerant controller for uncertain steer-by-wire systems using the delta operator. *IEEE/ASME Trans. Mechatron.* **2018**, *23*, 2587–2598. [CrossRef]
35. Hang, P.; Chen, X. Towards Autonomous Driving: Review and Perspectives on Configuration and Control of Four-Wheel Independent Drive/Steering Electric Vehicles. *Actuators* **2021**, *10*, 184. [CrossRef]
36. Du, H.; Zhang, N.; Dong, G. Stabilizing vehicle lateral dynamics with considerations of parameter uncertainties and control saturation through robust yaw control. *IEEE Trans. Veh. Technol.* **2010**, *59*, 2593–2597.
37. Ding, N.; Taheri, S. A modified Dugoff tire model for combined-slip forces. *Tire Sci. Technol.* **2010**, *38*, 228–244. [CrossRef]
38. Huang, C.; Huang, H.; Hang, P.; Gao, H.; Wu, J.; Huang, Z.; Lv, C. Personalized trajectory planning and control of lane-change maneuvers for autonomous driving. *IEEE Trans. Veh. Technol.* **2021**, *70*, 5511–5523. [CrossRef]
39. Hang, P.; Xia, X.; Chen, X. Handling stability advancement with 4WS and DYC coordinated control: A gain-scheduled robust control approach. *IEEE Trans. Veh. Technol.* **2021**, *70*, 3164–3174. [CrossRef]
40. Hang, P.; Chen, X.; Wang, W. Cooperative control framework for human driver and active rear steering system to advance active safety. *IEEE Trans. Intell. Veh.* **2021**, *6*, 460–469. [CrossRef]

41. Ma, J.; Cheng, Z.; Zhang, X.; Tomizuka, M.; Lee, T.H. Alternating direction method of multipliers for constrained iterative LQR in autonomous driving. *arXiv* **2020**, arXiv:2011.00462.
42. Ma, J.; Cheng, Z.; Zhang, X.; Tomizuka, M.; Lee, T.H. Optimal decentralized control for uncertain systems by symmetric Gauss–Seidel semi-proximal ALM. *IEEE Trans. Autom. Control* **2021**, *66*, 5554–5560. [CrossRef]

sensors

MDPI

Article

Generalized Single-Vehicle-Based Graph Reinforcement Learning for Decision-Making in Autonomous Driving

Fan Yang [1], Xueyuan Li [1,*], Qi Liu [1], Zirui Li [1,2] and Xin Gao [1]

[1] School of Mechanical Engineering, Beijing Institute of Technology, Beijing 100081, China; yangfanbitdb@163.com (F.Y.); 3120195257@bit.edu.cn (Q.L.); 3120195255@bit.edu.cn (Z.L.); 13403627345@163.com (X.G.)

[2] Department of Transport and Planning, Faculty of Civil Engineering and Geosciences, Delft University of Technology, Stevinweg 1, 2628 CN Delft, The Netherlands

* Correspondence: lixueyuan@bit.edu.cn

Abstract: In the autonomous driving process, the decision-making system is mainly used to provide macro-control instructions based on the information captured by the sensing system. Learning-based algorithms have apparent advantages in information processing and understanding for an increasingly complex driving environment. To incorporate the interactive information between agents in the environment into the decision-making process, this paper proposes a generalized single-vehicle-based graph neural network reinforcement learning algorithm (SGRL algorithm). The SGRL algorithm introduces graph convolution into the traditional deep neural network (DQN) algorithm, adopts the training method for a single agent, designs a more explicit incentive reward function, and significantly improves the dimension of the action space. The SGRL algorithm is compared with the traditional DQN algorithm (NGRL) and the multi-agent training algorithm (MGRL) in the highway ramp scenario. Results show that the SGRL algorithm has outstanding advantages in network convergence, decision-making effect, and training efficiency.

Keywords: autonomous driving; decision-making; graph convolution; deep reinforcement learning

Citation: Yang, F.; Li, X.; Liu, Q.; Li, Z.; Gao, X. Generalized Single-Vehicle-Based Graph Reinforcement Learning for Decision-Making in Autonomous Driving. *Sensors* **2022**, 22, 4935. https://doi.org/10.3390/s22134935

Academic Editors: Yafei Wang, Chao Huang, Peng Hang, Zhiqiang Zuo and Bo Leng

Received: 24 May 2022
Accepted: 28 June 2022
Published: 29 June 2022

1. Introduction

In autonomous driving, the decision-making system is mainly used to produce advanced actions of vehicles, such as lane changing, acceleration, braking, and so on. Tactical decision-making for autonomous driving is challenging due to the diversity of environments, the uncertainty in the sensor information, and the complex interaction with other road users [1,2]. Traditional vehicle trajectory modeling and tracking control methods, such as genetic algorithms, neural networks, and their optimizations, have played a positive role in the research of decision-making [3].

The operational space of an autonomous vehicle (AV) can be diverse and vary significantly. Due to this, formulating a rule-based decision-maker for selecting driving maneuvers may not be ideal [4]. With the development of deep learning, the domain of reinforcement learning (RL) has become a robust learning framework now capable of learning complex policies in high-dimensional environments [5]. Therefore, using reinforcement learning to solve decision-making problems has gradually become the mainstream of research. Carl-Johan Hoel et al. introduce a method based on deep reinforcement learning for automatically generating a general-purpose decision-making function [6]. They trained an RL agent to handle a truck–trailer combination's speed and lane change decisions in a simulated environment. Hongbo Gao et al. solved sequential decision optimization problems based on the inverse reinforcement learning algorithm, and the proposed method was verified in terms of efficiency [7]. Some algorithms for planning and decision-making are based on analyzing driver behavior [8,9].

The realization of the decision-making is based on the understanding and analysis of high-dimensional environmental information. However, the traditional reinforcement learning algorithm only has a good capacity for decision-making based on low-dimension features [10]. It will have the problem of insufficient understanding in the face of more complex scenario information. The deep neural network (DNN) has a strong ability to learn representations and for the generalization of matching patterns from high-dimensional data. Therefore, deep reinforcement learning (DRL) algorithms are effective in tasks requiring feature representation and policy learning, e.g., autonomous driving decision-making [11]. Using the functional approximation ability of a deep neural network (DNN), an intelligent controller integrating artificial intelligence technologies such as deep learning (DL) and reinforcement learning (RL) is designed to maintain and avoid obstacles in lanes [12–14], decision-making [15–21], longitudinal control [22], merger maneuvers [23], human-like driving strategies [24–26], and other large-scale autonomous driving control tasks. Yingjun Ye et al. put forward a framework for decision-making training and learning. It consists of a deep reinforcement learning (DRL) training program and a high-fidelity virtual simulation environment [27]. Compared with the DQN algorithm based on a value function, the deep deterministic policy gradient (DDPG) algorithm based on an action policy can solve the continuity problem of the action space. Haifei Zhang et al. used the DDPG algorithm to solve the control problem of automatic driving based on a reasonable reward function, deep convolution network, and exploration policy [28].

Interaction between vehicles in common public transport scenarios is necessary and pervasive. However, there are relatively few studies on how autonomous vehicles interact in public environments with reinforcement learning. Realizing coordination between vehicles in a shared environment is challenging due to the unique feature of vehicular mobility, which makes it infeasible to apply the existing reinforcement learning methods directly. Chao Yu et al. proposed using a dynamic coordination graph to model the continuously changing topology during vehicles' interactions and developed two basic learning approaches to coordinate the driving maneuvers for a group of vehicles [29].

Cooperative vehicle–infrastructure systems and automatic driving technology are developing rapidly [30]. The graph neural network (GNN) has gained increasing popularity in various domains, including social network analysis [31,32]. GNN can extract the relational data representations and generate useful node embeddings on the node features and the features from neighboring nodes. The interactions between the ego vehicle and other surrounding vehicles can also be represented by the dynamic potential field (DPF) and embedded in the gap acceptance model to ensure safety and personalization during driving [33]. Jiqian Dong et al. proposed a novel deep reinforcement learning (DRL)-based approach combining the graphic convolution neural network (GNN) and deep Q network (DQN), namely the graphic convolution Q network, as the information fusion module and decision processor [34]. The proposed model can aggregate the information obtained by collaborative perception and output collaborative lane change decisions for multiple vehicles. Even in the case of highly dynamic and partially observed mixed traffic, the intention can be satisfied.

However, the above multi-agent training-based GNN reinforcement learning (MGRL) has the following problems in the actual verification process (the scenario of highway ramp exiting):

1. Multi-agent training simultaneously increases the computational network complexity, resulting in a higher overall training time cost. Therefore, each parameter modification requires a more prolonged verification time, which is not conducive to the development and adjustment of the algorithm.
2. The reward and punishment offset each other in multi-agent overall training, resulting in poor network convergence in the training process. Due to the mutual influence of multiple agents' reward values, it cannot accurately evaluate the current state, resulting in the very unstable fluctuation of the loss curve.

3. Through the test of the final training model, the final task success rate of the GCQ algorithm is maintained at around 50%, which cannot meet the basic driving needs of vehicles.
4. The GCQ decision-making model stays in the lateral lane change and cannot control the longitudinal behavior of the vehicle at the same time. This is an incomplete driving control model, which leads to a low success rate, serious collisions, and low traffic efficiency.

Given the above problems, this paper proposes an improved single-agent GNN-based RL algorithm (SGRL algorithm). This paper has the following contributions:

1. Based on retaining interactive feature extraction (GNN), the trained object is transferred from multi-agent to single-agent. Through the internal processing of the network model, the training results of single-agent can be applied to the application scenarios of multi-agent. This training method can significantly reduce the time cost of model training and eliminate the interaction of rewards and punishments between agents to improve the training effect.
2. An inductive reward function is designed to improve the convergence speed of the model. The reward function incorporates driving intention, collision, lane change frequency, and vehicle speed into the calculation. The trained model simultaneously performs well in terms of task success rate, safety, driving stability, and traffic efficiency.
3. The dimension enhancement of the action space is realized by changing the network structure. Adding vehicle longitudinal velocity control gives the model a more robust control ability for task success rate, safety, and traffic efficiency.
4. In the training process, the real-time screening of stored data can improve the training speed. The random generation mechanism of vehicles' number, type, and position in the training scenario can improve the model's generalization ability and avoid overfitting.

The paper is organized as follows. Section 2 introduces the proposed SGRL algorithm in detail. Section 3 shows the model training and testing of the SGRL algorithm and comparison algorithms (MGRL and NGRL). Section 4 shows the results of training and testing and analyzes the comparison. Finally, Section 5 derives the conclusions and proposes future improvement directions.

2. Method

The proposed SGRL methods are used to model the Markov Decision Process (MDP). The agents can explore the environment by observing states, taking actions, and receiving rewards, as shown in Figure 1.

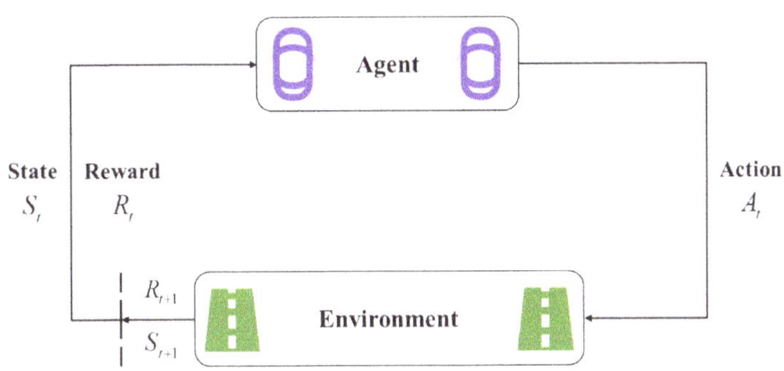

Figure 1. Markov Decision Process.

The implementation core of the SGRL method is based on the preprocessing of input data, the structure of the deep neural network, and the setting of the output end.

2.1. Network Input

In each time step in the training process, the vehicles in the scenario can be divided into human-driven vehicles (HVs) and autonomous vehicles (AVs). At each step t, the input of the training model can be divided into two parts: feature matrix X_t and correlation matrix C_t. The state $s_t = (X_t, C_t)$. Concerning the feature matrix X_t, the necessary basic information based on highway scenarios is included. The total number of human-driven vehicles (HVs) and autonomous vehicles (AVs) is set to $N = N_{HV} + N_{AV}$. Then, the specification of the matrix is $N \times 8$. The eight characteristic parameters of each vehicle describe the speed, lateral position, longitudinal position, and destination information, denoted as $x_i = (v_i, p_i, l_i, I_i)$, To serialize multiple data at the same scale, the parameters are set as follows:

- $v_i = \frac{v_{i-current}}{v_{\max}}$ is the relative speed, where $v_{\max} = \max(v_{\max-vehicle}, v_{\max-highway})$ is the maximum speed for the current vehicle.
- $p_i = \frac{x_i}{l_{highway}}$ is the relative longitudinal position normalized by the entire length of the highway. (The algorithm belongs to local planning. Its application scenarios are specific ramps and expressway sections within short distances, so the normalization of location information is more favorable for network computing.)
- l_i is the lateral lane position for the vehicle i. The position of the vehicle is represented by three-bit binary coding. For example, vehicles in the rightmost lane are denoted as $[1,0,0]$, the middle lane $[0,1,0]$ and the leftmost lane $[0,0,1]$.
- I_i is the destination intention feature for the vehicle i. The data type is similar to l_i. The three destinations are expressed as $[1,0,0]$ (merge out from the first ramp), $[0,1,0]$ (merge out from the second ramp) and $[0,0,1]$ (go straight along the highway).

Based on GNN, we can introduce a correlation matrix C_t to calculate the relationship between vehicle nodes. Considering that the sensors installed on autonomous vehicles have fixed sensing ranges, only vehicles within the sensor sensing range are recorded in the correlation matrix. C_t is a square matrix with the specification $N \times N$. The row data i of the matrix represent the relationship between vehicle i and other vehicles through binary data. The value C_{ij} of the vehicle j within the set distance of the target vehicle i will be set to 1, as shown in Figure 2.

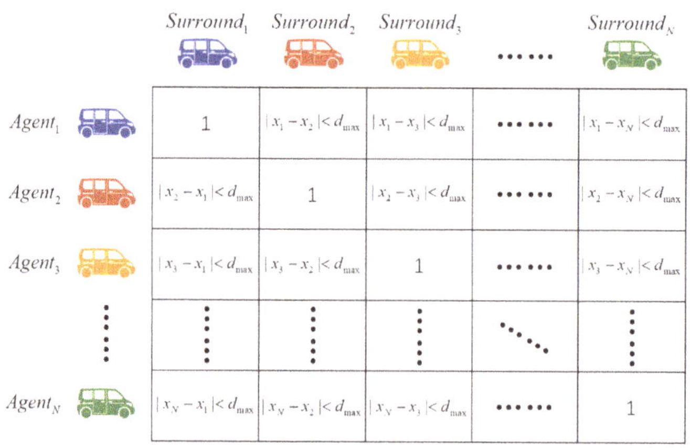

Figure 2. The schematic diagram of the correlation matrix setting. Each row in the matrix represents the correlation between a vehicle and all other vehicles, specifically the logical values of 0 and 1, where 0 represents the actual distance greater than the set distance, and 1 represents the actual distance less than the set distance.

Considering that the number of vehicles in the scene is dynamic, this situation has been considered when setting the feature matrix and the correlation matrix. The vehicles in the environment are divided into HVs and AVs, located in the feature matrix's upper and lower parts. When the number of vehicles is less than N, the positions in the matrix are occupied by 0.

To ensure the authenticity of the simulation process, all vehicles appear and disappear successively in the scenario, so the feature information and correlation information obtained at the step *t* cannot reach the predetermined matrix size. To ensure the standard calculation of GNN, matrix data filling is needed. Matrix segmentation prevents data confusion and error correspondence caused by random packing, as shown in Figure 3.

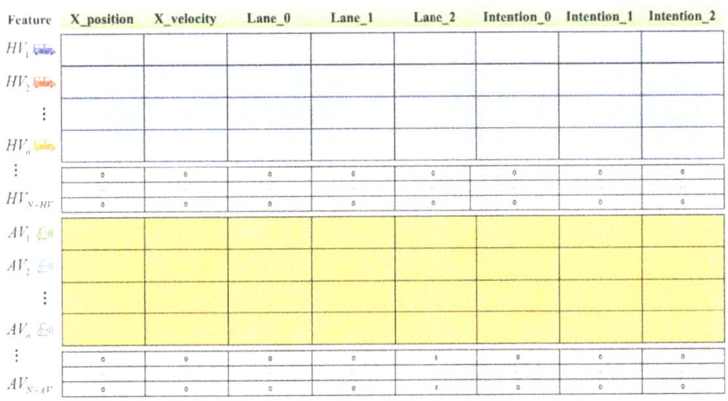

Figure 3. Matrix segmentation diagram. Each row in the matrix represents the characteristic information of a vehicle, and the matrix is divided into the upper and lower parts to separate AV and HV. The green part indicates that the vehicle is not currently in the scenario.

2.2. Network Structure

In the whole network structure, the function of graph convolution is to obtain the interaction information between vehicles. The function of the full connection layer is to parse the matrix information. Each newly obtained matrix in the network needs to be input into the full connection layer for recording and information analysis. In addition, the number of nodes in each layer also plays a decisive role in the model's performance. The optimal number of nodes is selected through comparative experiments.

Data records for comparative tests are shown in Table 1.

Table 1. Training effect for different numbers of nodes.

N of Nodes	Training Time for 1000 Episodes	Convergence Effect
32	1.5179 h	Poor (large fluctuation)
64	2.6438 h	Acceptable (occasional large fluctuations)
128	**3.4756 h**	**Good (small fluctuation)**
256	6.0987 h	Good (small fluctuation)
512	10.9542 h	Good (small fluctuation)

At each step t, the feature matrix X_t is first fed to a Fully Connected Network (FCN) encoder ψ and then the matrix $H_t \in {}^{N\times 128}$ is obtained (Equation (1)). The original eight characteristic values will be recoded to 128 values. FCN will be executed twice consecutively.

$$H_t = \psi(X_t) \in {}^{N\times 128} \tag{1}$$

Next is the calculation of graph convolution. The input of the convolution function is the newly obtained feature matrix H_t and correlation matrix C_t. The function output matrix K_t has the same specification as the original matrix H_t (Equation (2)).

$$K_t = \chi(H_t, C_t) \in \mathbb{R}^{N\times 128} \tag{2}$$

The original feature information and the new information obtained by graph convolution are fused by matrix stitching online as the total input of the subsequent neural network η.

The feature data density changes at each stage are as follows:

- Fully Connected Network (FCN): $Dense(8) \rightarrow Dense(128) \rightarrow Dense(128)$;
- Graph Neural Network (GNN): $Dense(128) \rightarrow Dense(128) \rightarrow Dense(128)$;
- Q Network: $Dense(256) \rightarrow Dense(128) \rightarrow Dense(128)$;
- Output: $Dense(128) \rightarrow Dense(33)$.

The overall structure of the SGRL algorithm is shown in Figure 4.

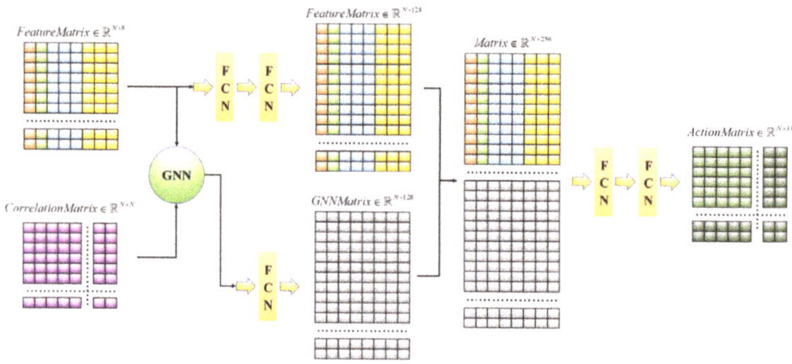

Figure 4. The overall structure diagram of the model. FCN represents the full connection layer, and GNN represents the graph neural network.

2.3. Network Output

Since the vehicle longitudinal control model built into the simulation environment is still rule-based, it cannot incorporate the interaction between vehicles into the control model. The SGRL algorithm fuses longitudinal and lateral control into the same model by increasing the output dimension, which significantly simplifies the complexity of the control process and solves the coupling problem of two directions.

The SGRL algorithm is trained based on DQN, so the output action space can only be discrete. For AVs, the longitudinal control is mainly reflected in the acceleration, and the lateral control is primarily reflected in the lane change direction. The improvement of the output action resolution is conducive to improving the control sensitivity, but it also increases the computational complexity and reduces the control frequency. In the algorithm of this paper, a compromise solution is selected. The longitudinal control is set to 11 discrete values in the interval $[-5, 5]$, and the lateral control is set to three actions: keeping, turning left and turning right. Thus, the output matrix of the model is set as $A_t \in \mathbb{R}^{N\times 33}$, as shown in Figure 5.

Action space correspondence relationship		Longitudinal Acceleration										
		-5	-4	-3	-2	-1	0	1	2	3	4	5
Lane Change Action	**left**	**0**	**3**	**6**	**9**	**12**	**15**	**18**	**21**	**24**	**27**	**30**
	keep	**1**	**4**	**7**	**10**	**13**	**16**	**19**	**22**	**25**	**28**	**31**
	right	**2**	**5**	**8**	**11**	**14**	**17**	**20**	**23**	**26**	**29**	**32**

Figure 5. Model action space diagram. The matrix row direction divides the longitudinal acceleration into discrete values, and the matrix column direction represents the lateral lane change of the vehicle.

2.4. *Reward Function*

The setting of the reward function needs clear guidance and a strong correlation with training objectives. In the SGRL algorithm, the task success rate, security, and traffic efficiency should be considered simultaneously. The corresponding reward values are set as intention reward, crash reward, and speed reward.

2.4.1. Intention Reward

To guide autonomous vehicles to complete driving tasks from the corresponding ramps out of the highway, the SGRL algorithm constructs reward gradients for different lanes, as shown in Figure 6. Merge_0 and merge_1, respectively, refer to the autonomous vehicles that need to drive away from ramp_0 and ramp_1 according to the task requirements. In the simulation experiment, the driving route of all autonomous vehicles is entirely determined by the reinforcement learning controller. Therefore, vehicles belonging to the merge_0 category will not necessarily exit from ramp_0; that is, they cannot complete the driving task.

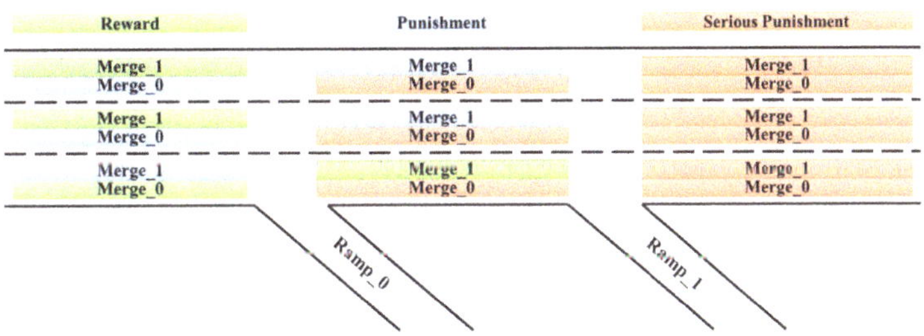

Figure 6. Intention reward gradient diagram. The task completion of autonomous vehicles is seen as a factor in judging the quality of the current reinforcement learning model, which is manifested in the reward values that can be obtained when each vehicle is in different driving sections and lanes. The strip area represents the reward types that the corresponding vehicles can obtain from the area. Green represents reward ($1 \times R_{I-Base}$), blue represents punishment ($-1 \times R_{I-Base}$), and orange represents serious punishment ($-2 \times R_{I-Base}$). R_{I-Base} is set to 1 for normalization.

To ensure practical guidance for all locations, the reward value R_{I-t} for the fixed area is set to a constant value. To ensure the interaction between various rewards and to pass it to the training process, it is necessary to pay attention to the consistency of the numerical scale when setting R_{I-t}. Moreover, the R_{I-t} needs to distinguish between positive and negative values, which is more conducive to the training. In addition, if the autonomous vehicle completes the task, it can obtain greater R_{I-t}, and if the task fails, it will also obtain a greater negative R_{I-t}.

2.4.2. Crash Reward

The safety of autonomous driving is the basis of all other characteristics. The simulation platform SUMO can detect collisions at step t and output the number $N_{collision-t}$ of vehicles involved in collisions. The collision reward is calculated according to $N_{collision-t}$ (Equation (3)).

$$R_{C-t} = -N_{collision-t}/2 \tag{3}$$

2.4.3. Speed Reward

To control the consistency of the reward scale, the speed reward value R_{S-t} needs to be normalized. $v_{\max} = \max(v_{\max-vehicle}, v_{\max-highway})$ is the max speed for the AV. To ensure the positive and negative values of R_{S-t}, the calculation is shown as Equation (4).

$$R_{S-t} = \frac{v_{i-t}}{v_{\max}} - 0.3 \tag{4}$$

2.4.4. Total Reward

The general method to calculate the total reward is a weighted summation of each component. Direct addition will lead to mutual coverage of rewards, resulting in poor training effects. The SGRL algorithm uses a new aggregation method, as shown in Equation (5).

$$R_t = \begin{cases} \omega_I \times R_{I-t} \times R_{S-t} + \omega_C \times R_{C-t} & (R_{I-t} > 0) \\ \omega_I \times R_{I-t} + \omega_C \times R_{C-t} & (R_{I-t} \leq 0) \end{cases} \tag{5}$$

This calculation method takes the task success rate and safety as the primary considerations and considers traffic efficiency simultaneously. The problem of vehicle parking for intention rewards can be completely avoided. Moreover, the weight relationship between rewards can be adjusted by parameters ω_I and ω_C. The parameter settings of this paper are $\omega_I = 1, \omega_C = 2$.

2.5. Model Training and Testing

The most prominent feature of the SGRL algorithm is training for a single agent, but the obtained model can be applied to a multi-agent environment. For trained agents, the vehicles around them can be considered the same category—surrounding vehicles. Therefore, the work done by the GNN-based reinforcement learning model can be interpreted as planning and decision-making based on the characteristics and relationships of the target agent and its surrounding vehicles. The model obtained by single-agent training can be directly transplanted to other agents in the same scenario, as shown in Figure 7.

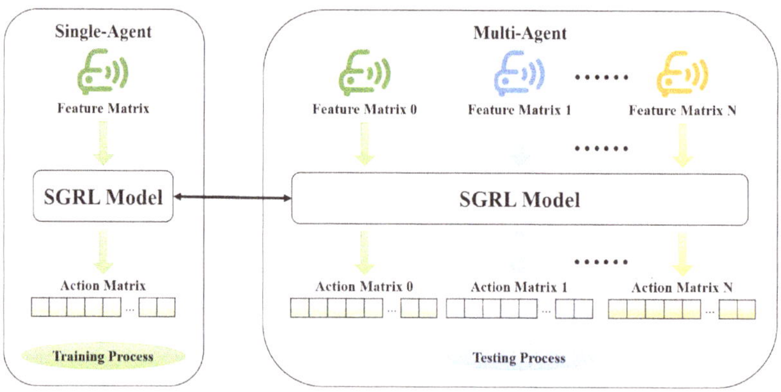

Figure 7. Reinforcement learning model transplant diagram.

To prevent the overfitting of the reinforcement learning process, we randomize the distribution of training vehicles and the task of autonomous vehicles in each episode based on fixed rules, as shown in Figure 8.

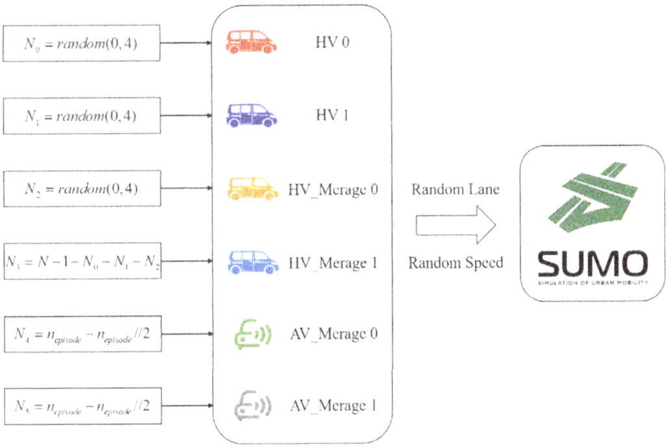

Figure 8. Scenario random setting diagram.

Algorithm 1 shows the detailed steps of training.

Algorithm 1 SGRL Q Learning Steps

Initialize the reply memory R to capacity N
Initialize the weights for the SGRL Net $(\psi, \chi, \eta \ldots\ldots)$
Jointly Current Network $\hat{Q}_\theta$ and Target Network $\hat{Q}_t = \hat{Q}_\theta$
Warming up
For step $t = 1$ to T_0(warming up steps) **do**
  Choose random action for agent i: $a_{t-r} = np.random.choice(np.arrange(3), N)$
  Get the transition (s_t, a_t, r_t, s_{t+1})
  Store in the buffer
Training step
For step $t = T_0 + 1$ to T(total steps) **do**
  With the probability e choose random action for agent i: a_{t-r}
  With the probability $1 - e$ do:
    State decoding: $(X_t, C_t) = s_t$
    Double FCN: $H_{t-1} = \psi_0(X_t) \in \mathbb{R}^{N \times 128}$; $H_{t-2} = \psi_1(H_{t-1}) \in \mathbb{R}^{N \times 128}$
    GNN + FCN: $K_t = \chi(H_{t-2}, C_t) \in \mathbb{R}^{N \times 128}$; $K_{t-1} = \psi_2(K_t) \in \mathbb{R}^{N \times 128}$
    Feature Stitching: $F_t = (H_{t-2}, K_{t-1}) \in \mathbb{R}^{N \times 256}$
    Double FCN: $F_{t-1} = \psi_3(F_t) \in \mathbb{R}^{N \times 128}$; $F_{t-2} = \psi_4(F_{t-1}) \in \mathbb{R}^{N \times 128}$
    Compute Q values: $\hat{Q}_\theta(s_t) = \psi_5(F_{t-3}) \in \mathbb{R}^{N \times 33}$
    Select $a_t^* = \arg\max \hat{Q}_\theta(s_t)$
  Execute a_t^* and get a new state s_{t+1}
  Store in the buffer
  Set $s_t = s_{t+1}$
Training model at training step
Sample a batch from the buffer and calculate the Q target:

$$Q_{target} = \begin{cases} r_t + \gamma \max\limits_a \hat{Q}_\theta(s_t) & done = 0 \\ r_t & done = 1 \end{cases}$$

Get average $Loss$
Update parameters θ of the model $\hat{Q}_\theta$
Updating Target Net every n steps
$\hat{Q}_{target} = \hat{Q}_\theta$

Algorithm 2 shows the detailed steps of testing.

Algorithm 2 SGRL Testing Steps

Initialize the simulation environment
Testing step
For step $t = 1$ to T(total test steps) **do**
  State decoding: $(X_t, C_t) = s_t$
  For AV $i = 1$ to N_{AV} **do**
    Move other AV's features to the back of HV: $X_{t0} = \lambda(X_t) \in \mathbb{R}^{N\times 8}$
    Calculate Q based on trained model: $\hat{Q}_\theta(s_t) = \pi_{SGRL}(X_{t0}, C_t) \in \mathbb{R}^{N\times 33}$
    Select $a^*_{t-i} = \arg\max \hat{Q}_\theta(s_t)$
  Get the action matrix $a^*_t = (a^*_1, a^*_2, \ldots\ldots a^*_{N_{AV}})$
  Execute a^*_t and get a new state s_{t+1}
  Set $s_t = s_{t+1}$

3. Simulation

3.1. Baseline Models

Two baseline models are introduced for comparative analysis in the simulation: the traditional DQN algorithm (NGRL) and the multi-agent training algorithm (MGRL). In the NGRL algorithm, the GNN part is removed, and the correlation matrix is used as input by splicing with the feature matrix.

The model of the MGRL algorithm is consistent with the SGRL proposed in this paper in terms of network structure, but its training process is based on multi-agent environment interaction.

By comparing the three models, the specific effects of the GNN structure and single-agent training of SGRL can be effectively analyzed. The simulation scenario is shown in Figure 9.

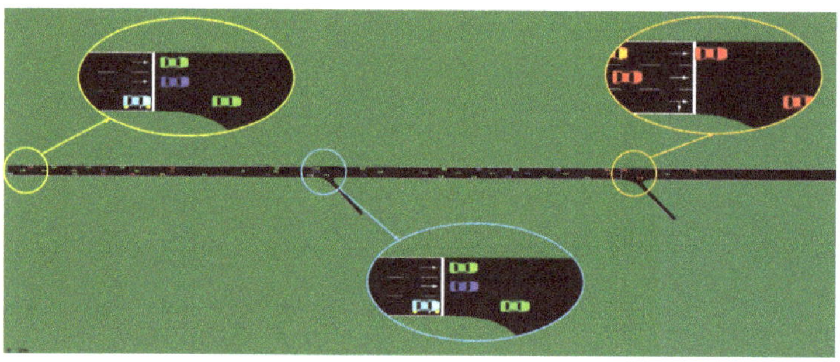

Figure 9. Simulation scenario diagram.

3.2. Simulator Parameters

The simulation scenario is a long highway with three lanes. There are exit ramps at one-third and two-thirds of the total length respectively. The speed limit for the whole road is set as 20 m/s (76 km/h) for all the vehicles. The AVs and HVs are put into the scenario at the probability of 0.1 and 0.4 from the left side of the road. And the initial speed and lane position are random.

There are six types of vehicles in the scenario, including HVs that travel straight through the highway, vehicles (HVs and AVs) that want to exit from ramp_0 and vehicles (HVs and AVs) that want to exit from ramp_1. The simulation environment controls the driving of HVs, and the AVs are completely controlled by the reinforcement learning model SGRL in real time.

The number of vehicles in the experiment is set as shown in Table 2.

Table 2. Number setting of experimental vehicles.

Algorithm Type		N of Vehicles		
		AVs		HVs
		Merge_0	Merge_1	
Training Process	SGRL	1		19
	MGRL	5	5	10
	NGRL	5	5	10
Testing Process	SGRL	5	5	10
	MGRL	5	5	10
	NGRL	5	5	10

4. Results and Discussion

4.1. Training Results

All three models were trained for 1000 episodes. The symbolic data of the training process are the reward, average Q value and loss value. Average rewards are obtained by averaging rewards against steps. The specific changes are shown in Figures 10–13.

For the comparison of reward values in the training process, under the same reward value calculation, the reward values and average reward values of the SGRL algorithm can converge faster and have better final convergence.

The changing trend of the Q value and loss value of the SGRL algorithm in the training process is consistent with the learning process. According to the loss results, SGRL has a faster convergence speed.

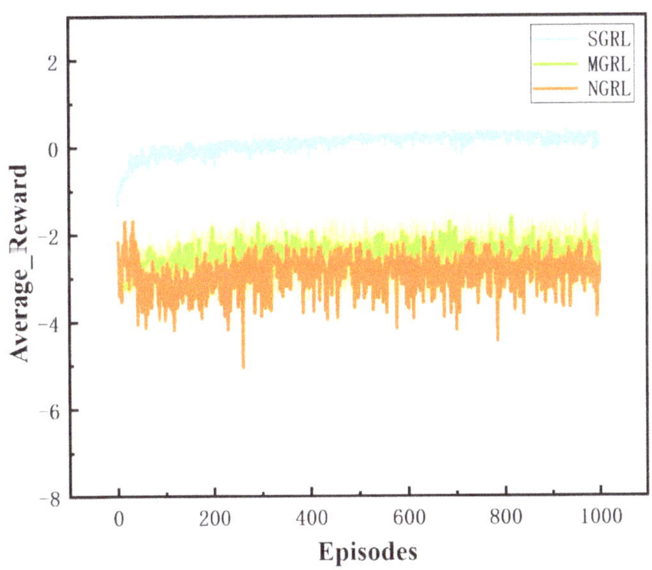

Figure 10. Diagram of average reward. This value is the average of each step reward value.

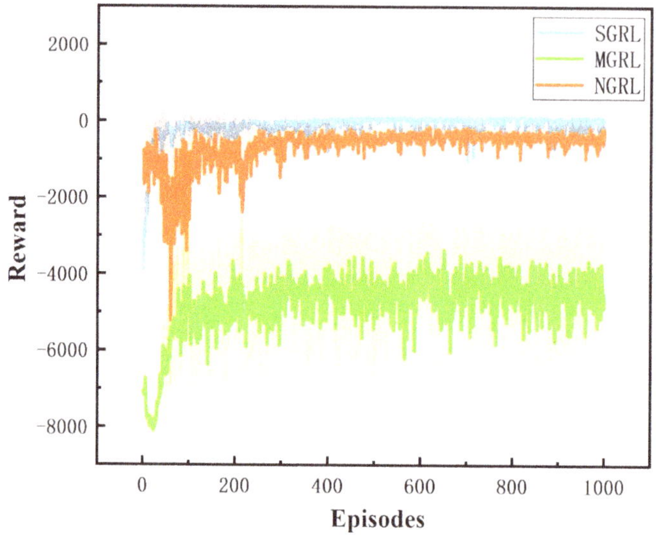

Figure 11. Diagram of reward. This value is the accumulation of each single step reward value.

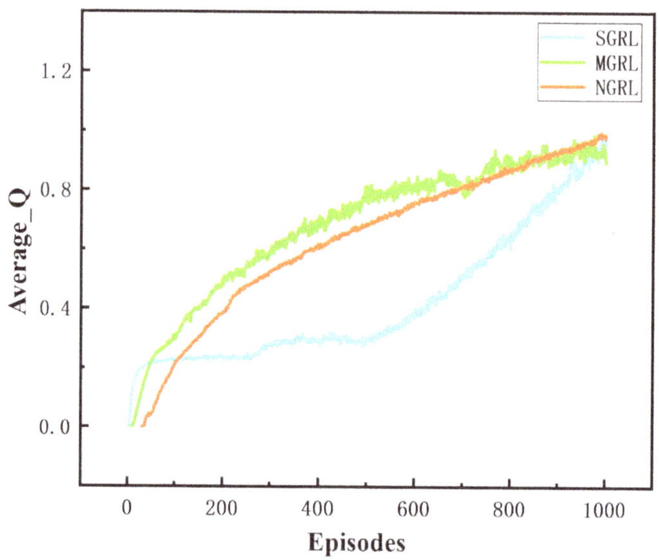

Figure 12. Diagram of average Q. This value is a training mark value in reinforcement learning.

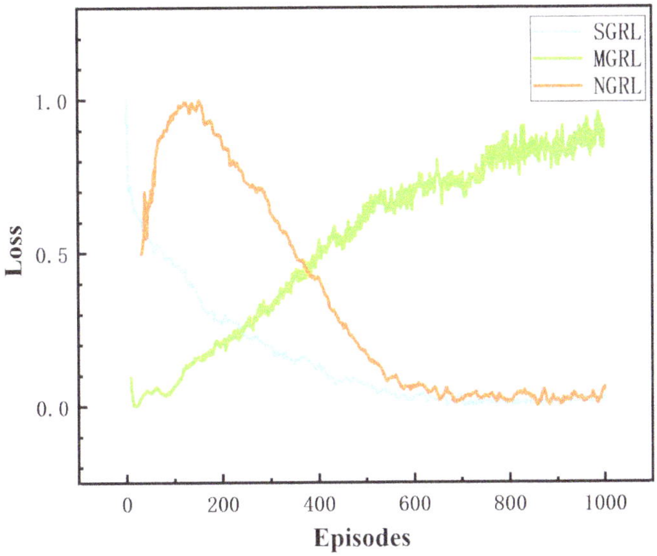

Figure 13. Diagram of loss. This value represents the difference between the real network and the ideal network.

To compare the task success rate, security, and traffic efficiency, the average velocity, number of collisions, success rate, and average steps per episode must be collected. The results of the above training data are shown in Figures 14–17.

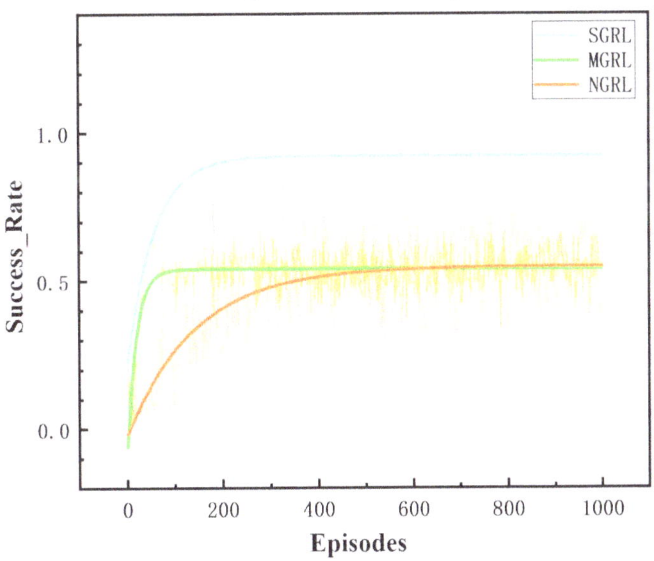

Figure 14. Diagram of success rate. This value is obtained by the ratio of the number of vehicles completing the task (entering the corresponding ramp) to the total number.

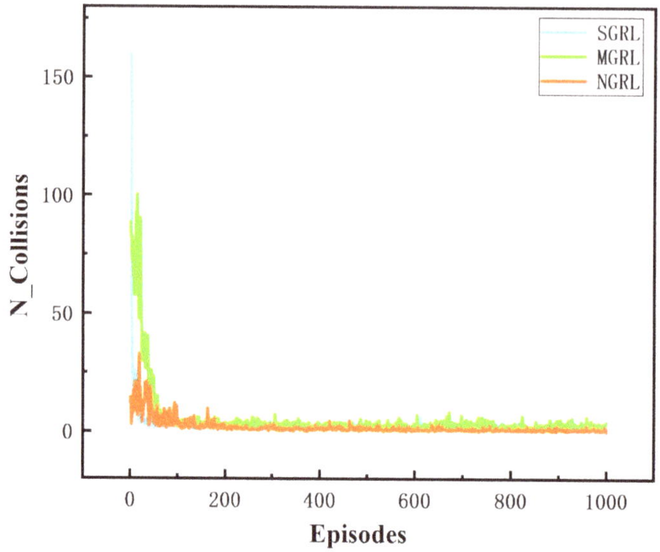

Figure 15. Diagram of collisions. This value is the number of collisions between vehicles obtained by real-time detection in the simulation scenario.

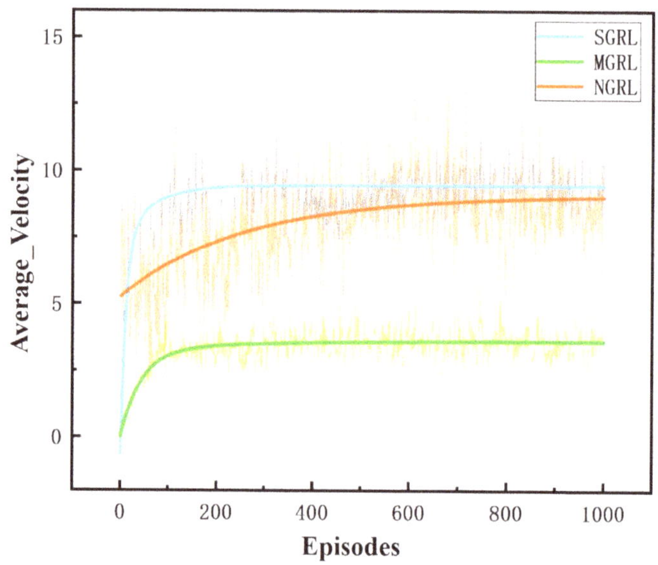

Figure 16. Diagram of average velocity. This value is the average velocity of all AVs in the scenario.

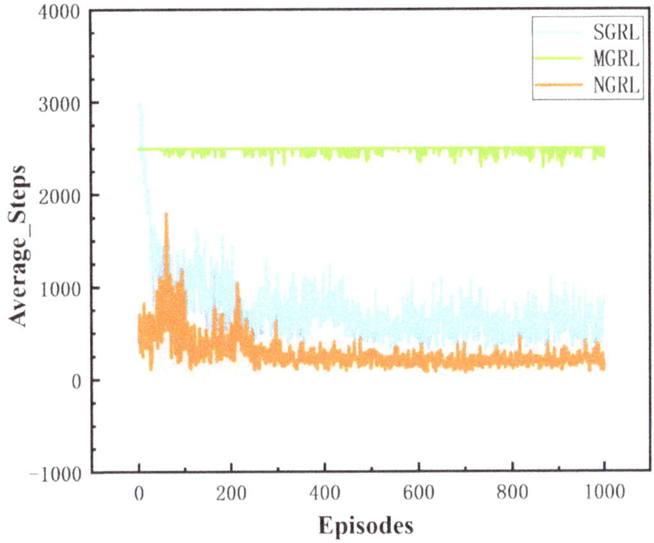

Figure 17. Diagram of average steps. This value is the number of steps experienced at the end of each episode.

The SGRL algorithm has obvious advantages regarding task success rate and average vehicle speed. In terms of collision times and average training step length, SGRL also meets driving safety requirements.

In addition, SGRL has apparent advantages in training efficiency under the premise of the same training episodes. The specific hardware parameters and training time are shown in Table 3. The comparison of the training time is shown in Table 4.

Table 3. Computer hardware information.

Item	Type
CPU	Intel I9 10980XE
GPU	NVIDIA RTX3090(24G)
RAM	Crucial DDR4 3200MHz 32G × 4
SSD	SAMSUNG 970 EVO Plus 1T × 2
OS	Ubuntu 20.04

Table 4. Training time statistics (ten experiments for each algorithm; time unit is hours).

	SGRL	MGRL	NGRL
1	3.335665	66.40787	3.251174
2	3.687935	78.94237	4.812508
3	3.782653	71.37449	3.737191
4	3.763133	82.16632	4.360045
5	3.38374	66.463	3.259155
6	3.286891	72.25948	3.890604
7	3.874574	67.38896	3.840993
8	3.043649	67.68016	3.668797
9	3.368947	65.51202	3.017189
10	3.832845	69.72951	4.072374
Mean	3.536003	70.79242	3.791003

4.2. Testing Results

The trained model needs to be verified by the test process. To fully verify the algorithm's effectiveness, we adjust the total length of the road under the premise of the same traffic flow (20 vehicles per episode). The three algorithms are tested on 1000 m, 750 m and 500 m roads to simulate different traffic flow and congestion levels.

Each test process includes 1000 episodes. In the test process, the reward value can still be used as an essential evaluation of model performance. The simulation results of reward value and average reward value are shown in Figures 18 and 19.

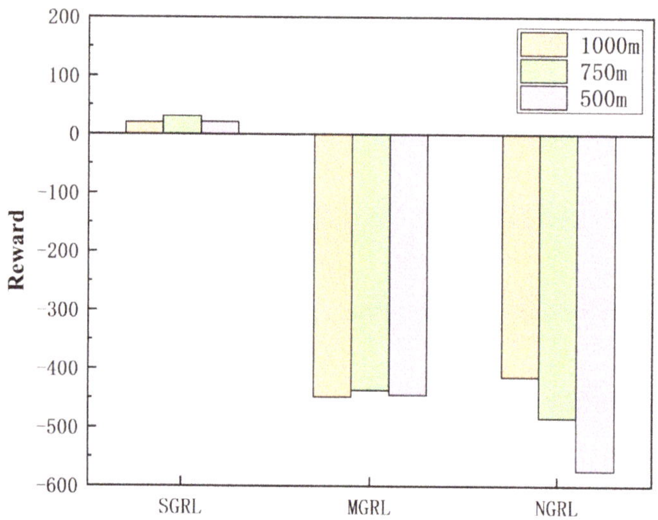

Figure 18. Diagram of testing reward. This value is the average of the total reward value for each episode in the test.

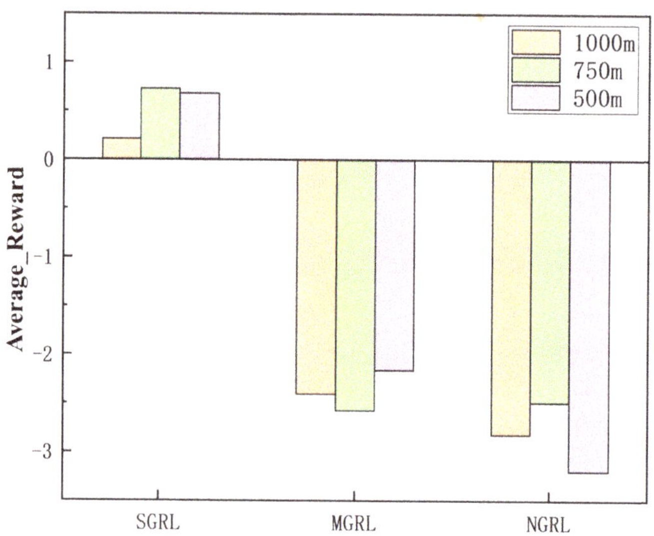

Figure 19. Diagram of testing average reward. This value is the average of the average reward value of each episode in the test.

For the most complex and congested 500 m highway scenario, the longitudinal motion spatial distribution of the three algorithms throughout the test cycle is as shown in Figure 20.

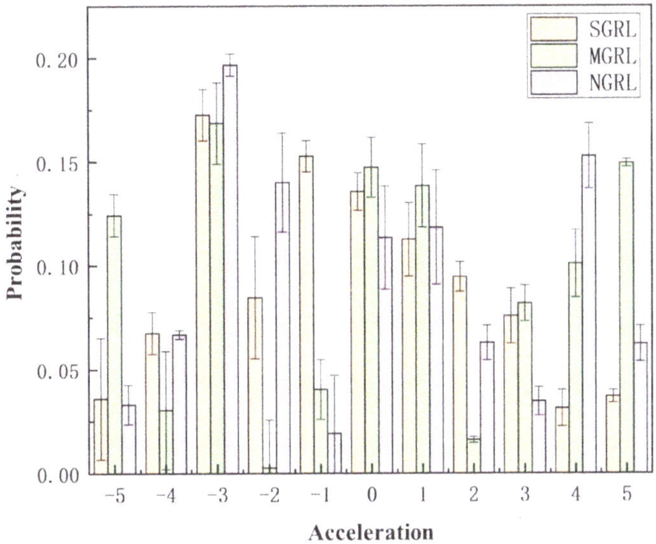

Figure 20. Spatial distribution diagram of longitudinal movement. Based on the frequency statistics of each action output in each testing process, the probability distribution of the longitudinal action can be obtained through probability calculation. The data in the figure are obtained from the average of ten repeated experiments.

The longitudinal control of autonomous vehicles will become more and more complex with the increase in road congestion, so the test results of the 500 m scene are the optimal reference. The test results show that the action output of the SGRL algorithm is mainly concentrated near 0, and the probability of large acceleration is acceptable.

To compare the task success rate, security, and traffic efficiency, the average velocity, number of collisions, success rate, and average step per episode must be collected.

The data of different methods are listed in Table 5, and the mean of the above data is shown in Figures 21–24.

Table 5. Performance comparison for different models.

Model	Road Length	Average_V	N_Collisions	Success_Rate	Average_Steps
SGRL	1000 m	**9.55588**	1.66888	**0.94068**	601.1014
	750 m	**8.45047**	**1.73714**	**0.92385**	494.9505
	500 m	**7.50548**	**2.06911**	**0.91493**	307.8586
MGRL	1000 m	3.60231	2.90205	0.54129	2478.409
	750 m	3.33545	2.96804	0.53864	1836.932
	500 m	3.20753	3.02719	0.47095	1334.704
NGRL	1000 m	8.92665	**1.23373**	0.53529	**204.3344**
	750 m	7.16045	1.83906	0.52708	**166.5145**
	500 m	6.70472	2.31418	0.4839	**111.9219**

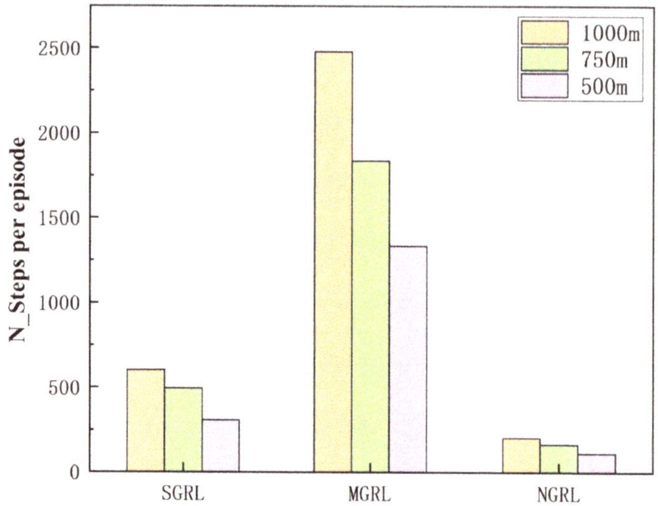

Figure 21. Diagram of testing average steps. This value is the number of steps experienced at the end of each episode.

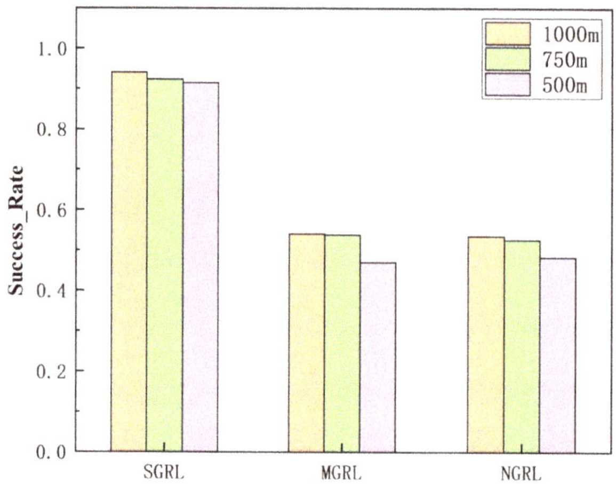

Figure 22. Diagram of testing success rate. This value is obtained by the ratio of the number of vehicles completing the task (entering the corresponding ramp) to the total number.

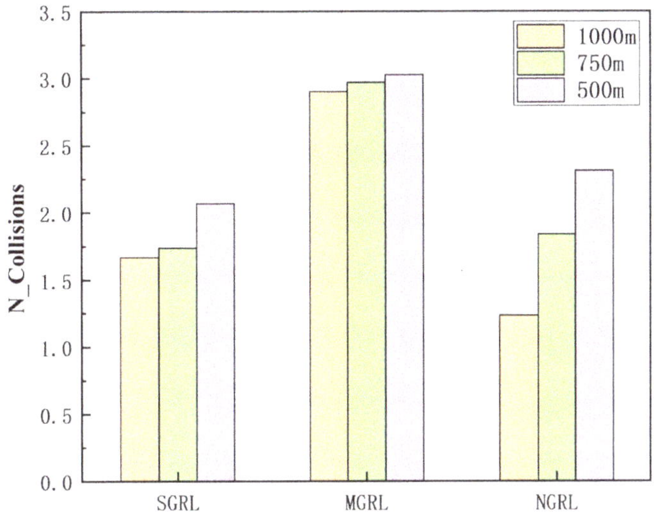

Figure 23. Diagram of testing collisions. This value is the number of collisions between vehicles obtained by real-time detection in the simulation scenario.

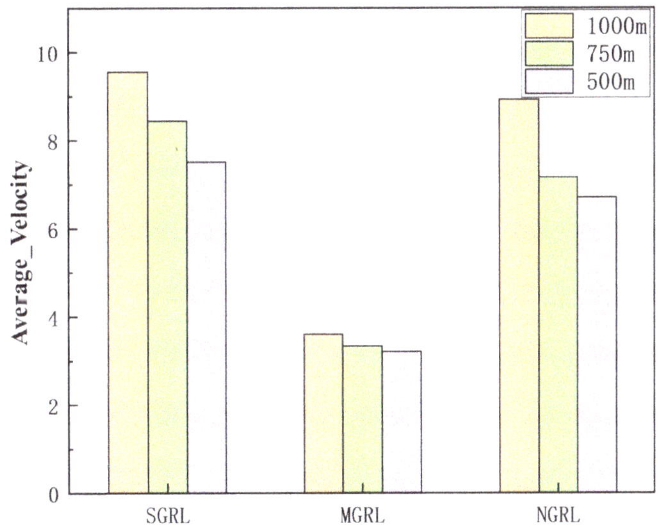

Figure 24. Diagram of testing average velocity. This value is the average velocity of all AVs in the scenario.

It can be seen from the figure and table that SGRL has apparent advantages in task success rate and average velocity. In terms of the number of collisions and the test steps, although the SGRL value is not the best, it is consistent with the best value.

4.3. Results Discussion

It can be seen that the SGRL algorithm has outstanding advantages over the MGRL algorithm and the NGRL algorithm. The SGRL algorithm can converge faster during the training process and achieve better data performance. In the testing process, the SGRL algorithm has outstanding data performance in terms of task success rate, average vehicle speed, and security.

The MGRL algorithm directly sums the reward value of all autonomous vehicles in the process of reward value calculation, so there is a phenomenon in which the excellent performance of vehicle behavior and the poor performance of vehicle behavior offset each other, which is not conducive to the adequate updating of parameters, and also causes the final convergence speed to be slow, and thus the convergence effect is not good.

The NGRL algorithm lacks the calculation consideration of the interaction process between vehicle individuals. Therefore, in the decision-making process, due to the relatively simple understanding of the environment, it cannot obtain sufficient data support, so the data performance is poor.

The SGRL algorithm considers and solves the above problems and optimizes the network structure. According to the comparison of data, it can be verified that the improvement of SGRL is obviously effective.

5. Conclusions

This paper proposes a generalized single-vehicle-based graph neural network reinforcement learning algorithm (SGRL algorithm). This algorithm combines GNN with deep reinforcement learning to solve the vehicle planning problem in the scenario of highway driving out of the ramp. The SGRL model is trained for a single agent and can be tested in multi-agent scenarios. At the same time, the algorithm sets up an improved reward function to provide a clear direction for training.

Comparing the three algorithms, the conclusions are as follows:

- Firstly, a training mode for single-agent training extended to multi-agent scenarios is proposed and verified in terms of training effectiveness and performance. The algorithm improves the analytical ability of the DRL by increasing the number of network nodes, thereby increasing the control dimension of vehicle decision-making to longitudinal and lateral dimensions.
- Secondly, the proposed SGRL algorithm simplifies the training mode and improves the training efficiency without affecting the training effect. This helps to adjust complex parameters and reduce time costs.
- Thirdly, SGRL is more sufficient in the training process to achieve a better convergence effect. The fluctuation is the smallest after the training data are stable. This shows that the proposed SGRL algorithm has outstanding training ability and is more suitable for decision-making based on reinforcement learning in multi-agent scenarios.
- Finally, the newly designed reward function effectively solves the problem of mutual influence between longitudinal and lateral control. SGRL can achieve higher task success rates and average velocity in the training and testing process. This shows that the new reward function, the training method for a single agent, and the incorporation of GNN effectively improve the decision performance of the model.

In future research, the continuity of model action space can be added to this algorithm, which will effectively improve the driving fluency of vehicles. In addition, the relationship between multiple agents in the scenario should not be limited to physical characteristics: decision-making, driving intention, and task priority can also be incorporated into the calculation process.

Author Contributions: Conceptualization, F.Y. and Q.L.; Data curation, F.Y.; Methodology, F.Y. and Z.L.; Project administration, X.L.; Resources, X.L.; Software, F.Y., Q.L. and X.G.; Supervision, X.L.; Validation, F.Y.; Visualization, F.Y.; Writing—original draft, F.Y.; Writing—review & editing, X.L. and Q.L. All authors have read and agreed to the published version of the manuscript.

Funding: This research received no external funding.

Conflicts of Interest: The authors declare no conflict of interest.

References

1. Hoel, C.J.; Driggs-Campbell, K.; Wolff, K.; Laine, L.; Kochenderfer, M.J. Combining Planning and Deep Reinforcement Learning in Tactical Decision Making for Autonomous Driving. *IEEE Trans. Intell. Veh.* **2020**, *5*, 294–305. [CrossRef]
2. Liu, Q.; Li, Z.; Yuan, S.; Zhu, Y.; Li, X. Review on Vehicle Detection Technology for Unmanned Ground Vehicles. *Sensors* **2021**, *21*, 1354. [CrossRef] [PubMed]
3. Peng, T.; Su, L.; Zhang, R.; Guan, Z.; Zhao, H.; Qiu, Z.; Zong, C.; Xu, H. A new safe lane-change trajectory model and collision avoidance control method for automatic driving vehicles. *Expert Syst. Appl.* **2020**, *141*, 112953. [CrossRef]
4. Nageshrao, S.; Tseng, H.E.; Filev, D. Autonomous highway driving using deep reinforcement learning. In Proceedings of the 2019 IEEE International Conference on Systems, Man and Cybernetics (SMC), Bari, Italy, 6–9 October 2019; pp. 2326–2331.
5. Kiran, B.R.; Sobh, I.; Talpaert, V.; Mannion, P.; Sallab, A.A.A.; Yogamani, S.; Perez, P. Deep Reinforcement Learning for Autonomous Driving: A Survey. *IEEE Trans. Intell. Transp. Syst.* **2021**, *23*, 4909–4926. [CrossRef]
6. Hoel, C.J.; Wolff, K.; Laine, L. Automated speed and lane change decision making using deep reinforcement learning. In Proceedings of the 2018 21st International Conference on Intelligent Transportation Systems (ITSC), Maui, HI, USA, 4–7 November 2018; pp. 2148–2155.
7. Gao, H.; Shi, G.; Xie, G.; Cheng, B. Car-following method based on inverse reinforcement learning for autonomous vehicle decision-making. *Int. J. Adv. Robot. Syst.* **2018**, *15*. [CrossRef]
8. Li, Z.; Gong, J.; Lu, C.; Li, J. Personalized Driver Braking Behavior Modeling in the Car-Following Scenario: An Importance-Weight-Based Transfer Learning Approach. *IEEE Trans. Ind. Electron.* **2022**, *69*, 10704–10714. [CrossRef]
9. Lu, C.; Hu, F.; Cao, D.; Gong, J.; Xing, Y.; Li, Z. Transfer learning for driver model adaptation in lane-changing scenarios using manifold alignment. *IEEE Trans. Intell. Transp. Syst.* **2019**, *21*, 3281–3293. [CrossRef]
10. Zhao, D.B.; Shao, K.; Zhu, Y.H.; Li, D.; Wang, C.H.J.C.T. Review of deep reinforcement learning and discussions on the development of computer Go. *Control Theory Appl.* **2016**, *33*, 701–717.
11. Wang, J.; Zhang, Q.; Zhao, D.; Chen, Y. Lane Change Decision-making through Deep Reinforcement Learning with Rule-based Constraints. In Proceedings of the 2019 International Joint Conference on Neural Networks (IJCNN), Budapest, Hungary, 14–19 July 2019; pp. 1–6. [CrossRef]
12. Li, Y.; Chen, S.; Ha, P.; Dong, J.; Steinfeld, A.; Labi, S. Leveraging Vehicle Connectivity and Autonomy to Stabilize Flow in Mixed Traffic Conditions: Accounting for Human-driven Vehicle Driver Behavioral Heterogeneity and Perception-reaction Time Delay. *arXiv* **2020**, arXiv:2008.04351.
13. Gong, C.; Li, Z.; Lu, C.; Gong, J.; Hu, F. A comparative study on transferable driver behavior learning methods in the lane-changing scenario. In Proceedings of the 2019 IEEE Intelligent Transportation Systems Conference (ITSC), Auckland, New Zealand, 27–30 October 2019; pp. 3999–4005.
14. Sallab, A.; Abdou, M.; Perot, E.; Yogamani, S.J.E.I. Deep Reinforcement Learning framework for Autonomous Driving. *Electron. Imaging* **2017**, *2017*, 70–76. [CrossRef]
15. Noh, S. Decision-Making Framework for Autonomous Driving at Road Intersections: Safeguarding Against Collision, Overly Conservative Behavior, and Violation Vehicles. *IEEE Trans. Ind. Electron.* **2019**, *66*, 3275–3286. [CrossRef]
16. Liu, Q.; Li, X.; Yuan, S.; Li, Z. Decision-Making Technology for Autonomous Vehicles: Learning-Based Methods, Applications and Future Outlook. In Proceedings of the 2021 IEEE International Intelligent Transportation Systems Conference (ITSC), Indianapolis, IN, USA, 19–22 September 2021; pp. 30–37. [CrossRef]
17. Schwarting, W.; Alonso-Mora, J.; Rus, D. Planning and Decision-Making for Autonomous Vehicles. *Annu. Rev. Control. Robot. Auton. Syst.* **2018**, *1*, 187–210. [CrossRef]
18. Li, L.; Ota, K.; Dong, M. Humanlike Driving: Empirical Decision-Making System for Autonomous Vehicles. *IEEE Trans. Veh. Technol.* **2018**, *67*, 6814–6823. [CrossRef]
19. Xu, X.; Zuo, L.; Li, X.; Qian, L.; Ren, J.; Sun, Z. A Reinforcement Learning Approach to Autonomous Decision Making of Intelligent Vehicles on Highways. *IEEE Trans. Syst. Man Cybern. Syst.* **2019**, *50*, 3884–3897. [CrossRef]
20. Zhang, Z.; Jiang, Q.; Wang, R.; Song, L.; Zhang, Z.; Wei, Y.; Mei, T.; Yu, B. Research on Management System of Automatic Driver Decision-Making Knowledge Base for Unmanned Vehicle. *Int. J. Pattern Recognit. Artif. Intell.* **2019**, *33*, 1959013. [CrossRef]
21. Duan, J.; Li, S.E.; Guan, Y.; Sun, Q.; Cheng, B.J.I.I.T.S. Hierarchical reinforcement learning for self-driving decision-making without reliance on labelled driving data. *IET Intell. Transp. Syst.* **2020**, *14*, 297–305. [CrossRef]
22. Cheng, X.; Jiang, R.; Chen, R. Simulation of decision-making method for vehicle longitudinal automatic driving based on deep Q neural network. In Proceedings of the 2020 the 7th International Conference on Automation and Logistics (ICAL), Beijing, China, 22–24 July 2020; pp. 12–17.
23. Wang, P.; Chan, C.; Fortelle, A.d.L. A Reinforcement Learning Based Approach for Automated Lane Change Maneuvers. In Proceedings of the 2018 IEEE Intelligent Vehicles Symposium (IV), Changshu, China, 26–30 June 2018; pp. 1379–1384. [CrossRef]
24. Forster, Y.; Hergeth, S.; Naujoks, F.; Beggiato, M.; Krems, J.F.; Keinath, A. Learning to use automation: Behavioral changes in interaction with automated driving systems. *Transp. Res. Part F Traffic Psychol. Behav.* **2019**, *62*, 599–614. [CrossRef]
25. Biondi, F.; Alvarez, I.; Jeong, K.A. Human–Vehicle Cooperation in Automated Driving: A Multidisciplinary Review and Appraisal. *Int. J. Hum. Comput. Interact.* **2019**, *35*, 932–946. [CrossRef]
26. Li, Z.; Gong, C.; Lu, C.; Gong, J.; Lu, J.; Xu, Y.; Hu, F. Transferable driver behavior learning via distribution adaption in the lane change scenario. In Proceedings of the 2019 IEEE Intelligent Vehicles Symposium (IV), Paris, France, 9–12 June 2019; pp. 193–200.

27. Ye, Y.; Zhang, X.; Sun, J.J.T.R.P.C.E.T. Automated vehicle's behavior decision making using deep reinforcement learning and high-fidelity simulation environment. *Transp. Res. Part C Emerg. Technol.* **2019**, *107*, 155–170. [CrossRef]
28. Zhang, H.; Xu, J.; Qiu, J.; Bashir, A.K. An Automatic Driving Control Method Based on Deep Deterministic Policy Gradient. *Wireless Commun. Mob. Comput.* **2022**, *2022*, 7739440. [CrossRef]
29. Yu, C.; Wang, X.; Xu, X.; Zhang, M.; Ge, H.; Ren, J.; Sun, L.; Chen, B.; Tan, G. Distributed Multiagent Coordinated Learning for Autonomous Driving in Highways Based on Dynamic Coordination Graphs. *IEEE Trans. Intell. Transp. Syst.* **2020**, *21*, 735–748. [CrossRef]
30. Yuan, S.; Zhao II, P.; Zhang III, Q. Research on automatic driving technology architecture based on cooperative vehicle-infrastructure system. In Proceedings of the International Conference on Artificial Intelligence, Virtual Reality, and Visualization (AIVRV 2021), Sanya, China, 19–21 November 2021; Volume 12153, pp. 111–117.
31. Li, Z.; Gong, J.; Lu, C.; Yi, Y. Interactive Behavior Prediction for Heterogeneous Traffic Participants in the Urban Road: A Graph-Neural-Network-Based Multitask Learning Framework. *IEEE/ASME Trans. Mechatron.* **2021**, *26*, 1339–1349. [CrossRef]
32. Li, Z.; Lu, C.; Yi, Y.; Gong, J. A hierarchical framework for interactive behaviour prediction of heterogeneous traffic participants based on graph neural network. *IEEE Trans. Intell. Transp. Syst.* **2021**, 1–13. [CrossRef]
33. Huang, C.; Lv, C.; Hang, P.; Xing, Y. Toward Safe and Personalized Autonomous Driving: Decision-Making and Motion Control With DPF and CDT Techniques. *IEEE/ASME Trans. Mechatron.* **2021**, *26*, 611–620. [CrossRef]
34. Dong, J.; Chen, S.; Ha, P.; Li, Y.; Labi, S. A DRL-based Multiagent Cooperative Control Framework for CAV Networks: A Graphic Convolution Q Network. *arXiv* **2020**, arXiv:2010.05437.

sensors

MDPI

Article

Multi-Agent Decision-Making Modes in Uncertain Interactive Traffic Scenarios via Graph Convolution-Based Deep Reinforcement Learning

Xin Gao [1], Xueyuan Li [1,*], Qi Liu [1,*], Zirui Li [1,2], Fan Yang [1] and Tian Luan [1]

1 School of Mechanical Engineering, Beijing Institute of Technology, Beijing 100080, China; 3120210298@bit.edu.cn (X.G.); 3120195255@bit.edu.cn (Z.L.); yangfanbitdb@163.com (F.Y.); 3220200285@bit.edu.cn (T.L.)

2 Department of Transport and Planning, Faculty of Civil Engineering and Geosciences, Delft University of Technology, Stevinweg 1, 2628 CN Delft, The Netherlands

* Correspondence: lixueyuan@bit.edu.cn (X.L.); 3120195257@bit.edu.cn (Q.L.)

Abstract: As one of the main elements of reinforcement learning, the design of the reward function is often not given enough attention when reinforcement learning is used in concrete applications, which leads to unsatisfactory performances. In this study, a reward function matrix is proposed for training various decision-making modes with emphasis on decision-making styles and further emphasis on incentives and punishments. Additionally, we model a traffic scene via graph model to better represent the interaction between vehicles, and adopt the graph convolutional network (GCN) to extract the features of the graph structure to help the connected autonomous vehicles perform decision-making directly. Furthermore, we combine GCN with deep Q-learning and multi-step double deep Q-learning to train four decision-making modes, which are named the graph convolutional deep Q-network (GQN) and the multi-step double graph convolutional deep Q-network (MDGQN). In the simulation, the superiority of the reward function matrix is proved by comparing it with the baseline, and evaluation metrics are proposed to verify the performance differences among decision-making modes. Results show that the trained decision-making modes can satisfy various driving requirements, including task completion rate, safety requirements, comfort level, and completion efficiency, by adjusting the weight values in the reward function matrix. Finally, the decision-making modes trained by MDGQN had better performance in an uncertain highway exit scene than those trained by GQN.

Keywords: multi-mode decision-making; connected autonomous vehicles; reward function matrix; uncertain highway exit scene; GQN; MDGQN

Citation: Gao, X.; Li, X.; Liu, Q.; Li, Z.; Yang, F.; Luan, T. Multi-Agent Decision-Making Modes in Uncertain Interactive Traffic Scenarios via Graph Convolution-Based Deep Reinforcement Learning. *Sensors* **2022**, *22*, 4586. https://doi.org/10.3390/s22124586

Academic Editors: Chao Huang, Yafei Wang, Peng Hang, Zhiqiang Zuo and Bo Leng

Received: 23 May 2022
Accepted: 15 June 2022
Published: 17 June 2022

1. Introduction

Artificial intelligence is in its golden development age due to the exponential increase in data production and the continuous improvements in computing power [1]. Autonomous driving is one of the main uses. A comprehensive autonomous driving system integrates sensing, decision-making, and motion-controlling modules [2–4]. As the "brains" of connected autonomous vehicles (CAVs) [5], the decision-making module formulates the most reasonable control strategy according to the state feature matrix transmitted by the sensing module, the vehicle state, and the cloud transmission information [6]. Moreover, it sends the determined control strategy to the motion-controlling module, including high-level behavior and low-level control requirements [7,8]. It is crucial to complete autonomous driving tasks safely and efficiently by making reasonable decisions based on other modules [9].

In an uncertain interactive traffic scenario, the driving environment has rigorous dynamic characteristics and high uncertainty, and the influences of driving behaviors of different traffic participants will be transmitted continuously [10]. On the level of transportation overall, all traffic participants need to cooperate efficiently [11]. At the traffic participant level, individuals need to judge the risk factors and make appropriate decisions sensitively based on dynamic scene changes [12]. In [13], a Gaussian mixture model and important weighted least squares probability classifier were combined and used for scene modeling. That model can identify the braking strength levels of new drivers under the condition of insufficient driving data. The key to participating in traffic scenarios for CAVs is that each CAV needs to generate appropriate and cooperative behavior to match human vehicles and other CAVs. Therefore, CAVs urgently demand efficient and accurate multi-agent decision-making technology to effectively handle the interactions between different traffic participants.

The current multi-agent decision-making technologies mainly focus on deep reinforcement learning (DRL) due to the excellent performance of DRL in high-dimensional dynamic state space [14]. The keys of DRL in uncertain interactive traffic scenarios can be summarized as follows: (1) Efficient modeling of interactive traffic scenes and accurate representation of state features. (2) Generating reasonable and cooperative decision-making behaviors based on uncertain scene changes and individual task requirements. The design of the reward function is an essential part of the DRL application. Concretizing and numericizing task objectives realizes the communication between objectives and algorithms. Therefore, the design of the reward function determines whether the agent can generate reasonable and cooperative decision-making behaviors [15]. In addition, the accurate modeling of the interactive traffic environment and representation of state characteristics are the requirements for agents to generate expected behaviors.

In studies of traffic scenarios using the DRL method, researchers found that in uncertain interactive traffic scenarios, sparse reward problems lead to agents having a lack of effective information guidance [16], making the algorithm difficult to converge. In order to solve the sparse reward problem, researchers divide the original goal into different sub-goals and give reasonable rewards or punishments. In [17], the DDPG was adopted to settle the autonomous braking problem. The reward function was split into three parts to solve the problems: braking too early, braking too late, and braking too quickly. In [18], the reward function was divided into efficiency and safety rewards to train the comprehensive safety and efficiency decision model. In addition, considering the changes in driving task requirements and scene complexity, some studies have given different weights to the sub-reward functions based on the decomposition of the reward function, so as to train different decision-making modes. In [19], the reward function was divided into sub-reward functions based on security, efficiency, and comfort. The system can realize different control targets by adjusting the weight values in the reward function.

In the decision-making research involving uncertain interactive traffic scenarios, the DRL method only takes the individual characteristics of each vehicle as the input. It ignores the interactive influence of transitivity between vehicles. This will result in CAVs not generating reasonable and cooperative behavior, which may reduce total traffic efficiency and the occurrence of traffic accidents. Graph representation can accurately describe the interactions between agents, representing the relationship between vehicles in uncertain interactive traffic scenarios. Therefore, some researchers focused on the graph reinforcement learning (GRL) method and modeled the interaction with graph representation [20]. In [21], a hierarchical GNN framework was proposed and combined with LSTM to model the interactions of heterogeneous traffic participants (vehicles, pedestrians, and riders) to predict their trajectories. The GRL method combines GNN and DRL: the features of interactive scenes are processed by GNN, and cooperative behaviors are generated by the DRL framework [22]. In [23], the traffic network was modeled by dynamic traffic flow probability graphs, and a graph convolutional policy network was used in reinforcement learning.

This paper proposes an innovative, dynamic reward function matrix, and various decision-making modes can be trained by adjusting the weight coefficient of the reward function matrix. Additionally, the weight coefficient is further set as a function of reward, forming an internal dynamic reward function. In the traffic environment adopted in this paper, the randomness and interactions between HVs and CAVs are strengthened by using some human vehicles making uncertain lane changes. Two GRL algorithms are used in this paper, GQN and MDGQN. Finally, we report a simulation based on the SUMO platform and a comparative analysis from various perspectives, such as reward functions, algorithms, and decision-making modes. The schematic diagram of the designed framework is shown in Figure 1.

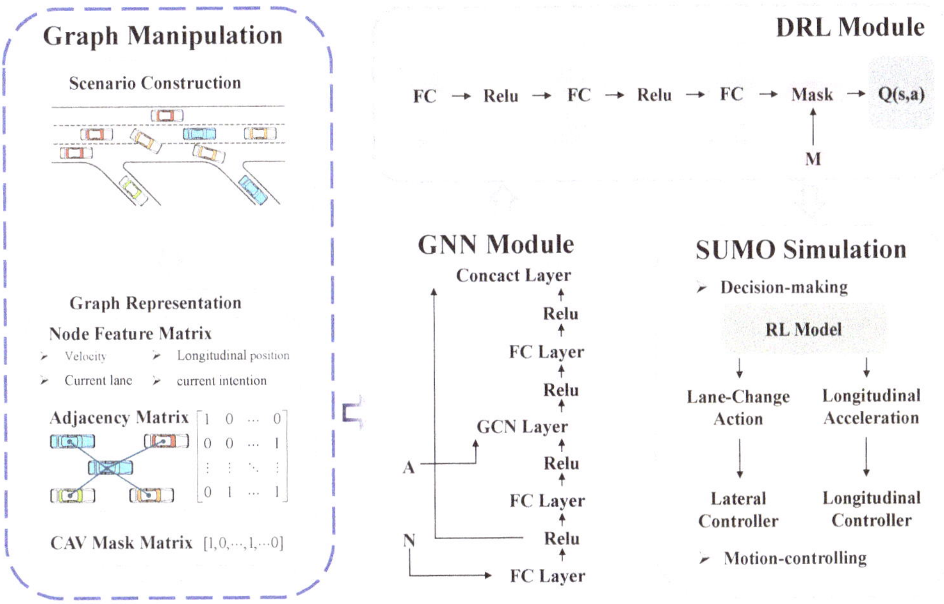

Figure 1. The schematic diagram of the designed framework. Letters N, A, and M represent node feature matrix, adjacency matrix, and CAV mask matrix respectively.

To summarize, the contributions of this paper are as follows:

1. Innovative dynamic reward function matrix: we propose a reward function matrix including a decision-weighted coefficient matrix, an incentive-punished-weighted coefficient matrix, and a reward–penalty function matrix. By adjusting the decision-weighted coefficient matrix, the decision-making modes of different emphases among driving task, traffic efficiency, ride comfort, and safety can be realized. Based on the premise that the incentive-punished function matrix separates the reward and the penalty, the optimization of individual performance can be achieved by adjusting the incentive and punishment ratio of each sub-reward function. In addition, the weight coefficient of the incentive-punished-weighted coefficient matrix is further set as a function of reward functions, which can reduce the impact of proportional adjustment on important operations, such as collision rate.
2. Adjust the parameters to train multiple decision-making modes: We compare the proposed reward function matrix with the traditional reward function under the same conditions. By adjusting the parameters of the decision-weighted coefficient matrix and the incentive-punished-weighted coefficient matrix, we can achieve aggressive or conservative incentive and punitive decision-making modes, respectively. Specifically, the four decision-making modes trained in this paper are the aggressive incentive (AGGI), aggressive punishment (AGGP), conservative incentive (CONI), and conservative punishment (CONP).
3. Modeling of interactive traffic scene and evaluation of decision-making modes: We designed a highway exit scene with solid interactions between CAVs and human vehicles (HVs) and adopted two algorithms to verify their differences. Addition-

ally, we also propose a set of indicators to evaluate the performance of driverless decision-making and used them to verify the performance differences among various algorithms and decision-making modes.

This article is organized as follows. Section 2 introduces the problem formulation. Section 3 introduces the methods used. Section 4 proposes the reward function matrix. Section 5 describes and analyzes the simulation results. Section 6 summarizes this paper and gives the future development directions.

2. Problem Formulation

2.1. Scenario Construction

As is shown in Figure 2, a 3-lane highway containing two exits was designed, and the three lanes were sorted from bottom to top as first, second, and third lanes. All the vehicles (HVs and CAVs) enter the road segment from the left of "Highway 0". The color of vehicle indicates its type and intention. In this paper, the vehicles were set as follows for safety principles and actual human driving rules:

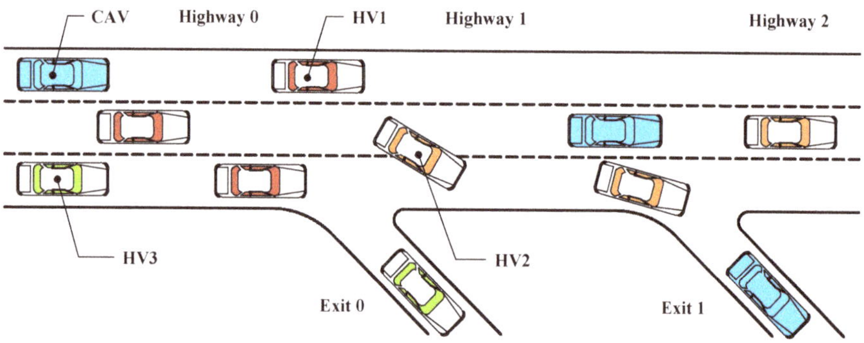

Figure 2. The highway exit scene with solid interactions between CAVs and HVs.

- A white body with a colored roof represents an HV, and an all-blue vehicle represents a CAV;
- The HV with the red roof (HV1) is set to appear in three lanes randomly and can only go out from "Highway 2";
- The HV with the orange roof (HV2) is set to emerge from the second lane and can go out from "Exit 1" when it is in the first lane and go out from "Highway 2" when it is in the other lanes;
- The HV with the green roof (HV3) is set to only appear from the first lane and can only go out from "Exit 0";
- When there is no car at the longitudinal distance of 10 m in the lane to be changed, the HV2 will be switched to the right;
- The CAV is set to appear in three lanes randomly and can go out from "Highway 2", "Exit 0", and "Exit 1".

The interaction between CAVs and HVs is enhanced through the above setting. The setting of HV1 enhances the uncertainty of the scene considerably, increasing the requirements for CAVs' decision-making. Straightening away from "Highway 2" is the most straightforward choice with the lowest collision risk, owing to CAVs' least lane change decisions. In addition, due to the uncertain lane change behavior of HV2, it is difficult for CAVs to explore "Exit 1" safely, and the collision risk of leaving the highway from "Exit 1" is higher than it is for other exits. Specific scenario setting parameters are shown in Table 1:

Table 1. Specific scenario setting parameters.

Parameter	Value	Parameter	Value
Length of "Highway 0"	150 m	Number of HV1	20
Length of "Highway 1"	200 m	Number of HV2	10
Length of "Highway 2"	150 m	Number of HV3	15
Time step	0.1 s	Number of CAVs	15
Initial velocity of HV1	24m/s	Probability of HV1 appearing every 0.1 s	0.18
Initial velocity of HV2	22 m/s	Probability of HV2 appearing every 0.1 s	0.08
Initial velocity of HV3	20 m/s	Probability of HV3 appearing every 0.1 s	0.12
nitial velocity of CAVs	20 m/s	Probability of CAVs appearing every 0.1 s	0.12

2.2. *State Representation*

The scenario is modeled as an undirected graph. Each vehicle in this scenario is regarded as the node of the graph, and the interaction between vehicles is considered the edge of the graph. Three matrices can represent the state space: a node features matrix X_t, an adjacency matrix A_t, and a CAV mask matrix M_t; each of them are described in the following.

Node Features Matrix: The vehicle's features are speed, position, location, and intention, denoted as $[V_i, Y_i, L_i, I_i]$. The node features matrix represents the features of each vehicle in the constructed scenario, which can be described as follows:

$$N_t = \begin{bmatrix} [V_1, Y_1, L_1, I_1] \\ [V_2, Y_2, L_2, I_2] \\ \cdots \\ [V_i, Y_i, L_i, I_i] \\ \cdots \\ [V_n, Y_n, L_n, I_n] \end{bmatrix} \quad (1)$$

where $V_i = v_{i-actual} / v_{\max}$ denotes the ratio of actual longitudinal velocity to maximum longitudinal velocity. $Y_i = y_{i-actual} \Big/ L_{highway}$ denotes the percentage of actual longitudinal position of the total length of the highway. L_i denotes the one-hot encoding matrix of the current lane of vehicles. I_i denotes the vehicle's one-hot encoding matrix of current intention (change to left lane, change to right lane, and go straight).

Adjacency matrix: In this paper, the interactions between vehicles are embodied in the information sharing between vehicles, expressed by the adjacency matrix. The calculation of the adjacency matrix is based on three assumptions:

- All CAVs can share information in the constructed scenario;
- Information cannot be shared between HVs;
- Vehicles can share information with themselves $a_{ii} = 1$.
- All CAVs can share information with HVs in their sensing range.

The derivation of the adjacency matrix is as follows:

$$A_t = \begin{bmatrix} a_{11} & a_{12} & \cdots & & \cdots & a_{1n} \\ a_{21} & a_{22} & \cdots & & \cdots & a_{2n} \\ \vdots & \vdots & \ddots & & & \vdots \\ & & & a_{ij} & & \\ \vdots & \vdots & & & \ddots & \vdots \\ a_{n1} & a_{n2} & \cdots & & \cdots & a_{nn} \end{bmatrix} \quad (2)$$

where a_{ij} denotes the edge value of the ith vehicle and the jth vehicle. $a_{ij} = 1$ denotes that the ith vehicle and the jth vehicle share information; $a_{ij} = 0$ denotes that the ith vehicle and the jth vehicle share no information.

CAV mask matrix: CAV mask is used to filter out the embeddings of HVs after the GCN fusion block. The specific mathematical expression is as follows:

$$M_t = [m_1, m_2, \cdots, m_i, \cdots m_n] \tag{3}$$

where m_i = 0 or 1. If the ith vehicle is controlled by the GRL algorithm, $m_i = 1$; otherwise, $m_i = 0$.

2.3. Action Space

At each time step, each CAV has a discrete action space representing potential actions to be executed at the next time step, as follows:

$$a_i = \{a_{lane-change}, a_{acceleration}\} \tag{4}$$

where $a_{lane-change}$ indicates that the lane change action can be taken, including a left lane change, right lane change, or straight line; $a_{acceleration}$ denotes the discrete acceleration that CAV can take, and its value is equalized by interval $[-8\,\mathrm{m}\cdot\mathrm{s}^{-2}, 5\,\mathrm{m}\cdot\mathrm{s}^{-2}]$ at $0.5\,\mathrm{m}\cdot\mathrm{s}^{-2}$.

3. Methods

This section describes the principles of the methods used, including graph convolutional neural networks, Q-learning, GQN, and MDGQN.

3.1. Graph Convolutional Neural Network

GCN is a neural network model that directly encodes graph structure. The goal is to learn a function of features on a graph. A graph can be represented by $G = (V, E)$ in theory, where V denotes the set of nodes in the graph, the number of nodes in the graph is denoted by N, and E denotes the set of edges in the graph. The state of G is considered a tuple of 3 matrices of information: feature matrix X, adjacency matrix A, and degree matrix D.

- The adjacency matrix A is used to represent connections between nodes;
- The degree matrix D is a diagonal matrix, and $D_{ii} = \sum_j A_{ij}$;
- The feature matrix X is applied to represent node features, $X \in R^{N\times F}$, where F represents the dimensions of the feature.

GCN is a multi-layer graph convolution neural network, a first-order local approximation of spectral graph convolution. Each convolution layer only deals with first-order neighborhood information, and multi-order neighborhood information transmission can be realized by adding several convolution layers.

The propagation rules for each convolution layer are as follows [24]:

$$Z = g(H, A) = \sigma(\hat{D}^{-1/2}\hat{A}\hat{D}^{-1/2}H^{(l)}W^{(l)} + b) \tag{5}$$

where $\hat{A} = A + I_N$ is the adjacency matrix of the undirected graph G with added self-connections. I_N is the identity matrix; $\hat{D}_{ii} = \sum_j \hat{A}_{ij}$ and $W^{(l)}$ are layer-specific trainable weight matrices. $\sigma(\cdot)$ denotes an activation function, such as the $\mathrm{Re}LU(\cdot) = \max(0, \cdot)$. $H^{(l)} \in R^{N\times D}$ is the matrix of activations in the lth layer, $H(0) = X$.

3.2. Deep Q-Learning

Q-learning [25] is a value-based reinforcement learning algorithm. Q_t is the expectation that $Q(s_t, a_t)$ can obtain benefits by taking action $a_t(a_t \in A_t)$ under the state of $s_t(s_t \in S_t)$ at time t. The environment will feed back the corresponding reward R_t according to the agent's action. Each time step produces a quadruplet (s_t, a_t, r_t, s_{t+1}), and it is stored in the experience replay. Deep Q-learning [26] replaces the optimal action–value function with the deep neural network $Q(s, a, \omega)$. The following is a description of the principle of the algorithm.

The predicted values of DQN can be calculated according to the given four-tuple (s_t, a_t, r_t, s_{t+1}):

$$\hat{q}_t = Q(s_t, a_t, \omega) \tag{6}$$

The TD target can be calculated based on the actual observation reward r_t:

$$\hat{y}_t = r_t + \gamma \cdot \max_{a \in \mathcal{A}} Q(s_{t+1}, a, \omega) \tag{7}$$

DQN updates network parameters according to the following formula:

$$\omega \leftarrow \omega - \alpha \cdot (\hat{q}_t - \hat{y}_t) \cdot \nabla_\omega Q(s_t, a_t, \omega) \tag{8}$$

where α represents the learning rate.

3.3. Graph Convolutional Q-Network (GQN)

As described in Section 3.2, Q-learning uses a Q-Table to store the Q value of each state–action pair. However, if the state and action space are high-dimensionally continuous, there will be the curse of dimensionality; that is, with a linear increase in dimensions, the calculation load increases exponentially. GQN [27] replaces the optimal action–value function $Q_\star(s, a)$ with the graph convolutional neural network $Q(s, a, \theta)$:

$$Q(s, a, \theta) = \rho(Z, a) \tag{9}$$

where Z is the node embeddings output from graph convolution layer. ρ represents the neural network block, including the fully connected layer. θ is the aggregation of all the weights.

The specific training process of GQN is the same as DQN [26]. Firstly, the predictive value of GQN can be calculated according to the four-tuple (s_t, a_t, r_t, s_{t+1}) sampled from the experience replay:

$$\tilde{q}_t = Q(s_t, a_t, \theta_{now}) \tag{10}$$

However, since the Q-network uses the same estimate to update itself, it will cause bootstrapping and lead to deviation propagation. Therefore, another neural network can be used to calculate the TD target, called the target network $Q(s, a, \theta_{now}^-)$. Its network structure is precisely the same as that of the Q-network, but the parameter θ_{now}^- is different from θ_{now}. Selecting the optimal action and forward propagation of the target network:

$$a^\star = \arg\max_{a \in \mathcal{A}} Q(s_{t+1}, a_t, \theta_{now}^-) \tag{11}$$

$$\tilde{q}_{t+1} = Q(s_{t+1}, a^\star, \theta_{now}^-) \tag{12}$$

Calculation of TD target $\hat{y}_t$ and TD error δ_t:

$$\hat{y}_t = r_t + \gamma \cdot \tilde{q}_{t+1} \tag{13}$$

$$\delta_t = \tilde{q}_t - \hat{y}_t \tag{14}$$

The gradient $\nabla_\theta Q(s_t, a_t, \theta_{now})$ is calculated by the backpropagation of a Q-Network, and the parameter of the Q-Network is updated by gradient descent:

$$\theta_{new} \leftarrow \theta_{now} - \alpha \cdot \delta_t \cdot \nabla_\theta Q(s_t, a_t, \theta_{now}) \tag{15}$$

where θ_{new} is the parameter of the updated Q-Network. α represents the learning rate.

Finally, GQN adopts the weighted average of the two networks to update the target network parameters:

$$\theta_{new}^- \leftarrow \tau \cdot \theta_{new} + (1 - \tau) \cdot \theta_{now}^- \tag{16}$$

where τ represents soft update rate.

3.4. Multi-Step Double Graph Convolutional Q-Network (MDGQN)

MDGQN further adopts double Q-learning and a multi-step TD target algorithm based on GQN. As shown in Section 3.3, the target network cannot completely avoid bootstrapping, since the parameters of the target network are still related to the Q-Network. The double Q-learning algorithm is improved based on the target network. The Q-learning with the target network uses the target network to select the optimal action and calculate the Q value of the optimal action. However, the double Q-learning selects the optimal action according to the Q-Network and uses the target network to calculate the Q value of the optimal action. Equation (11) is modified as follows:

$$a^{\star} = \underset{a \in \mathcal{A}}{\arg\max}\, Q(s_{t+1}, a_t, \theta_{\text{now}}) \tag{17}$$

The multi-step TD target algorithm can balance the significant variance caused by Monte Carlo and the significant deviation caused by bootstrapping. Equation (13) is modified as follows:

$$\hat{y}_t = \sum\nolimits_{i=0}^{m-1} \gamma^i r_{t+i} + \gamma^{\text{m}} \cdot \tilde{q}_{t+\text{m}} \tag{18}$$

where $\hat{y}_t$ is called the m-step TD target.

4. Reward Functions

The design of the reward function is an important criterion and goal of the DRL training process. Based on four aspects—the results the of the driving task, traffic efficiency, ride comfort, and safety—the reward function is divided into four blocks. Further, we propose a reward function matrix including a decision-weighted coefficient matrix, an incentive-punished-weighted coefficient matrix, and a reward–penalty function matrix.

$$\begin{aligned} r &= tr(A_{k1}A_{k2}A_r) \\ &= tr\left[\begin{bmatrix} k_r & & & \\ & k_e & & \\ & & k_c & \\ & & & k_s \end{bmatrix} \begin{bmatrix} k_{rI} & & & \\ k_{rP} & & & \\ & k_{eI} & & \\ & k_{eP} & & \\ & & k_{cI} & \\ & & k_{cP} & \\ & & & k_{sI} \\ & & & k_{sP} \end{bmatrix}^T \begin{bmatrix} r_{result-I} & & & \\ r_{result-P} & & & \\ & r_{efficiency-I} & & \\ & r_{efficiency-P} & & \\ & & r_{comfort-I} & \\ & & r_{comfort-P} & \\ & & & r_{safe-I} \\ & & & r_{safe-P} \end{bmatrix}\right] \end{aligned} \tag{19}$$

Specific parameters are described below:

1. A_{k1} is the decision-weighted coefficient matrix. k_r, k_e, k_c, and k_s denote the weights of reward and penalty functions based on the results of the driving task, traffic efficiency, ride comfort, and safety.
2. A_{k2} is the incentive-punished-weighted coefficient matrix. k_{rI}, k_{eI}, k_{cI}, and k_{sI} denote the weights of reward functions based on the results of the driving task, traffic efficiency, ride comfort, and safety, respectively. k_{rP}, k_{eP}, k_{cP}, and k_{sP} denote the weights of penalty functions based on the results of the driving task, traffic efficiency, ride comfort, and safety, respectively.
3. A_r is a reward–penalty function matrix. $r_{result-I}$, $r_{efficiency-I}$, $r_{comfort-I}$, and r_{safe-I} are reward functions based on the results of the driving task, traffic efficiency, ride comfort, and safety, respectively. $r_{result-P}$, $r_{efficiency-P}$, $r_{comfort-P}$, and r_{safe-P} are penalty functions based on the results of the driving task, traffic efficiency, ride comfort, and safety, respectively.

By adjusting the weight coefficient of the decision-weighted coefficient matrix and incentive-punished-weighted coefficient matrix, DRL can train different goal-oriented decision-making modes. In the decision-making process of autonomous vehicles, the upper control module can choose different decision-making modes according to different needs. In order to select a decision-making model with excellent comprehensive performance and strong contrast, we conducted multiple sets of experiments, and some experimental data were put in Appendix A. This paper determined four decision modes: AGGI, AGGP, CONI, and CONP, by adjusting the parameters in Tables 2 and 3.

Table 2. Weight coefficient of the decision-weighted coefficient matrix.

Parameters	Conservative	Aggressive
The weight of task k_r	0.2	0.25
The weight of efficiency k_e	0.2	0.35
The weight of comfort k_c	0.25	0.15
The weight of safety k_s	0.35	0.25

Table 3. Weight coefficient of the incentive-punished-weighted coefficient matrix.

Parameters	Incentive	Punished
The weight of incentive sub-reward function k_{rI0} , k_{eI0} , k_{cI0} , k_{sI0}	0.6	0.4
The weight of punished sub-reward function.k_{rP0} , k_{eP0} , k_{cP0} , k_{sP0}	0.4	0.6

Since the change of weight coefficient will dilute some essential rewards or punishments, this paper improves the reward function against this defect. The weight coefficient of the incentive-punished-weighted coefficient matrix is further set as a functional of reward functions, which forms an internal dynamic reward function. The specific formula is as follows:

$$\begin{cases} k_{rI} = k_{rI0} \cdot \exp\left((r_{result-P} + r_{comfort-P}) / 100{,}000 \right) \\ k_{eI} = k_{eI0} \cdot \exp\left((r_{result-P} + r_{comfort-P}) / 80{,}000 \right) \\ k_{sI} = k_{sI0} \cdot \exp\left((r_{result-P} + r_{comfort-P}) / 150{,}000 \right) \end{cases} \tag{20}$$

Based on [3], the specific reward functions and penalty functions were designed. Firstly, we designed the corresponding reward function and penalty function based on the results of the driving tasks. The independent variable of the reward function is the number of CAVs and HV2 reaching destinations, which aims to train decisions that can assist HVs in completing driving tasks. The penalty function is designed based on collisions.

$$r_{result-I} = 300(n_{r1} + n_{r2}) \tag{21}$$

$$r_{result-P} = -60{,}000, if\ collision \tag{22}$$

where n_{r1} is the number of CAVs leaving from "Exit 1". n_{r2} is the number of CAVs leaving from "Exit 1".

To train the decision-making model to improve traffic efficiency, this paper divides the speed interval of CAVs into three parts. The corresponding reward and penalty functions were designed to curb speeding, encourage high-speed driving, and punish low-speed blocking for these three-speed ranges. In order to make CAVs faster and more stable to explore the optimal speed, we used the exponential function to design a soft reward function [28].

$$r_{efficiency-I} = \exp\left[\frac{6 \times (v_y - v_{y\min})}{v_{y\max}} \right], if\ v_{y\min} \le v_y \le v_{y\max} \tag{23}$$

$$r_{efficiency-P} = \begin{cases} -\exp(v_y - v_{y\max}), if\ \ v_y > v_{y\max} \\ -\exp\left(1 + \frac{v_{y\min} - v_y}{3}\right), if\ v_y < v_{y\min} \end{cases} \tag{24}$$

where v_y is the velocity of the CAV. $v_{y\max}$ represents the maximum velocity allowed by the current lane; its value is 25 $\text{m} \cdot \text{s}^{-1}$, or 15 $\text{m} \cdot \text{s}^{-1}$.

In order to improve the ride comfort of all vehicles in this traffic section, the corresponding reward function and penalty function are designed based on the acceleration and lane change times of all vehicles.

$$r_{comfort-I} = \exp(3) \times n_{c1} \tag{25}$$

$$r_{comfort-P} = -2000 \times n_{c2} - \exp\left(\frac{m}{2}\right) \quad (26)$$

where n_{c1} is the number of vehicles with acceleration of $[-2\ \mathrm{m \cdot s^{-2}}, 2\ \mathrm{m \cdot s^{-2}}]$. n_{c2} is the number of vehicles with acceleration of $(-\infty, -4.5\ \mathrm{m \cdot s^{-2}}]$. m is the number of lane changes in this traffic section within 0.5 s.

Superior security performance is the premise of developing decision technology [29]. In [30], the length of CAVs' safety time is one of the most important factors affecting road safety. This paper introduces the safety time of the CAVs into the corresponding reward function. The definition of the safety time is as follows:

$$t_1 = t_{safe-follower} = \frac{y_{AV} - y_{follower}}{v_{follower} - v_{AV}} \quad (27)$$

$$t_2 = t_{safe-leader} = \frac{y_{leader} - y_{AV}}{v_{AV} - v_{leader}} \quad (28)$$

where y_{AV} is the longitudinal position of the CAV. y_{leader} and $y_{follower}$ are the longitudinal positions of the front and rear vehicles of CAV, respectively. v_{leader} and $v_{follower}$ are the longitudinal speeds of the front and rear vehicles of CAV, respectively.

A driving hazard diagram is proposed to represent the degree of danger of the vehicle's state based on safety time $t_{safe-follower}$ and $t_{safe-leader}$. As shown in Figure 3, three primary colors are used to represent the degrees of danger in this state. The red region represents collision accident danger, and the deeper the color, the greater the likelihood. The yellow area indicates that the vehicle needs to pay attention to the occurrence of a possible emergency. The green area indicates that the vehicle is in a safe state. By dividing Figure 3 into five categories, the sub-reward functions were designed. On the basis of the above principles, a security-based reward function is proposed for training security decisions. The formula is shown in Table 4.

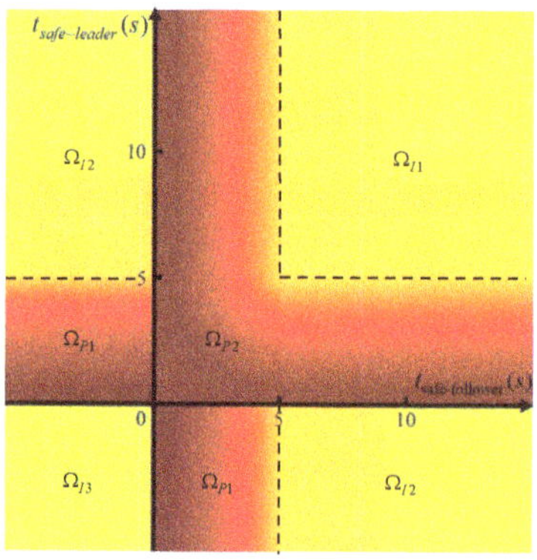

Figure 3. The driving hazard diagram.

Table 4. The security-based reward function.

Subfunction	Category	Calculation
r_{safe-I}	Ω_{I1}	$\max\left(\exp\left(\frac{\min(t_1,t_2)-5}{2.5}\right), \exp(2)\right)$
	Ω_{I2}	$\max\left(\exp\left(\frac{\max(t_1,t_2)-5}{2.5}\right), \exp(2)\right)$
	Ω_{I3}	$\exp(2)$
r_{safe-P}	Ω_{P1}	$-\exp(4 - \min(t_1, t_2))$
	Ω_{P2}	$-\exp(4 - \max(t_1, t_2))$

5. Experiments

5.1. Parameter Setting

In this section, we show the solid interaction scene designed through SUMO. Firstly, based on this scene, the proposed reward function is compared with the traditional reward function. The two algorithms were used to train four decision-making modes, and the differences between different algorithms and modes are compared. Finally, the performances of four decision-making modes based on the two algorithms are evaluated. The main parameters for algorithms are listed in Table 5.

Table 5. Parameters of the graph reinforcement learning.

Parameters	Value
Optimizer	Adam
Nonlinearity	Relu
Learning rate	0.005
Discount factor	0.99
Updating rate	0.05
Minibatch size	32
Multi-step	3

5.2. Evaluation Indexes

In order to further evaluate the test performance of each decision-making mode, four kinds of evaluation indexes are proposed, namely, efficiency index, safety index, comfort index, and task index.

1. Efficiency: The average longitudinal velocity of CAVs and all vehicles in this scenario is proposed and used to evaluate the traffic efficiency of CAVs and the comprehensive efficiency of total traffic flow under different decision modes. Their functions are defined as follows:

$$\bar{v}_{AVs} = \frac{\sum_{i=1}^{n}\sum_{j=1}^{m} v_{AV-ij}}{N} \tag{29}$$

$$\bar{v}s. = \frac{\sum_{i=1}^{n}\sum_{j=1}^{m} v_{ij}}{N_{all}} \tag{30}$$

where v_{AV-ij} represents the longitudinal velocity of the ith CAV detected at the jth time step. v_{ij} is the longitudinal velocity of the ith vehicle detected at the jth time step. N indicates the total number of CAVs' longitudinal velocities detected in all-time steps. N_{all} indicates the total number of all vehicles' longitudinal velocities detected in all-time steps.

2. Safety: Due to the presence of multiple CAVs in the test scenario, we define the collision rate as the probability of each CAV collision accident in each episode.

$$p_{collision} = \frac{N_{collision}}{N_{CAV}} \tag{31}$$

where $N_{collision}$ is the number of collisions in a single episode. N_{CAV} represents the number of CAVs.

3. Comfort: For the evaluation of comfort, we mainly studied the deceleration of all vehicles under different decision-making modes. All vehicles' total emergency braking times $N_{braking}$ in a single episode are defined as comfort evaluation indexes. It should be noted that the deceleration between $(-\infty, -4.5\ \mathrm{m \cdot s^{-2}}]$ is regarded as emergency braking.

4. Task: Based on the solid interaction between CAVs and HVs in the test scenario, the driving task completion rates of CAVs and HV2 are used as the task indexes. Their functions are defined as follows.

$$p_{CAV} = \frac{N_{CAV-Exit1}}{N_{CAV}} \quad (32)$$

$$p_{HV2} = \frac{N_{HV2-Exit1}}{N_{HV2}} \quad (33)$$

where $N_{CAV-Exit1}$ is the number of CAVs that left via "Exit 1". $N_{HV2-Exit1}$ indicates the number of HV2 that left via "Exit 1". N_{HV2} represents the number of HV2.

5.3. Validation of the Reward Function

The proposed reward function was used in GQN and MDGQN to train different decision-making modes while improving training efficiency. In [27], the traditional reward function only considers intention reward, speed reward, lane-changing penalty, and collision penalty. In this article, this traditional reward function is used as the baseline, and the reward of the CONP decision-making mode training process is compared. In order to allow them to make a fair comparison based on the maximum and minimum reward values they can achieve, the reward values in the training process are normalized.

As shown in Figure 4, using the proposed reward function for training can explore the maximum reward rapidly. The traditional reward function's training process experiences repeated reward reductions, and the maximum reward cannot be explored in the specified number of rounds. In addition, the training stability is greatly improved by comparing the reward fluctuation in the training process with the baseline. After the 200th episode, the average normalized reward values of Baseline-GCQ, CONP-GCQ, and CONP-MDGCQ were 0.7399, 0.9882, and 0.9886, respectively; and their variances were 0.0871, 0.0028, and 0.0049. Under the condition of using the GQN algorithm, the CONP decision-making model was 35.40% better than the average of baseline, and the variance is only 3.27% higher than that of the baseline. In summary, the proposed reward function can promote the fast convergence of the algorithm and greatly enhance the stability of the training process.

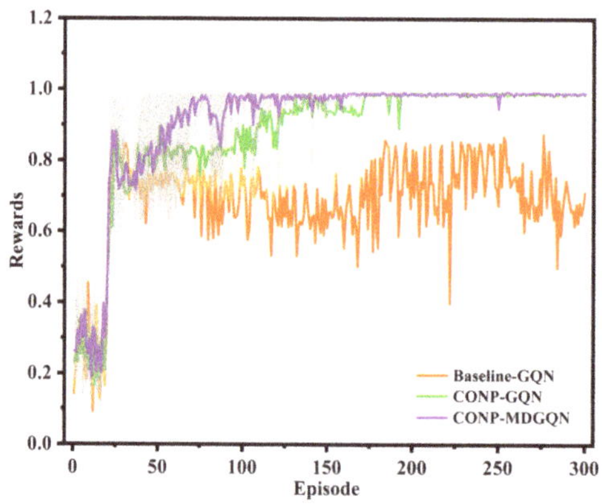

Figure 4. The comparison between the proposed reward function and baseline.

5.4. Training of Different Decision-Making Modes

As shown in Figure 5, under the four decision-making modes, the MDGQN algorithm converged faster than GQN, and the fluctuation of reward decline decreased significantly. This verified that MDGQN effectively alleviates the overestimation problem after using the multi-step TD target and double Q-learning algorithm to explore the optimal action more quickly. Based on MDGQN algorithm analysis, the AGGI decision-making mode converged in the 80th episode. In contrast, the AGGP decision-making mode did not fully converge until the 160th episode, which further verifies that the reward and punishment ratio of the reward function affects the convergence speed of the algorithm. In summary, by comparing the four decision-making modes, the aggressive decision-making mode uses more punishment than the conservative decision-making mode. The incentive decision-making mode can promote the convergence of the algorithm faster than the punished decision-making mode.

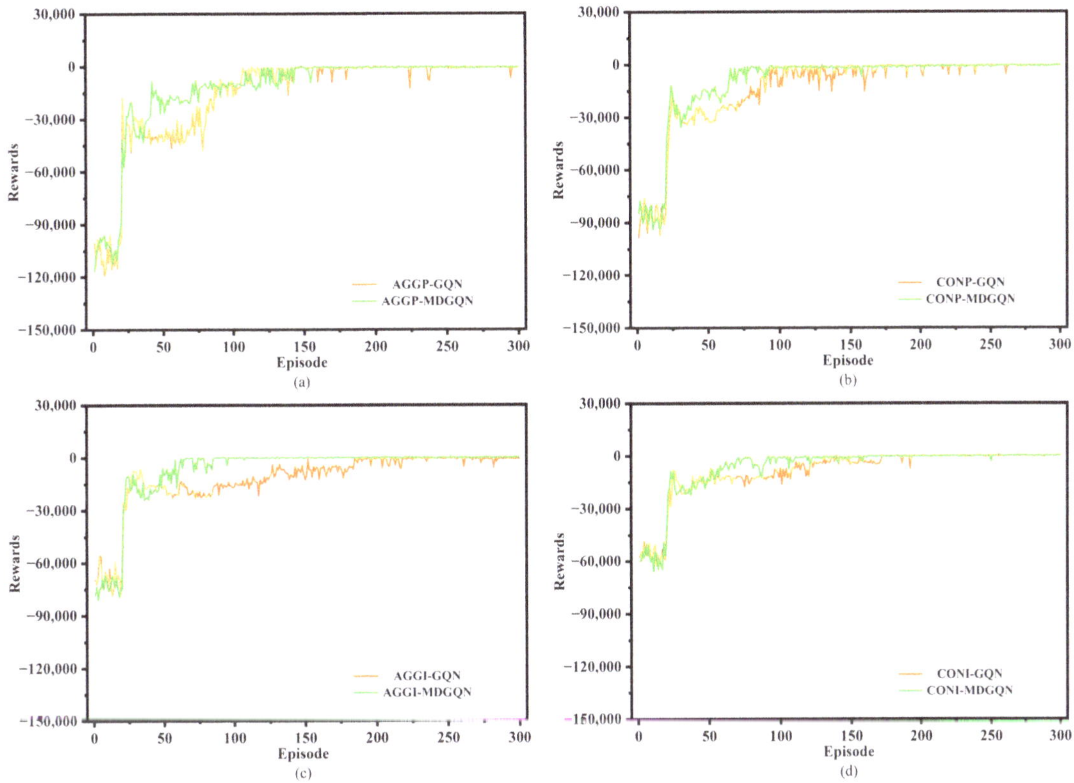

Figure 5. The training reward diagrams of four modes are compared and verified by the GQN and MDGQN algorithms. (**a**) The AGGP decision-making mode, (**b**) the CONP decision-making mode, (**c**) the AGGI decision-making mode, and (**d**) the CONI decision-making mode.

Furthermore, the training decision-making mode was tested, and the reward mean and variance are compared. As shown in Figure 6, MDGQN explored higher reward thresholds and averages than GQN in all four decision modes. Under AGGP, AGGI, CONP, and CONI decision-making modes, the corresponding test reward variances of MDGQN were 93.91, 80.26, 64.71, and 87.10; these were less than 108.51, 108.19, 93.78, and 97.01 for GQN. It can be concluded that the test stability of the CONP decision mode is the strongest, but the ability to distinguish algorithm differences is poor. AGGI is the decision-making mode that can best reflect the differences between algorithms, but the test stability decreases slightly.

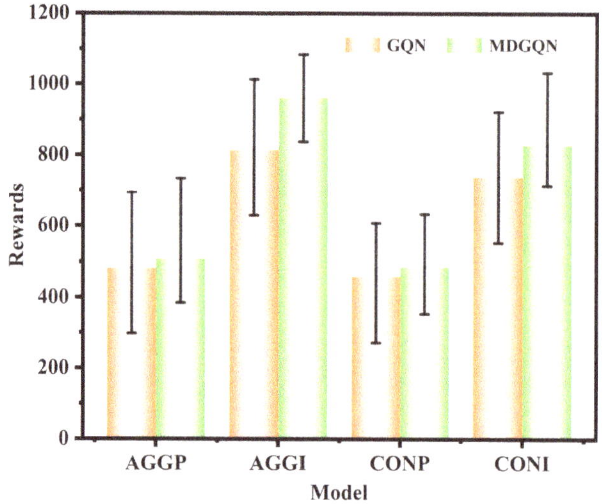

Figure 6. Test reward diagram of four decision-making modes. Orange represents the decision-making mode using the GQN algorithm, and green indicates the decision-making mode using the MDGQN algorithm. In the figure, the two ends of the error rod represent the maximum and minimum values of the test values.

5.5. Evaluation of Decision-Making Modes

Based on Figure 7, the two algorithms are first compared. In the four modes, the average speed of CAVs was better than that of GQN except for the CONI decision mode. MDGQN performed better than GQN in the speed of total traffic flow under the four modes. However, the proportion of CAVs' average speed increase in the total traffic flow's average speed increase is less than 1. In the CONI decision-making mode, the average speed of CAVs trained by MDGQN was lower than that of GQN, but the total traffic flow speed increased. This shows that based on the designed reward function, the MDGQN algorithm can obtain decisions that improve the total traffic efficiency by training CAVs. As shown in Table 6, MDGQN had a lower collision rate, fewer emergency braking times, and a higher arrival rate of CAVs and HV2 than GQN under the same decision-making mode, except for a slight increase in the number of emergency braking incidents under CONP decision-making mode. This can be analyzed from the corresponding data. Under the CONP decision-making mode, the p_{HV2} of MDGQN was 0.717, which was 18.12% higher than that of GQN. When the HV2 increases in "Exit 1", the uncertainty of the scene will enhance, increasing the number of emergency brake events. From the above results, it can be concluded that the comprehensive performance of the MDGQN algorithm is superior to that of GQN in this specific autonomous driving scene, and the MDGQN is more suitable for solid, interactive, uncertain scenes.

Additionally, the four decision-making modes are compared. Compared with the conservative decision-making mode, the aggressive decision-making mode had a higher average speed, a higher collision rate, more emergency braking incidents, and a higher vehicle arrival rate. Compared with the punished decision-making model, the average speed of the incentive decision-making mode was slightly higher, the collision rate was higher, the number of emergency braking incidents was higher, and the arrival rate of HV2 was further improved. Among the four decision-making modes, AGGI decision-making mode had the highest traffic efficiency; CAVs and HV2 had the highest arrival rate, but their collision rate and number of emergency braking incidents were are also the highest. The collision rate and emergency braking incidents of CONP decision mode were the lowest, but the traffic efficiency was the lowest, and the arrival rates of CAVs and HV2 were the lowest. Those performance differences are consistent with the weight coefficient allocation of the proposed reward function matrix. This fully proves that the proposed reward function matrix can effectively train various decision-making modes to adapt to autonomous driving scenarios.

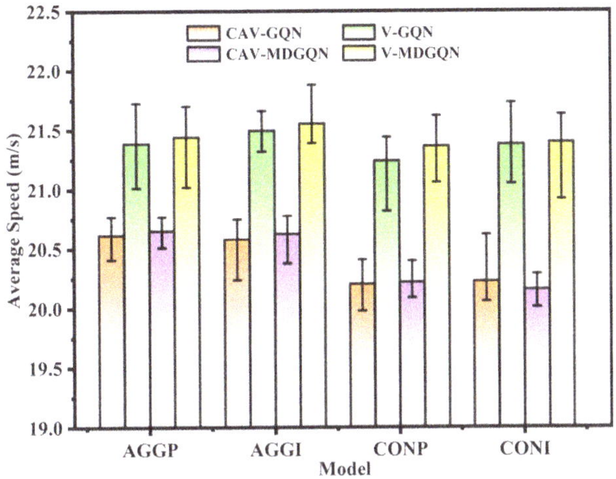

Figure 7. Average longitudinal velocity of CAVs and total traffic flow under four decision-making modes. The CAV-GQN means the average longitudinal speed of CAVs under the training of the GQN algorithm. The CAV-MDGQN represents the average longitudinal velocity of CAVs under the MDGQN algorithm training. The V-GQN indicates the average longitudinal speed of the total traffic flow under the training of the GQN algorithm. The V-MDGQN indicates the average longitudinal velocity of total traffic flow under MDGQN training. The two ends of the error rod indicate the maximum and minimum values of the test values.

Table 6. Test evaluation results.

Modes	Algorithm	$p_{collision}$	$N_{braking}$	p_{CAV}	p_{HV2}
AGGP	GQN	0.0111	3.1	0.991	0.763
	MDGQN	0.0067	1.867	1	0.813
AGGI	GQN	0.0133	3.567	1	0.93
	MDGQN	0.0022	2.533	1	0.937
CONP	GQN	0.0045	1.667	0.987	0.607
	MDGQN	0.0	1.9	0.996	0.717
CONI	GQN	0.0111	2.5	1	0.857
	MDGQN	0.0045	2.3	1	0.83

6. Conclusions

We proposed a reward function matrix, including a decision-weighted coefficient matrix, an incentive-punished-weighted coefficient matrix, and a reward–penalty function matrix. By adjusting the weight coefficient of the reward function matrix, various decision-making modes can be trained. We trained four decision-making modes, namely, AGGI, AGGP, CONI, and CONP; and the GQN algorithm and the MDGQN algorithm based on GQN improvement were used for verification by comparison. A large number of simulation results proved the following three conclusions. Firstly, the proposed reward function can promote the fast convergence of the algorithm and greatly improve the stability of the training process. Taking the CONP decision-making mode as an example, the average normalized reward value after the 200th round was 35.40% higher than that of the baseline, and the variance was only 3.27% greater than that of the baseline. Secondly, the comprehensive performance of the MDGQN algorithm is superior to that of GQN. Under the four decision-making modes, the averages and variances of test reward values of MDGQN are better than those of GQN. In terms of driving performance, MDGQN performs better than GQN, except that the number of brakes increases slightly in CONP decision-making mode. Finally, the proposed reward function matrix can effectively train various decision-making modes to adapt to different autonomous driving scenarios. With an increase in incentive weight, the comparison effect of the algorithm is more obvious, but the security will decrease. In our future work, we will further study the interactions of autonomous vehicles and decision mode switching.

Author Contributions: Conceptualization , X.G. and Q.L.; methodology, X.G.; software, Q.L. and Z.L.; validation, X.G., Q.L. and X.L.; formal analysis, X.G.; investigation, F.Y.; resources, T.L.; data curation, X.G.; writing—original draft preparation, X.G.; writing—review and editing, Q.L., X.L. and Z.L.; visualization, X.G.; supervision, X.L.; project administration, X.L.; funding acquisition, X.L. All authors have read and agreed to the published version of the manuscript.

Funding: This research received no external funding.

Institutional Review Board Statement: Not applicable.

Informed Consent Statement: Not applicable.

Data Availability Statement: Not applicable.

Conflicts of Interest: The authors declare no conflict of interest.

Appendix A

In the training process of decision-making mode, we adjusted different weight coefficients to find a mode with high comprehensive performance. To better validate the effectiveness of the proposed reward function matrix at training different decision patterns, we list four other sets of parameter settings, as shown in the Table A1.

Table A1. Parameter settings.

Modes	k_r	k_e	k_c	k_s	k_I0	k_P0
(a)	0.1	0.1	0.2	0.6	0.6	0.4
(b)	0.4	0.4	0.05	0.15	0.6	0.4
(c)	0.25	0.35	0.15	0.25	0.75	0.25
(d)	0.2	0.2	0.25	0.35	0.7	0.3

The simulation results are shown in Table A2. From this we can conclude that too much of an increase in the proportion of safety will lead to a decline in the completion rate of the driving tasks, especially CAVs' loss of the driving task objectives. Excessive increases in the proportions of the driving tasks and efficiency can lead to a rapid decline in safety. An increase in incentive rewards will lead to a decrease in security, and higher task completion rate and efficiency. Based on the above results, four groups of reasonable parameters were selected, and four decision models, CONP, CONI, AGGP, and AGGI, were trained.

Table A2. Test evaluation results.

Modes	$\bar{v}$	$p_{collision}$	$N_{braking}$	p_{CAV}	p_{HV2}
(a)	20.872	0.033	1.967	3.733	10.967
(b)	21.747	1.367	7.833	9.733	15
(c)	21.581	0.333	4.367	9.633	15
(d)	21.41	0.233	2.8	8.933	15

References

1. Stoma, M.; Dudziak, A.; Caban, J.; Droździel, P. The Future of Autonomous Vehicles in the Opinion of Automotive Market Users. *Energies* **2021**, *14*, 4777. [CrossRef]
2. Liu, Q.; Li, X.; Yuan, S.; Li, Z. Decision-Making Technology for Autonomous Vehicles Learning-Based Methods, Applications and Future Outlook. In Proceedings of the IEEE International Intelligent Transportation Systems Conference, Indianapolis, IN, USA, 19–22 September 2021.
3. Chen, L.; He, Y.; Wang, Q.; Pan, W.; Ming, Z. Joint optimization of sensing, decision-making and motion-controlling for autonomous vehicles: A deep reinforcement learning approach. *IEEE Trans. Veh. Technol.* **2022**, *71*, 4642–4654. [CrossRef]
4. Cadena, C.; Carlone, L.; Carrillo, H.; Latif, Y.; Scaramuzza, D.; Neira, J.; Reid, I.; Leonard, J.J. Past, Present, and Future of Simultaneous Localization and Mapping: Toward the Robust-Perception Age. *IEEE Trans. Robot.* **2016**, *32*, 1309–1332. [CrossRef]
5. Liu, Q.; Yuan, S.; Li, Z. A Survey on Sensor Technologies for Unmanned Ground Vehicles. In Proceedings of the 2020 3rd International Conference on Unmanned Systems, Harbin, China, 27–28 November 2020. [CrossRef]
6. Badue, C.; Guidolini, R.; Carneiro, R.V.; Azevedo, P.; Souza, A. Self-driving cars: A survey. *Expert Syst. Appl.* **2020**, *165*, 113816. [CrossRef]
7. Yu, Y.; Lu, C.; Yang, L.; Li, Z.; Gong, J. Hierarchical Reinforcement Learning Combined with Motion Primitives for Automated Overtaking. In Proceedings of the 2020 IEEE Intelligent Vehicles Symposium (IV), Las Vegas, NV, USA, 19 October–13 November 2020.
8. Kłosowski, G.; Gola, A.; Amila, T. Computational Intelligence in Control of AGV Multimodal Systems. *IFAC-PapersOnLine* **2018**, *51*, 1421–1427. [CrossRef]
9. Liu, Q.; Li, Z.; Yuan, S.; Zhu, Y.; Li, X. Review on Vehicle Detection Technology for Unmanned Ground Vehicles. *Sensors* **2021**, *21*, 1354. [CrossRef]
10. Bouton, M.; Nakhaei, A.; Fujimura, K.; Kochenderfer, M.J. Cooperation-aware reinforcement learning for merging in dense traffic. In Proceedings of the 2019 IEEE Intelligent Transportation Systems Conference (ITSC), Auckland, New Zealand, 27–30 October 2019; pp. 3441–3447.
11. Caban, J.; Nieoczym, A.; Dudziak, A.; Krajka, T.; Stopková, M. The Planning Process of Transport Tasks for Autonomous Vans–Case Study. *Appl. Sci.* **2022**, *12*, 2993. [CrossRef]
12. Nieoczym, A.; Caban, J.; Dudziak, A.; Stoma, M. Autonomous vans - the planning process of transport tasks. *Open Eng.* **2020**, *10*, 18–25. [CrossRef]
13. Li, Z.; Gong, J.; Lu, C.; Li, J. Personalized Driver Braking Behavior Modelling in the Car-following Scenario: An Importance Weight-based Transfer Learning Approach. *IEEE Trans. Ind. Electron.* **2022**, *69*, 10704–10714. [CrossRef]
14. Schwarting, W.; Alonso-Mora, J.; Rus, D. Planning and Decision-Making for Autonomous Vehicles. *Annu. Rev. Control Robot. Auton. Syst.* **2018**, *1*, 187–210. [CrossRef]
15. Matignon, L.; Laurent, G.J.; Fort-Piat, N.L. *Reward Function and Initial Values: Better Choices for Accelerated Goal-Directed Reinforcement Learning*; Springer: Berlin/Heidelberg, Germany, 2006.
16. Ou, X.; Chang, Q.; Chakraborty, N. Simulation study on reward function of reinforcement learning in gantry work cell scheduling. *J. Manuf. Syst.* **2019**, *50*, 1–8. [CrossRef]
17. Fu, Y.; Li, C.; Yu, F.R.; Luan, T.H.; Zhang, Y. A Decision-Making Strategy for Vehicle Autonomous Braking in Emergency via Deep Reinforcement Learning. *IEEE Trans. Veh. Technol.* **2020**, *69*, 5876–5888. [CrossRef]
18. Wang, H.; Gao, H.; Yuan, S.; Zhang, F.; Li, K. Interpretable Decision-Making for Autonomous Vehicles at Highway On-Ramps with Latent Space Reinforcement Learning. *IEEE Trans. Veh. Technol.* **2021**, *70*, 8707–8719. [CrossRef]
19. Chen, Q.; Zhao, W.; Li, L.; Wang, C.; Chen, F. ES-DQN: A Learning Method for Vehicle Intelligent Speed Control Strategy under Uncertain Cut-in Scenario. *IEEE Trans. Veh. Technol.* **2022**, *71*, 2472–2484. [CrossRef]
20. Peng, Y.; Tan, G.; Si, H.; Li, J. DRL-GAT-SA: Deep reinforcement learning for autonomous driving planning based on graph attention networks and simplex architecture. *J. Syst. Archit.* **2022**, *126*, 102505. [CrossRef]
21. Li, Z.; Lu, C.; Yi, Y.; Gong, J. A Hierarchical Framework for Interactive Behaviour Prediction of Heterogeneous Traffic Participants Based on Graph Neural Network. *IEEE Trans. Intell. Transp. Syst.* **2021**, 1–13. [CrossRef]
22. Jiang, J.; Dun, C.; Huang, T.; Lu, Z. Graph Convolutional Reinforcement Learning. *arXiv* **2018**, arXiv:1810.09202.
23. Peng, H.; Du, B.; Liu, M.; Liu, M.; He, L. Dynamic Graph Convolutional Network for Long-Term Traffic Flow Prediction with Reinforcement Learning. *Inf. Sci.* **2021**, *578*, 401–416. [CrossRef]
24. Kipf, T.N.; Welling, M. Semi-Supervised Classification with Graph Convolutional Networks. *arXiv* **2016**, arXiv:1609.02907.
25. Watkins, C.J.C.H. Learning from Delayed Rewards. Ph.D. Thesis, Kings College University of Cambridge, Cambridge, UK, 1989.
26. Mnih, V.; Kavukcuoglu, K.; Silver, D.; Rusu, A.A.; Veness, J.; Bellemare, M.G.; Graves, A.; Riedmiller, M.; Fidjel, A.K.; Ostrovski, G.; et al. Human-level control through deep reinforcement learning. *Nature* **2015**, *518*, 529–533. [CrossRef]
27. Dong, J.; Chen, S.; Ha, P.; Li, Y.; Labi, S. A DRL-based Multiagent Cooperative Control Framework for CAV Networks: A Graphic Convolution Q Network. *arXiv* **2020**, arXiv:2010.05437.
28. Li, G.; Yang, Y.; Li, S.; Qu, X.; Lyu, N.; Li, S.E. Decision making of autonomous vehicles in lane change scenarios: Deep reinforcement learning approaches with risk awareness. *Transp. Res. Part Emerg. Technol.* **2022**, *134*, 103452. [CrossRef]

29. Khayyam, H.; Javadi, B.; Jalili, M.; Jazar, R.N. Artificial Intelligence and Internet of Things for Autonomous Vehicles. In *Nonlinear Approaches in Engineering Applications: Automotive Applications of Engineering Problems;* Jazar, R.N., Dai, L., Eds.; Springer International Publishing: Cham, Switzerland, 2020; pp. 39–68. [CrossRef]
30. Tarkowski, S.; Rybicka, I. Distraction of the Driver and Its Impact on Road Safety. *Transp. Res. Procedia* **2020**, *44*, 196–203. [CrossRef]

sensors MDPI

Article

Local Path Planning of Autonomous Vehicle Based on an Improved Heuristic Bi-RRT Algorithm in Dynamic Obstacle Avoidance Environment

Xiao Zhang [1], Tong Zhu [2,*], Lei Du [1], Yueqi Hu [3] and Haoxue Liu [1]

[1] School of Automobile, Chang'an University, Xi'an 710064, China
[2] College of Transportation Engineering, Chang'an University, Xi'an 710064, China
[3] School of Vehicle Engineering, Xi'an Aeronautical Institute, Xi'an 710077, China
* Correspondence: zhutong@chd.edu.cn; Tel.: +86-2982334729

Abstract: The existing variants of the rapidly exploring random tree (RRT) cannot be effectively applied in local path planning of the autonomous vehicle and solve the coherence problem of paths between the front and back frames. Thus, an improved heuristic Bi-RRT algorithm is proposed, which is suitable for obstacle avoidance of the vehicle in an unknown dynamic environment. The vehicle constraint considering the driver's driving habit and the obstacle-free direct connection mode of two random trees are introduced. Multi-sampling biased towards the target state reduces invalid searches, and parent node selection with the comprehensive measurement index accelerates the algorithm's execution while making the initial path gentle. The adaptive greedy step size, introducing the target direction, expands the node more effectively. Moreover, path reorganization minimizes redundant path points and makes the path's curvature continuous, and path coherence makes paths between the frames connect smoothly. Simulation analysis clarifies the efficient performance of the proposed algorithm, which can generate the smoothest path within the shortest time compared with the other four algorithms. Furthermore, the experiments on dynamic environments further show that the proposed algorithm can generate a differentiable coherence path, ensuring the ride comfort and stability of the vehicle.

Keywords: autonomous vehicle; local path planning; Bi-RRT; path reorganization; path coherence

Citation: Zhang, X.; Zhu, T.; Du, L.; Hu, Y.; Liu, H. Local Path Planning of Autonomous Vehicle Based on an Improved Heuristic Bi-RRT Algorithm in Dynamic Obstacle Avoidance Environment. *Sensors* **2022**, *22*, 7968. https://doi.org/10.3390/s22207968

Academic Editor: Chao Huang

Received: 20 September 2022
Accepted: 17 October 2022
Published: 19 October 2022

1. Introduction

The intelligent transportation system is a real-time, accurate, efficient, and comprehensive transportation management system that plays a role in various directions [1]. It can effectively improve road capacity, reduce traffic accidents, improve transportation efficiency, and alleviate traffic congestion [2,3]. Meanwhile, it can also reduce energy consumption and improve environmental pollution [4,5]. Therefore, it has become the future development direction of the transportation system and has attracted more and more attention from all countries. As a component, the autonomous vehicle plays an essential role in the intelligent transportation system. It consists of an environmental perception layer, a path planning layer, and a path tracking control layer, and the study of path planning has always been a core problem. Commonly, path planning refers to efficiently finding a collision-free and feasible path from a starting point to a target point in a workspace [6–8]. In practical usage, the quality of the planned path will directly affect the vehicle's driving performance, so how to plan a passable path that can be tracked is very important for autonomous vehicles.

Scholars have carried out much research on path planning, and new path-planning algorithms are constantly emerging and developing. In the previous studies, five common categories of path planning algorithms can be found: geometric algorithms [9,10], graph search algorithms [11,12], intelligent bionic algorithms [13,14], the artificial potential field algorithm [15], and sampling-based search algorithms. Sampling-based search algorithms, including the rapidly exploring random tree (RRT) and the probability roadmap, have effectively solved many challenging planning problems, especially in complex environments [16,17]. In path planning, the basic RRT algorithm is widely used for actuators with nonholonomic constraints because of the advantages, including probability completeness, low computational cost, and no need to model search space [18–20]. However, the basic RRT only focuses on finding a path, with less regard to the convergence speed, the search efficiency, and the path optimality [21–23]. To overcome the basic RRT's shortcomings, some scholars made many improvements. The biased RRT uses target-biased search to form an extended mode of nonrandom sampling, thus improving planning efficiency [24]. The bidirectional RRT (Bi-RRT) can simultaneously generate two trees from the starting point and the target point to explore the search space, improving the algorithm's search efficiency [25]. The RRT-connect, a Bi-RRT version fusing a greedy function, generates two trees from the starting point and the target point, respectively, which reduces the search space and accelerates the convergence speed of the algorithm [26]. The RRT* uses new steps, including reselecting the parent node and rewiring the neighboring nodes of the newly inserted node to change search mode, thus generating a path with the optimal or approximate optimal length [27]. These algorithms improve the performance in planning speed and path length, respectively. However, they do not take the steering constraints of the wheels into account, resulting in them not being applied to the path planning of the autonomous vehicle directly.

When local practical environments are partially known or dynamic, supposing that some unknown or dynamic obstacles occupy the pre-generated global path at a certain moment, the autonomous vehicle will collide with the obstacles while tracking the path. Therefore, to avoid dynamic obstacles, the autonomous vehicle needs to regenerate a feasible path in real-time according to the environmental information obtained from its perception module. Park et al. constructed an algorithm combining the A* method and the artificial potential field method to solve online local path planning problems in the campus environment, guaranteeing real-time performance and the shortest path generation [28]. Chen et al. use a two-layered path planning model structure consisting of the modified Bi-RRT based on the steering constraint and a vector field histogram-guided polynomial planning method to plan a safe and smooth path meeting the real-time requirement [29]. Ge et al. utilized the resultant force of the potential field, the separating axis theorem, and the cubic B-spline to improve the Bi-RRT* and take the vehicle constraints into account, resulting in obtaining the smoothest path by taking the shortest time in practice the complicated environment [30]. Qi et al. utilized a modified RRT* to obtain an initial path, regard the state tree structure as prior knowledge, and design an approach to choose the best node among several candidates to regenerate the path quickly, resulting in planning a path avoiding dynamic obstacles [31]. Zou et al. proposed a path-planning algorithm based on RRT and reinforcement learning optimization, which can generate a smooth and steady path in complex and unknown environments without collision with obstacles [32]. Li et al. presented a real-time RRT-based path planning strategy consisting of a pre-processing RRT path planner and a real-time planner, which can modify the path rather than regenerate a path to avoid the unknown moving obstacle [33]. Peng et al. introduced a new way to choose candidate nodes, incremental step size, and the rapidly exploring random vines with a trajectory parameter space to form a kinematically constrained RRT-based path planning algorithm, which can find collision-free and kinematically feasible paths in various environments, such as dense environments and environments with narrow passages [34]. Wen et al. employed environmental knowledge to guide the planning procedure of the optimal RRT* method to propose a heuristic dual sampling domain reduction-based optimal RRT scheme including a layered online path planning framework in accordance with the model predictive control method, which outperforms traditional reduction schemes in terms of improving the execution efficiency of RRT* and is more reliable [35]. Niu et al. proposed a global dynamic path planning method based on an improved A* algorithm and combine it with the dynamic window method to improve the real-time performance of the dynamic obstacle avoidance of the intelligent vehicle [36]. Wu et al. utilized the genetic algorithm to optimize the path length and turning angle to obtain a short and smooth path [37]. The above path planning algorithms have improved the length and smoothness of the path and can be applied in a dynamic environment. However, they seldomly consider the curvature consistency of paths in multiple frames which refers to the fact that there is no sudden change between the path planned in the current frame and the path planned in the previous frame.

Furthermore, path optimization can effectively reduce the control difficulty of autonomous vehicles with nonholonomic constraints. Ge et al. used the cubic B-spline directly to optimize the path, resulting in the planned path with more turns [30]. Lu et al. utilized Dubins curves to generate a path, but the curvature of the generated path is discontinuous [38]. Yang et al. used path pruning to delete unnecessary path nodes without considering the included angles between line segments between path nodes, resulting in excessive curvature of the final planned path [39]. Chen et al. adopted path pruning based on inserted points to solve the initial path without considering the influence of inserted points on the path length, causing the length of the final planned path may not be optimal [29]. Thus, developing a path optimization method that can consider both path pruning and smoothing is necessary.

Therefore, in this article, an improved heuristic Bi-RRT path planning algorithm is proposed to solve local path planning problems of the dynamic environment by considering path length and the continuity of path curvature between frames. The improved heuristic Bi-RRT algorithm has contributed to node sampling, node selection, node extension, the interconnection mode of two trees, path organization, and the coherence of path curvature between frames.

1. The multiple-sampling states plus a guided method biased towards the target point are designed to reduce the blind growth of the random trees, and the node extension mechanism integrating the greedy algorithm, namely the adaptive greedy step size considering the target direction, can effectively accelerate the growth of two random trees.
2. The nearest node selection mechanism considering the kinematic constraint of the vehicle and the target state is put forward to reduce the effect of random sampling on path smoothness and speed up the growth of the random trees.
3. An amplifying vehicle constraint considering the driver's driving habit is introduced to make the vehicle move more safely, and the obstacle-free direct connection mode of two trees is introduced to further accelerate the execution of the algorithm.
4. A path reorganization process is designed to optimize the initial path to decrease the length of the final planned path to the maximum extent while ensuring path smoothness.
5. A novel path coherence method considering the inter-frame correlation of paths is used to ensure the curvature continuity of the path and make the vehicle be controlled easily and move more steadily.

The article is organized as follows: Section 2 discusses the Bi-RRT path planning algorithm, the simplified vehicle model, and the differences in path planning between front and back frames. The improved heuristic Bi-RRT algorithm is presented in Section 3. Section 4 presents simulation experiments to demonstrate the effectiveness and practicability of the proposed algorithm. Section 5 presents the discussion about its performance, and conclusions are provided thereafter.

2. Problem Statements

The basic bi-RRT algorithm is briefly described in this section, and its shortcomings are pointed out. A brief introduction of the vehicle model points out that the turning radius constraint of the vehicle needs to be considered in the path planning process. Furthermore, the influence of the difference between the path planning results of the front and rear frames in the dynamic path planning process on vehicle driving is described.

2.1. Basic Bi-RRT

The Bi-RRT is a variant of the basic RRT, which changes the expansion mode of the algorithm. That is, two random trees are constructed from the initial state and the target state, respectively. In each cycle, a random tree is first expanded to generate a new tree node, and then another random tree also starts to generate a new tree node, making two random trees expand towards each other. The two random trees expand alternately until the nodes of the two random trees meet. The searching schematic diagram of the basic Bi-RRT is shown in Figure 1.

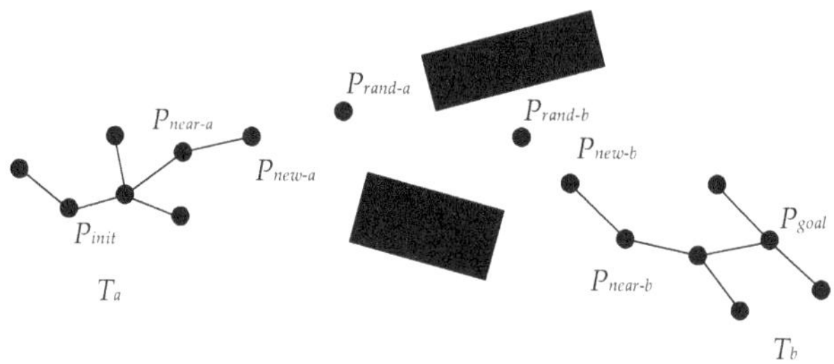

Figure 1. Schematic diagram of basic Bi-RRT expansion.

Algorithm 1 shows the basic Bi-RRT algorithm. Once initialized, the basic Bi-RRT algorithm conducts its iterative circle by selecting a random point P_{rand} from the configuration space using the sampling function *Random_State* ( ) (Line 3). The algorithm then determines a near tree node P_{near} by the function *Nearest_Neighbor* ( ) and obtains a new tree node P_{new} by the function *Extend* ( ) (Lines 4–5). If there are no obstacles between P_{near} and P_{new}, the new tree node P_{new} is added to the random tree T_a, and the nearest tree node $P_{nearest}$ from the random tree T_b is found by the function *Nearest_Neighbor* ( ) (Lines 6–8). The iterative circle terminates if the distance between P_{new} and $P_{nearest}$ is less than $l_{threshold}$ (Lines 9–11). Otherwise, T_a and T_b are swapped, and the procedures mentioned above are executed on the random tree T_b again (Line 12). Additionally, then, a path is generated by the function *Get_Path T* ( ) (Line 16).

Algorithm 1: Build Bi − RRT (P_{init} ,P_{goal})

1: T_a (P_{init}); T_b (P_{goal});
2: **while** 1 **do**
3: $P_{rand} \leftarrow$ *Random_State* ( );
4: $P_{near} \leftarrow$ *Nearest_Neighbor* (P_{rand}, T_a);
5: $P_{new} \leftarrow$ *Extend* (P_{near}, P_{rand});
6: **if** *Collision_Free* (P_{near}, P_{new}) **then**
7: T_a.*Add* (P_{new}), T_a.*Add* (P_{near}, P_{new})
8: $P_{nearest} \leftarrow$ *Nearest_Neighbor* (P_{new} , T_b);
9: **if** *Dis* tan *ce* (P_{new}, $P_{nearest}$) $< l_{threshold}$ **then**
10: *Return T* (T_a, T_b)
11: *break*
12: **else** *Swap* (T_a, T_b)
13: **end if**
14: **end if**
15: **end while**
16: *Path* $\leftarrow$ *Get_Path T* (T_a, T_b);

Algorithm 2 outlines the implementation procedure of the function *Get_Path T* ( ). Once the basic Bi-RRT algorithm completes the construction of two random trees, T_a and T_b, two path point sets, *path_a* and *path_b*, are defined, and the last added tree nodes of the two random trees are put into two sets, *path_a* and *path_b*, respectively (Lines 1–3).

Then, the two random trees are searched reversely according to indexes of parent nodes until their initial points are put into the path point sets, *path_a* and *path_b*, respectively (Lines 4–17). Finally, the path point set *path_a* is reversed and then combined with the path point set *path_b* to obtain a final path point set *path* (Lines 18–19).

Algorithm 2: Function *Get_Path T* (*Bi* − *T*)

```
1: Var path_a, path_b;
2: path_a.Add_Node (T_a.node_n);
3: path_b.Add_Node (T_b.node_n);
4: while 1 do
5:    i ← Index_pre_node (T_a);
6:    path_a.Back_Add_Node (T_a.node_i);
7:       if i = 1 then
8:          break
9:    end if
10: end while
11. while 1 do
12:    j ← Index_pre_node(T_b);
13:    path_b.Back_Add_Node (T_b.node_j);
14:    if j = 1 then
15:       break
16:    end if
17: end while
18: path_a ← reverse (path_a);
19: path ← path_a ∪ path_b;
20: Return path
```

The basic Bi-RRT algorithm simultaneously generates two random trees from the starting and target points and expands them in opposite directions, accelerating the convergence speed of the algorithm. However, the expansion mode of tree nodes still lacks directivity, the connective mode of two trees can be further improved, and the generated path is difficult to be directly tracked by the vehicle.

2.2. Vehicle Kinematical Model

Since the Bi-RRT is an incremental path planning algorithm, the kinematic vehicle model can be used to limit the expansion process of tree nodes to ensure the feasibility of the path. That is, the nonholonomic constraint of the vehicle should be considered when increasing new tree nodes. Because the sophisticated kinematic model is seldom available, a simplified theoretical motion model is provided, as shown in Figure 2. Assuming that the vehicle does not slip laterally, and the rear wheels do not steer, the vehicle kinematic model is expressed by Equation (1). Furthermore, the steering radius of the vehicle can be expressed by Equation (2).

$$\begin{cases} \dot{x} = v\cos\varphi \\ \dot{y} = v\sin\varphi \\ \dot{\varphi} = \frac{v\tan\delta_f}{L} \end{cases} \tag{1}$$

$$|k| \leq k_{\max} = \frac{1}{L}\tan\delta_{f_{\max}} = \frac{1}{R_{\min}} \tag{2}$$

where (x, y) represents the coordinate of the vehicle gravity center in the coordinate reference frame, v is the longitudinal speed of the vehicle, φ is the included angle between the vehicle main axis and the X axis, δ_f is the steering angle of the front wheels and $|\delta_f| \leq \delta_{f_{\max}}$, L is the wheelbase of the vehicle, k and R are the turning curvature and steering radius of the vehicle, respectively.

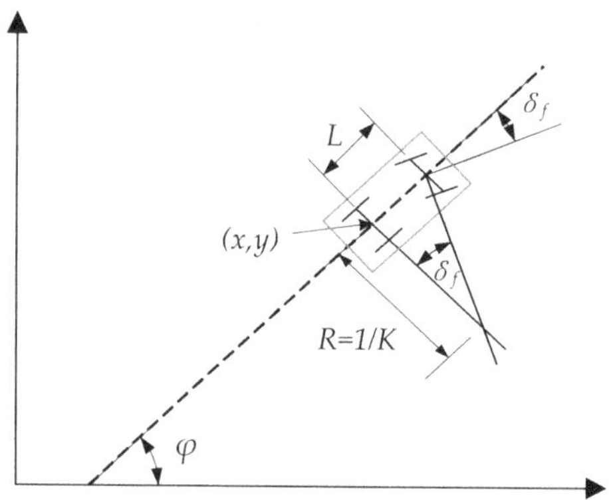

Figure 2. Schematic of vehicle kinematic model.

In order to consider the feasibility of newly generated path segments, the minimum steering radius constraint should be taken into account during the node extension procedure of Bi-RRT.

2.3. Path Planning Difference between Previous and Subsequent Frames

Because the length and smoothness of the path are calculated based on the environmental information collected in a certain frame, only considering the length and smoothness of the path cannot guarantee the steady driving of the autonomous vehicle. Suppose the path planned in the current frame deviates too far from the previous one. In that case, the driving stability of the autonomous vehicle will decline and even collide with the obstacle vehicle. It can be seen from Figure 3 that the path of the previous frame of the autonomous vehicle is on the left side of the obstacle, whereas the path of the current frame is on the right side of the obstacle. Because of the inconsistency of the path of the previous and subsequent frames, the autonomous vehicle may not avoid the obstacle and has the risk of collision with the obstacle. The dotted arrow may represent the actual driving direction of the autonomous vehicle.

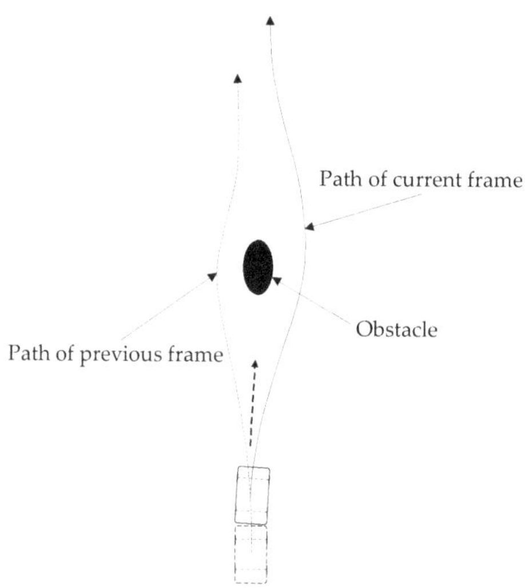

Figure 3. Path difference between current and previous frames.

Consequently, to prevent the difference in the paths generated by the previous and current frames of two adjacent planning cycles from influencing vehicle driving stability, it is necessary to consider the path information of the previous frame when planning the path in the current frame.

3. Improved Heuristic Bi-RRT Algorithm

Based on the above analysis, this section proposes an improved heuristic Bi-RRT algorithm for path planning in a dynamic obstacle avoidance environment. Figure 4 illustrates the model structure of the improved heuristic Bi-RRT. The input of the proposed model is a local driving environment and positioning information. The proposed algorithm model based on heuristic random sampling, heuristic nearest neighbor, heuristic extension, collision detection, direct connection detection, and path organization quickly generates a differentiable and collision-free path. Algorithm 3 shows the specific steps of the improved heuristic Bi-RRT algorithm.

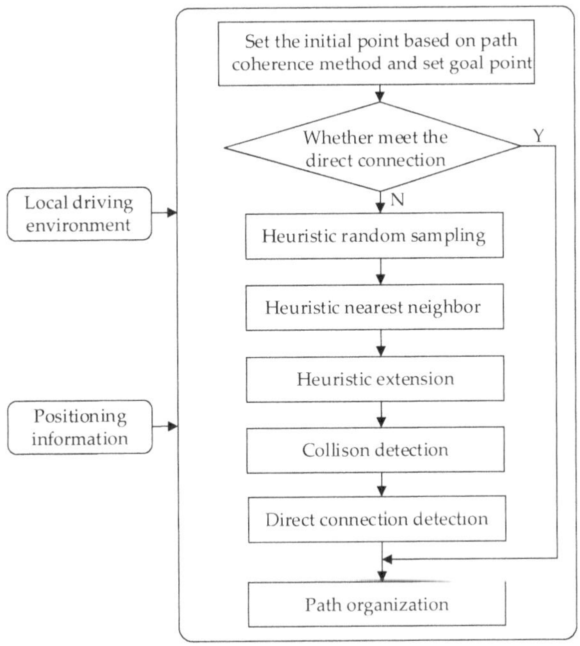

Figure 4. Model structure of the improved heuristic Bi-RRT.

The improved heuristic Bi-RRT algorithm obtains an initial point by the function *Current_Root* ( ) and initializes two random trees T_a and T_b in the same manner as they are in the basic Bi-RRT (Lines 1–2). Once the two random trees cannot be directly interconnected, the improved heuristic Bi-RRT begins its iterative processing by picking a random point in the feasible domain space through a sampling function *Heuristic_Random_Sampling* ( ) (Lines 4–5). Then, the parent node P_{near} from tree T_a is found by the function *Heuristic_Nearest_Neighbor* ( ), and the new tree node P_{new} is generated by the function *Heuristic_Extend* ( ) (Lines 6–7). If there are no obstacles between P_{near} and P_{new}, the new tree node P_{new} is added to the random tree T_a, and the nearest node $P_{nearest}$ from tree T_b is found (Lines 8–11). Subsequently, the iterative processing ends if there are no obstacles between $P_{nearest}$ and P_{new}, namely P_{judge2} and respectively (Lines 13–15). Otherwise, random trees T_b and T_b are swapped, and the procedures mentioned above are executed on the other random tree T_b again (Line 16). Afterward, an initial path is generated by the function

Get_Path T ( ) (Line 20). Path organization, including path node reconnection and path smoothing, processes the initial path to obtain a feasible path (Lines 21–22).

Algorithm 3: Build Improved Heuristic Bi-RRT (P_{init},P_{goal})

1: $P_{init} \leftarrow$ *Current_Root* ( );
2: T_a (P_{init});T_b (P_{goal});
3: $P_{judge1} \leftarrow P_{init}$; $P_{judge2} \leftarrow P_{goal}$;
4: **while** *Obstacle_Collision* (P_{judge1} , P_{judge2}) **do**
5: $P_{rand} \leftarrow$ *Heuristic_Random_Sampling* ( );
6: $P_{near} \leftarrow$ *Heuristic_Nearest_Neighbor*(P_{rand}, T_a);
7: $P_{new} \leftarrow$ *Heuristic_Extend* (P_{near}, P_{rand});
8: **if** *Collision_Free* (P_{near}, P_{new}) **then**
9: T_a.*Add*(P_{new}), T_a.*Add*(P_{near}, P_{new})
10: $P_{judge1} \leftarrow P_{new}$;
11: $P_{nearest} \leftarrow$ *Nearest_Neighbor* (P_{new}, T_b);
12: $P_{judge2} \leftarrow P_{nearest}$;
13: **if** *Collision_Free* (P_{judge1}, P_{judge2}) **then**
14: *Return T* (T_a, T_b)
15: *break*
16: **else** *Swap* (T_a, T_b)
17: **end if**
18: **end if**
19: **end while**
20: *Path* $\leftarrow$ *Get_Path T* (T_a, T_b);
21: *Path* $\leftarrow$ *Heuristic_Reconnection* (*Path*);
22: *Trajectory* $\leftarrow$ *Smoothing* (*Path*);

The improved heuristic Bi-RRT algorithm contains a set of heuristic methods to the benefit of path planning of the autonomous vehicle. The improvements in the connective mode of two random trees, node sampling, node selection, and node expansion based on the Bi-RRT framework are used to generate an initial path quickly. Additionally, then, path reorganization, including path reconnection and path smoothing, is employed to improve the quality of the initial path, making it suitable for tracking by the autonomous vehicle. The improved constraints containing the improved road environment and the improved vehicle constraint are conducive to generating a feasible path complying with the driver's driving habit. In addition, a novel path coherence method is introduced to make the generated path smoothly connected when planning a dynamic obstacle avoidance path. Specific methods of the algorithm are described as follows in detail.

3.1. Connective Mode of Two Random Trees

In order to further accelerate the running speed of the algorithm, the connective mode of two random trees can change from how the distance between two random trees is less than a certain distance threshold to the way of obstacle-free direct connection, as shown in Figure 5. After expanding the random tree T_a to obtain a new tree node P_{new-a}, the nearest node $P_{nearest}$ on the random tree T_b closest to the node P_{new-a} is calculated, and whether the area between P_{new-a} and $P_{nearest}$ is passable is checked. If there are no obstacles between P_{new-a} and $P_{nearest}$, P_{new-a} and $P_{nearest}$ are directly connected, and the initial path planning is complete. Otherwise, the algorithm continues to execute until the two random trees are connected successfully.

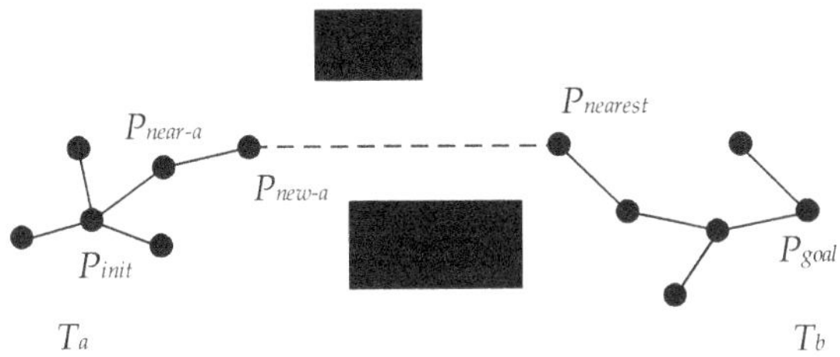

Figure 5. Obstacle-free direct connection of two random trees.

3.2. Heuristic Target Bias Sampling Method

The basic Bi-RRT usually adopts random searches in the global scope during the random sampling process, which will cause the generation of random points without guidance and too much unnecessary computation. As to this problem, a heuristic bias sampling strategy composed of a multiple-sampling method and a target bias method is adopted to make the random tree grow in a biased direction, enabling the initial state and the goal state to meet more effectively and faster. Equations (3) and (4) demonstrate how to calculate the random sampling state P_{rand}.

$$P_{randm} = Heuristic\_Random\_State\ (\ ) \tag{3}$$

$$P_{rand} = \begin{cases} P_{randm} & d_{\text{ob}} \leq d_{\text{threshold}} \\ P_{randm} + \chi \cdot \left(\dfrac{\overrightarrow{P_{randm}P_{\text{target}}}}{\left|\overrightarrow{P_{randm}P_{\text{target}}}\right|} \right) & d_{\text{ob}} > d_{\text{threshold}} \end{cases} \tag{4}$$

where P_{randm} is the random sampling state generated by the multiple-sampling method, P_{target} is the target point defined in each sampling process, and χ is the biased step size. d_{ob} and $d_{\text{threshold}}$ are the distance from the random point P_{randm} to the obstacle and the distance threshold from the obstacle, respectively.

3.2.1. Heuristic Random Sampling

With the knowledge of the initial and goal state, to make the sampling random point close to the target state, the *Multiple_Random_State* function generates several random points instead of one random point generated by the basic Bi-RRT algorithm in the free region using the *Random_State* function. The multiple-sampling point function does not include the probability of bias to the target, hence avoiding the local minimum problem. Additionally, then, the introduction of the *Nearest_To_Target* function is used to select the random state closer to the goal from several candidate sampling points. The random point out of these candidate sampling points closest to the target becomes the chosen random state P_{randm}, causing the Bi-RRT to search toward the target point. The number of random states to be selected was set to two after simulation results showed the best performance when the number of random states was limited to two.

3.2.2. Heuristic Target Bias

After obtaining the chosen random state P_{randm}, the target bias method is introduced. It can make the random point expand a step size χ along the direction from it to the target state to generate the random sampling point further closer to the target state, resulting in making the random tree grow more directionally and improving computational efficiency. When d_{ob} is less than $d_{\text{threshold}}$, if the generated random sampling point is still closer to the target state, it will make the newly generated tree node easy to collide with the obstacle, resulting in sampling failure and directly affecting the solution speed. Therefore, the target bias method introduces the distance threshold $d_{\text{threshold}}$, which is the projection distance of the semi-major axis of the expanded safety ellipse on the x axis of the coordinate reference frame. When $d_{\text{ob}} \leq d_{\text{threshold}}$, that is, the chosen random state P_{randm} is close to the obstacle, P_{randm} is regarded as the generated random sampling point P_{rand}, when $d_{ob} > d_{\text{threshold}}$, that is, the chosen random state P_{randm} is far from the obstacle, P_{rand} is generated from the target bias function. This strategy can maintain the balance between exploration and search speed.

Based on the basic Bi-RRT sampling function framework, the novel target bias sampling method changes the generation mode of the sampling point to make the generated sampling points more directional and effective, improving search efficiency and accelerating the algorithm's convergence. Algorithm 4 outlines the working of the function *Heuristic_Random_Sampling* ( ). The improved heuristic Bi-RRT obtains candidate sampling points through the sampling function *Random_State* ( ) and selects the sampling point with the minimum distance cost as the current sampling point P_{randm} (Lines 1–3). Then, according to the distance between the current sampling point P_{randm} and the obstacle, the current sampling point P_{randm} is processed by the heuristic target bias method to obtain the final sampling point P_{rand} (Lines 4–7). This procedure is performed in each sampling process.

Algorithm 4: Function *Heuristic_Random_Sampling* ( )

1: $P_{rand1} \leftarrow$ *Random_State* ( );
2: $P_{rand2} \leftarrow$ *Random_State* ( );
3: $P_{randm} \leftarrow$ *Nearest_To_Target* (P_{rand1}, P_{rand2});
4: **if** $d_{ob} \leq d_{threshold}$ **then**
5: $\quad P_{rand} \leftarrow P_{randm}$;
6: **else**
7: $\quad P_{rand} \leftarrow P_{randm} + \chi \cdot \left(\dfrac{\overrightarrow{P_{randm}P_{target}}}{\left|\overrightarrow{P_{randm}P_{target}}\right|} \right)$;
8: **end if**

3.3. Heuristic Parent Node Selection Method

In the basic Bi-RRT algorithm, the nearest tree node is measured by the Euclidean distance from the random point to the tree node, which may generate a polyline path with sharp included angles between connecting line segments of path nodes. Even if the polyline line path is smoothed, it cannot be successfully followed by vehicles. Because the vehicle has a minimum steering radius in actual motion, the Euclidean distance should not be the only factor to consider during each node selection process. As shown in Figure 6, the current vehicle state is a solid rectangle containing a yellow vehicle with three alternative driving states shown as dashed rectangles. In order to find the nearest driving state for the current vehicle, if the Euclidean distance is only considered, there is no doubt that the first driving state is the closest with a Euclidean distance of zero and the second driving state takes second place. However, the first and second driving states, also named in situ steering and lateral translation, are not possible for the vehicle due to the minimum steering radius. Hence, the third driving state is more reasonable.

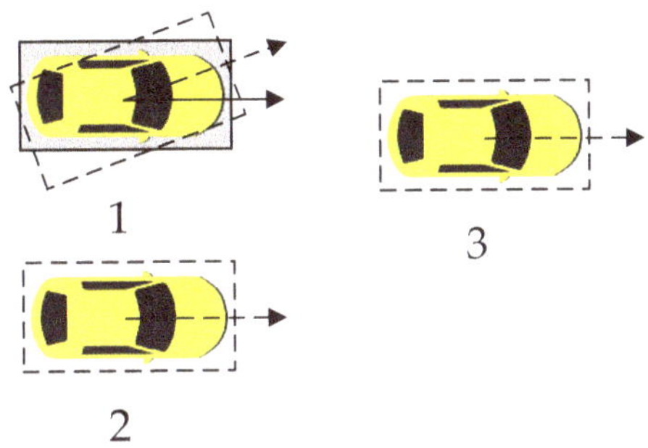

Figure 6. Different driving states.

Given the above analysis, the steering radius of the vehicle needs to be considered to choose the near tree node, namely the parent tree node. Therefore, a heuristic selection method of the parent tree node, namely a comprehensive measurement index considering the distance factor and angle factor, is introduced to make the generated path gentler and easily tracked by the vehicle. Furthermore, it can speed up the algorithm convergence and reduce the calculation time. Equations (5)–(11) show the calculation process when selecting the near tree node. The near tree node P_{near} is determined as the tree node corresponding to the maximum comprehensive measurement index.

$$C_M = \omega_1 \cdot C^1_{dis\tan ce} + \omega_2 \cdot C^1_{angle} \tag{5}$$

$$C_{dis\tan ce} = \xi_1 \cdot C_{dis\tan ce1} + \xi_2 \cdot C_{dis\tan ce2} \tag{6}$$

$$C_{dis\tan ce1} = \left| \overrightarrow{P_{tree}P_{rand}} \right| \tag{7}$$

$$C_{dis\tan ce2} = \left| \overrightarrow{P_{tree}P_{goal}} \right| \tag{8}$$

$$C_{angle} = \pi - \left(\cos^{-1} \left(\frac{\overrightarrow{P_iP_j} \cdot \overrightarrow{P_iP_{rand}}}{\left|\overrightarrow{P_iP_j}\right| \cdot \left|\overrightarrow{P_iP_{rand}}\right|} \right) \right) \tag{9}$$

$$C^1_{dis\tan ce} = \frac{C_{dis\tan ce\_max} - C_{dis\tan ce}}{C_{dis\tan ce\_max}} \tag{10}$$

$$C^1_{angle} = \frac{C_{angle\_max} - C_{angle}}{C_{angle\_max}} \tag{11}$$

where ω_1 and ω_2 are weighted coefficients of the distance index $C^1_{dis\tan ce}$ and the angle index C^1_{angle}, respectively. $C^1_{dis\tan ce}$ and C^1_{angle} are the normalized values of the distance $C_{dis\tan ce}$ and the angle C_{angle}, respectively. The diagrammatic presentation of the angle C_{angle} is shown in Figure 7. $C_{dis\tan ce}$ is the weighted sum of the distance $C_{dis\tan ce1}$ and the distance $C_{dis\tan ce2}$, ξ_1 and ξ_2 are the weighted coefficients of $C_{dis\tan ce1}$ and $C_{dis\tan ce2}$, respectively. $C_{dis\tan ce1}$ represents the distance from each tree node to the random sampling node, and $C_{dis\tan ce2}$ represents the distance from each tree node to the target state. The introduce of the distance $C_{dis\tan ce2}$ can make the selected near tree node have a trend close to the target state to increase the computational efficient.

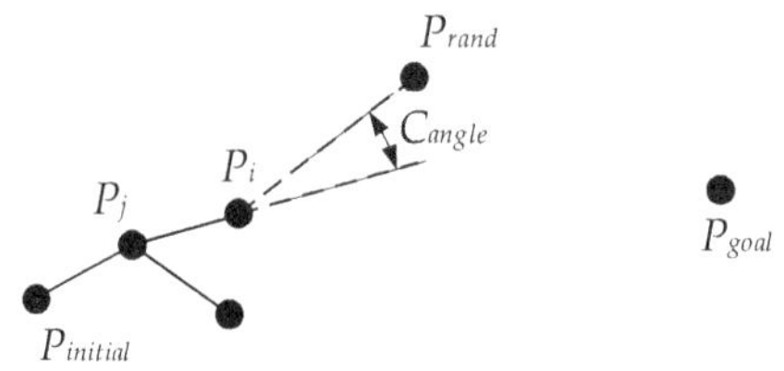

Figure 7. Calculation of the included angle.

The transition from Euclidean distance to the comprehensive measurement index changes the growing characteristic of the random tree. The introduction of the comprehensive measurement index hinders the pure expansion of the random tree growth towards the configuration region. On the contrary, it drives the tree growth towards the direction associated with the target state and the gentle trend. As a result, it improves the running speed of the algorithm to some extent, and there is a natural trade-off between quick tree growth and better path quality.

3.4. Heuristic Node Extension Method

Expanding the near tree node to the random point usually adopts a fixed step size in the basic Bi-RRT. The large fixed step size tends to make the new node difficult to extend in the surrounding area of the obstacle, especially dense obstacle areas, resulting in extension failure and reducing the growth rate of the random tree. The small fixed step size could lead to a slow convergence speed of the algorithm. In other words, the fixed step size has the disadvantages of low flexibility and low security. In addition, the expansion direction of the tree node is along the vector direction from P_{near} to P_{rand}. When utilizing the large step size to extend the parent tree node along the expansion direction of the tree node deviating from the target state, an invalid and unnecessary node could be generated, resulting in the low quality of the generated path, and affecting the running speed of the algorithm. Thus, to solve these two problems, a heuristic node extension method, namely the adaptive greedy step size, is introduced. 'Adaptive' is reflected in the pattern that the step size is dynamically and gradually changing rather than fixed, and 'greedy' is reflected in the pattern that the greater the extent of the node expansion direction approaching the direction of the target state, the larger the step size. Equations (12)–(16) show the calculation of the adaptive greedy step size, and its simple legend is shown in Figure 8.

$$\lambda_{adaptive\_greedy} = \begin{cases} \lambda & d_{near\_ob} < d_{threshold} \\ \begin{cases} (\eta_{\cos} + \sqrt{s}) \cdot \lambda & \beta < \frac{\pi}{2} \\ (1 - \eta_{\cos} + \sqrt{s}) \cdot \lambda & \beta \geq \frac{\pi}{2} \end{cases} & d_{near\_ob} \geq d_{threshold} \end{cases} \tag{12}$$

$$\eta_{\cos} = \left| \frac{\overrightarrow{V_{unit\_r}} \cdot \overrightarrow{V_{unit\_t}}}{\left|\overrightarrow{V_{unit\_r}}\right| \cdot \left|\overrightarrow{V_{unit\_t}}\right|} \right| \tag{13}$$

$$\beta = arc\cos\left(\frac{\overrightarrow{V_{unit\_r}} \cdot \overrightarrow{V_{unit\_t}}}{\left|\overrightarrow{V_{unit\_r}}\right| \cdot \left|\overrightarrow{V_{unit\_t}}\right|} \right) \tag{14}$$

$$\overrightarrow{V_{unit\_t}} = \frac{\overrightarrow{P_{near}P_{target}}}{\left|\overrightarrow{P_{near}P_{target}}\right|} \tag{15}$$

$$\overrightarrow{V_{unit\_r}} = \frac{\overrightarrow{P_{near}P_{rand}}}{\left|\overrightarrow{P_{near}P_{rand}}\right|} \tag{16}$$

where λ is the basic lower bound of the step size, $\overrightarrow{V_{unit\_t}}$ and $\overrightarrow{V_{unit\_r}}$ are unit vectors from P_{near} to P_{target} and from P_{near} to P_{rand}, respectively. φ is the included angle between $\overrightarrow{V_{unit\_t}}$ and $\overrightarrow{V_{unit\_r}}$, $\eta_{\cos}$ is the cotangent of φ, s is a constant as a regulating coefficient, and d_{near_ob} is the distance from P_{near} to the obstacle. Far away from the obstacle means $d_{nearest\_ob} \geq d_{threshold}$.

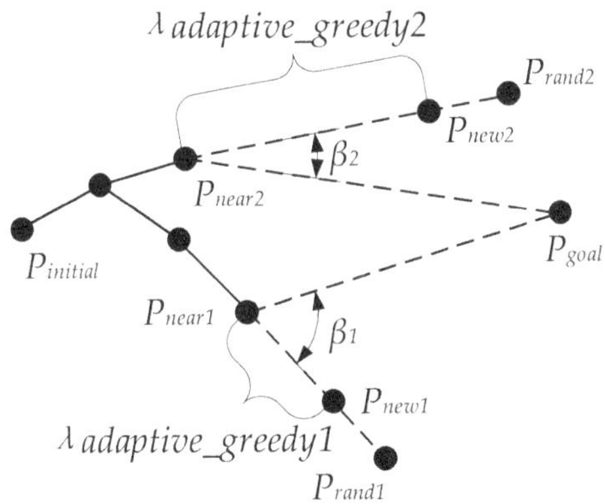

Figure 8. Adaptive greedy step size.

Compared with the traditional fixed step size, the adaptive greedy step size is no longer a constant. Its size is not only related to the distance from the obstacle but also to the included angle of the vector $\overrightarrow{P_{near}P_{rand}}$ towards the target state. In this way, when near an obstacle, a small step size can be used for the basic extension, making the search path more detailed and safer. When far from an obstacle, the step size can be greedily adjusted as the included angle changes, making the random tree adopt the larger step size to expand rapidly along the extension direction closer to the target state and reduce some blind expansions. As a result, the greedy mode of the step size accelerates the growth of the random tree and makes the growth of the path tend to approach the target state to a certain extent. It is exactly because of its dynamic adjustability that the adaptive greedy step size extension can often pass the obstacle detection when approaching the obstacle, unlike the fixed step size extension. Using the adaptive greedy step size can significantly improve the success rate of new node generation and accelerate overall search efficiency effectively.

3.5. Improved Constraints

Suppose a path can be successfully and effectively tracked by an autonomous vehicle. In that case, the path should not only meet the obstacle constraints but also comply with the driving characteristics of the actual driver, resulting in avoiding causing excessive tension and discomfort to the driver and passengers. The obstacle constraints include the road environment and the environmental constraint formed by the vehicle being regarded as an obstacle, namely, the vehicle constraint. Therefore, the road environment and the vehicle constraint are improved to obtain a feasible path meeting the driving habits of the actual driver.

3.5.1. Improved Road Environment

In order to generate a practical path, the path nodes need to meet the constraints of road environments when planning. The road environment restricts the planned path more effectively by considering the vehicle width to avoid collision between the vehicle and the road edge. Thus, the newly extended nodes need to meet the requirements of the following Equation (17), and the schematic diagram of the road environment is shown in Figure 9.

$$\begin{cases} P_{initial\_x} < T_{node\_x} < P_{goal\_x} \\ B_r + \frac{W_h}{2} < T_{node\_y} < B_l - \frac{W_h}{2} \end{cases} \tag{17}$$

where $P_{initial\_x}$ and P_{goal_x} are the x coordinates of the initial point and the goal point in the reference coordinate frame, respectively. B_l and B_r are the left and right boundaries of the road, and W_h is the width of the host vehicle.

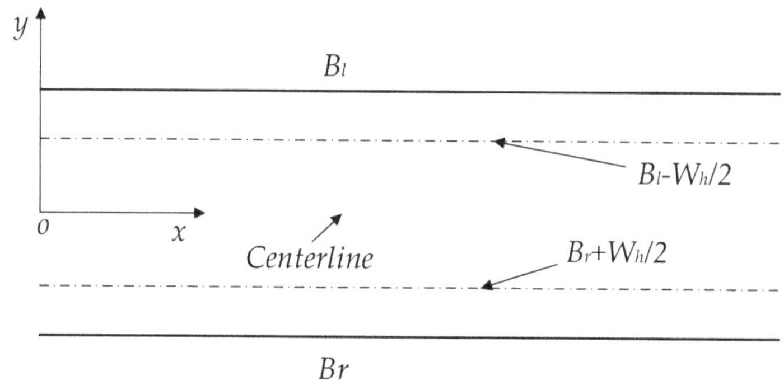

Figure 9. The improved road environment constraint.

3.5.2. Improved Vehicle Constraint

In order to express the constraint generated by the obstacle vehicle conveniently, a safety ellipse is introduced to envelop the obstacle vehicle. When considering the subjective comfort of the driver and passengers, a planned path needs to be able to avoid the obstacle vehicle in advance, that is, avoiding the generation of excessive path curvature when approaching the obstacle vehicle. In this case, the planned path can avoid causing discomfort to the driver and passengers and be easily tracked by the vehicle. Therefore, on the basis that half of the vehicle length is used as the semi-major axis of the safety ellipse, the advanced obstacle avoidance distance is added to the semi-major axis to ensure that the vehicle obstacle avoidance occurs at a long distance. Equation (18) expresses the condition that the newly expanded node needs to meet, and the schematic diagram of the vehicle constraint is shown in Figure 10.

$$\begin{cases} \frac{(x - x_{obstacle})^2}{\left[s_{f1}\cdot\left(d_{safe}+\frac{L_0}{2}\right)\right]^2} + \frac{(y - y_{obstacle})^2}{(s_{f2}\cdot W_o)^2} > 1 \\ d_{safe} = \frac{v^2}{2\cdot\eta\cdot g} \end{cases} \tag{18}$$

where (x, y) is the point on the connecting segment between the newly extended node P_{new} and its parent node P_{near}, (x_{ob}, y_{ob}) is the position of the obstacle vehicle, L_o and W_o are the length and width of the obstacle vehicle, respectively. s_{f1} and s_{f2} are the expansion coefficients of the semi-major axis and the semi-minor axis of the safety ellipse, respectively, v is the speed of the host vehicle, η is the friction coefficient, g is the acceleration of gravity, and d_{safe} is the advanced obstacle avoidance distance.

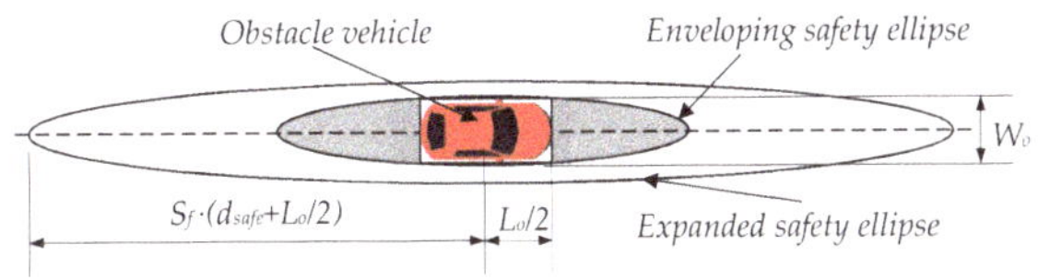

Figure 10. The improved vehicle constraint.

To sum up, during the process of the specific node extension, the improved constraints, including the improved road environment and the improved vehicle constraint, can make the planned path meet the actual driving requirement of the vehicle, namely, the path feasibility and the characteristics of avoiding the obstacle in advance.

3.6. Path Reorganization Method

Because the basic Bi-RRT algorithm adopts random sampling, the obtained path, also named the initial path, often has poor quality, mainly reflecting in containing too many unnecessary turn points and the discontinuity of path curvature. When tracking the initial path, the autonomous vehicle has to stop and change its driving direction so that it cannot drive smoothly and has unnecessary mechanical wear. Thus, a path reorganization method is needed to optimize the initial path, that is, to remove unnecessary turn points and smooth the path.

After obtaining an initial path by the function $Get\_Path\ T\ (Bi - T)$, as shown in Algorithm 2, the path reorganization method containing path node reconnection and path smoothing is used to process it. The path node reconnection is utilized to remove unnecessary turn nodes and insert path nodes to replace some necessary path nodes. As a result, it can make the included angles of the connecting line between path points meet the vehicle steering requirement, decreasing the control difficulty of the autonomous vehicle while reducing the path distance to the maximum extent. Additionally, then, the path obtained by the path node reconnection is a broken line composed of discrete path nodes; thus, it needs a further smoothing process to make the vehicle drive smoothly and steadily. Algorithm 5 describes the pseudocode of the path reorganization method.

3.6.1. Path Node Reconnection

A path obtained by the function $Get\_Path\ T\ (Bi - T)$ usually has poor connectivity due to the random attribute of the algorithm. Furthermore, there are many redundant turning segments in the path. As a result, it is often not continuously differentiable and infeasible. Thus, the path needs the path node reconnection to meet the prerequisite of path smoothing. Path node reconnection shown in lines 1–43 of Algorithm 5 is conducted to remove redundant path nodes and insert path nodes to replace some existing path nodes for obtaining a new path with maximum length reduction and no collision with obstacles. Moreover, it can ensure that the complementary angles of the included angles of line segments between path nodes are less than the steering angle of the front wheel. The function $Get\_Path\ T\ (Bi - T)$ is used to obtain a path node set S_0 of the initial path from the forward node of the goal node to the root node of the initial node (Line 2). The first path node is defined as the root node p_{root}, the second path node is defined as the parent node p_{parent}, and the third path node is defined as the current node $p_{current}$ (Line 3). A line segment connects the current node $p_{current}$ and the first subsequent path node from the set S_2. Additionally, then, connecting this current node $p_{current}$ and one of all subsequent path nodes from the set S_2 through a line segment successively in sequence is conducted until a collision occurs between the line segment and the obstacle (Lines 7–10). In this case, the path nodes between the line segment are removed, and the parent node of the path node that results in the collision with the obstacle is defined as the previous forward node $p_{forward\_last}$ (Line 11). Meanwhile, the complementary angle of the included angle between the vector $\overrightarrow{p_{current}p_{parent}}$ and the vector $\overrightarrow{p_{curren}p_{forward\_lastt}}$ is calculated. When the complementary angle is less than the steering angle constraint δ_f, p_{root} is redefined as the previous root node p_{root_last}, p_{parent} is redefined as the new root node p_{root}, $p_{current}$ is redefined as the new parent node p_{parent}, and $p_{forward\_last}$ is redefined as the new current node $p_{current}$ (Lines 12–13). On the contrary, a path node p_{insert} is inserted between the parent node p_{parent} and the previous forward node $p_{forward\_last}$ to replace the current node $p_{current}$ and meet the conditions that the complementary angle of the included angle between the vector $\overrightarrow{p_{parent}p_{insert}}$ and the vector $\overrightarrow{p_{parent}p_{root}}$ and the complementary angle of the included angle between the vector $\overrightarrow{p_{insert}p_{parent}}$ and the vector $\overrightarrow{p_{insert}p_{forward\_last}}$ are less than the steering angle constraint δ_f, and the line segment connecting the parent node p_{parent} and the inserted node p_{insert} and the line segment connecting the inserted node q_{insert} and the previous forward node $p_{forward\_last}$ do not collide with the obstacle (Lines 17–18). After that, p_{root} is redefined as the previous root node p_{root_last}, p_{parent} is redefined as the new root node p_{root}, p_{insert} is redefined as the new parent node p_{parent}, and $p_{forward\_last}$ is redefined as the new current node $p_{current}$ (Line 19). The above operations are applied to the remaining path nodes in the set S_2 in sequence until the path node p_{n-1} is found. Finally, suppose the complementary angle of the included angle between the line segment connecting the finally defined $p_{current}$ and the finally defined p_{parent} and the line segment connecting the finally defined $p_{current}$ and p_n is less than the steering angle constraint δ_f. In that case, p_n is added to the set S_1 (Lines 29–30). Otherwise, the final path node p_n is added to the set S_1 after meeting the conditions that the complementary angle of the included angle between the line segment connecting the finally defined p_{root} and the finally defined p_{root_last} and the line segment connecting the finally defined p_{root} and the inserted node p_{insert}, the complementary angle of the included angle between the line segment connecting the inserted node p_{insert} and the finally defined p_{root} and the line segment connecting the inserted node p_{insert} and the finally defined $p_{current}$, and the complementary angle of the included angle between the line segment connecting the finally defined $p_{current}$ and the inserted node p_{insert} and the line segment connecting the finally defined $p_{current}$ and the final node p_n are less than the steering angle constraint δ_f, and the line segment connecting the finally defined p_{root} and the inserted point p_{insert} and the line segment connecting the inserted point p_{insert} and the finally defined $p_{current}$ do not collide with the obstacle (Lines 31–43).

The path node reconnection process is illustrated in Figure 11 specifically. The solid black line is the initial path obtained by the function $Get\_Path\ T\ (Bi-T)$. The connecting line segments between the black nodes P_1, P_2, P_3, P_4, P_5, and P_6, which are obtained by the process of the path node reconnection, constitute a path to be smoothed in the next step, which is represented as the red solid line. The connecting line segments between the black nodes P_2, P_3, P_4, and P_5 do not intersect with the obstacle, removing the redundant nodes between them, namely, the green path nodes. $P_3{}'$, a path node from the initial path, can directly connect with P_2 and P_4. However, the complementary angle of the included angle between the line segment connecting $P_3{}'$ and P_2, and the line segment connecting $P_3{}'$ and P_4 is more than the steering angle constraint δ_f. As a result, a node, namely, node P_3, is inserted between P_2 and P_4 based on the constraint δ_f to replace the node $P_3{}'$ for obtaining a relatively gentle path with the maximum length reduction, namely the red path. Furthermore, the complementary angles of the included angles of line segments consisting of these black nodes are less than δ_f, that is, β_1, β_2, β_3, and β_4 are less than δ_f. Thus, the path node reconnection method can reduce the length of the initial path to the greatest extent and obtain a path convenient for further smoothing.

Algorithm 5: Function *path_reorganization* ( )

$Var\ S_0,\ S_1,\ S_2 :\ path$
$S_0(p_n \ldots, p_2, p_1, p_0) \leftarrow Get\_path\ T\ (Bi - T)$;
$p_{root} \leftarrow p_0$; $p_{parent} \leftarrow p_1$; $p_{current} \leftarrow p_2$;
$S_1 \leftarrow (p_{root}, p_{parent}, p_{current})$;
$S_2 \leftarrow (p_3 \ldots p_n)$;
while $p_{current}! = p_{n-1}$ **do**
  for *each node* $p_i \in S_2$ **do**
    if $Collision\_Free\ (p_{current}, p_i)$ **then**
      $p_{forward} \leftarrow p_i$;
    else
      $p_{forward\_last} \leftarrow p_{i-1}$;
      if $(\pi - Angle\ (\overrightarrow{p_{current}p_{parent}},\ \overrightarrow{p_{current}p_{forward\_last}})) < \delta_f \leq \delta_{f\_max}$ **then**
        $p_{root\_last} \leftarrow p_{root}$; $p_{root} \leftarrow p_{parent}$; $p_{parent} \leftarrow p_{current}$; $p_{current} \leftarrow q_{forward\_last}$;
        $S_1.Backward\_Add\_Node\ (p_{current})$;
      else
        while 1 **do**
          $p_{insert} \leftarrow Insert\_Node\ (p_{parent}, p_{forward\_last})$;
          if $(\pi - Angle\ (\overrightarrow{p_{parent}p_{insert}},\ \overrightarrow{p_{parent}p_{root}})) < \delta_f \leq \delta_{f\_max}$
and $(\pi - Angle\ (\overrightarrow{p_{insert}p_{parent}},\ \overrightarrow{p_{insert}p_{forward\_last}})) < \delta_f \leq \delta_{f\_max}$
and $Collision\_Free\ (p_{parent}, p_{insert})$
and $Collision\_Free\ (p_{insert}, p_{forward\_last})$ **then**
            $p_{root\_last} \leftarrow p_{root}$; $p_{root} \leftarrow p_{parent}$; $p_{parent} \leftarrow p_{insert}$; $p_{current} \leftarrow p_{forward\_last}$;
            $S_1.Backward\_Add\_Node\ (p_{insert})$;
            $S_1.Backward\_Add\_Node\ (p_{current})$;
            break
          end if
        end while
      end if
    end if
  end for
end while
if $(\pi - Angle\ (\overrightarrow{p_{current}p_{parent}},\ \overrightarrow{p_{current}p_n})) < \delta_f \leq \delta_{f\_max}$ **then**
$S_1.Backward\_Add\_Node\ (p_n)$;
else
while 1 **do**
  $p_{insert} \leftarrow Insert\_Node\ (p_{root}, p_{current})$;
  if $(\pi - Angle\ (\overrightarrow{p_{root}p_{root\_last}},\ \overrightarrow{p_{root}p_{insert}})) < \delta_f \leq \delta_{f\_max}$
    and $(\pi - Angle\ (\overrightarrow{p_{insert}p_{current}},\ \overrightarrow{p_{insert}p_{root}})) < \delta_f \leq \delta_{f\_max}$
   and $(\pi - Angle\ (\overrightarrow{p_{current}p_{insert}},\ \overrightarrow{p_{current}p_n})) < \delta_f \leq \delta_{f\_max}$
and $Collision\_Free\ (p_{root}, p_{insert})$
and $Collision\_Free\ (p_{insert}, p_{current})$ **then**
    $S_1.Backward\_Delete\_Node\ (p_{end})$;
    $S_1.Backward\_Delete\_Node\ (p_{end-1})$;
    $S_1.Backward\_Add\_Node\ (p_{insert})$;
    $S_1.Backward\_Add\_Node\ (p_{current})$;
    $S_1.Backward\_Add\_Node\ (p_n)$;
    break;
  end if
end while
end if
$trajectory \leftarrow Cubic\_Bspline\ (S_1)$;
Return *trajectory*

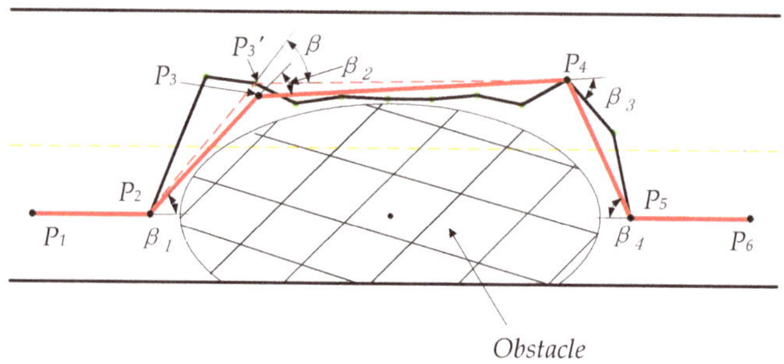

Figure 11. Path node reconnection processing.

3.6.2. Path Smoothing

After the path node reconnection process, a simplified path, the solid red line in Figure 11, is obtained, but the path contains some necessary turning nodes. As a result, the path is still not continuously differentiable such that it cannot be directly tracked by the vehicle. Thus, the obtained path should be further smoothed to make it be successfully tracked by the vehicle [40]. The cubic B-spline curve can move a control point to make the local modification without affecting the overall shape of the path and have a simple implementation and relatively low computational cost to ensure kinematic feasibility while avoiding an obstacle [41]. In addition, it is also often adopted to obtain a continuously differentiable cure [42]. Therefore, the cubic B-spline curve can be used to optimize the obtained path.

Supposing there are $m+1$ control points $P_i(i=0,1,\cdots,m)$, the cubic B-spline curve is expressed as

$$\begin{cases} P_{k,3}(t) = \sum_{i=0}^{3} P_{i+k} \cdot G_{i,3}(t) \quad t \in [0,1) \\ k = 0,1,\cdots m-3 \end{cases} \tag{19}$$

where the basic function $G^{i,3}(t)$ is defined as

$$\begin{cases} G_{0,3}(t) = \frac{-t^3+3t^2-3t+1}{6} \\ G_{1,3}(t) = \frac{3t^3-6t^2+4}{6} \\ G_{2,3}(t) = \frac{-3t^3+3t^2+3t+1}{6} \\ G_{3,3}(t) = \frac{t^3}{6} \end{cases} \tag{20}$$

In order to make the cubic B-spline curve start from P_0, tangent to the vector $\overrightarrow{P_0P_1}$, end at P_m, and tangent to the vector $\overrightarrow{P_mP_{m\text{-}1}}$, the control points P_{-1} and P_{m+1} are added to meet the conditions $P_{-1}+P_1=2P_0$ and $P_{m-1}+P_{m+1}=2P_m$.

The path node organization method can make the generated path continuously differentiable while reducing path length to the maximum extent. Meanwhile, the generated path does not have unnecessary steering and meets the tracking requirement of the vehicle.

3.7. Path Coherence Method

Due to the randomness of the algorithm, the difference in the planned paths between two adjacent frames could easily cause the vehicle to shake during driving, If the difference is too large, it may even cause the vehicle to collide with an obstacle. Hence, to solve this path planning problem in dynamic environments, the interframe path coherence method is introduced to guarantee the smooth connection of planned paths between frames. The interframe path coherence method refers to the need to consider the information of the planned path of the previous frame when planning a path in the current frame. Its main idea is that at the beginning of each new planning cycle, the root node P_{root_newly} of the new planning cycle is the position whose distance from the root node in the trajectory planned in the previous planning cycle is equal to R. At the same time, a point along the tangent line of the newly generated root node is selected as the new search starting node $P_{initial\_newly}$. Equations (21) and (22) express the calculation of the root node P_{root_newly}. Equation (23) shows how to calculate $P_{initial\_newly}$.

$$\begin{cases} \begin{cases} P_{k,3}(t) = \sum\limits_{i=0}^{3} P_{i+k} \cdot G_{i,3}(t) & t \in [0,1) \\ k = 0, 1, \cdots m-3 \end{cases} \\ (x - P_{root\_x})^2 + (y - P_{root\_y})^2 = R^2 \end{cases} \tag{21}$$

$$R = \rho \cdot v \tag{22}$$

$$\begin{cases} \vartheta = \tan^{-1}(B_{slope}) \\ P_{root\_newly} = P_{initial\_newly} + L_{skew}(\cos\vartheta, \sin\vartheta) \end{cases} \tag{23}$$

where P_{root} is the root node of the previous planning cycle, R is the offset distance of the root node P_{root}, and ρ is the coefficient of proportionality. R is more than the preview distance of the vehicle to ensure that the preview point is still on the trajectory of the previous frame. As a result, there is no need to make more tracking control adjustments. P_{root_newly} is the right intersection point of the cubic B-spline curve and the designed circle. B_{slope} is the slope of the cubic B-spline curve at the intersection point, ϑ is the included angle between the tangent line and the x axis at the intersection point, and L_{skew} is the offset distance along the direction of the tangent line. P_{root_newly} and $P_{initial\_newly}$ are introduced to ensure that the trajectory generated in the current frame is tangent to the tangent line at the intersection point P_{root_newly} and passes through the intersection point P_{root_newly}. Thereby, there are smooth connections in the interframe paths. As seen in Figure 12, the solid green line refers to the broken line formed by the path control points of the previous frame, and the solid black line is the broken line formed by the path control points of the current frame. The red curve represents the trajectory generated in the previous frame, the blue curve represents the trajectory generated in the current frame, and both curves are tangent to the vector $\overrightarrow{P_{root\_newly}P_{initial\_newly}}$. Hence, those two paths are smoothly connected at P_{root_newly}.

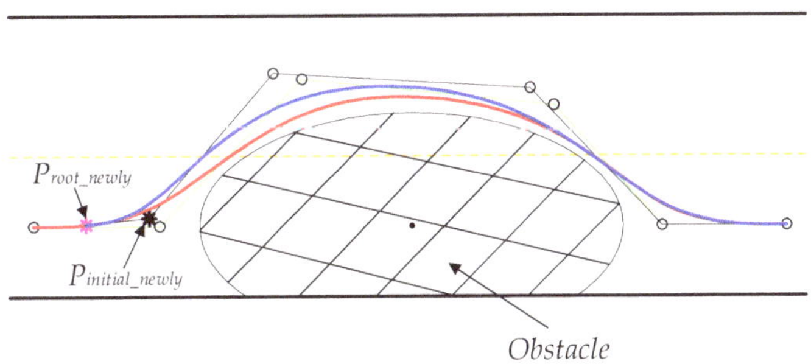

Figure 12. Path coherence illustration.

The path coherence method can significantly eliminate the difference between inter-frame trajectories and make the planned trajectory smooth and continuous, resulting in maintaining the vehicle's overall stability during driving.

4. Simulation Results and Analysis

4.1. Simulation Environment

In order to verify the performance of the improved heuristic Bi-RRT algorithm, two typical road scenarios, including a straight road scenario and a curved road scenario, as shown in Figure 13a,b, are separately taken into account. Road information and other parameters in the whole simulation are shown in Tables 1 and 2. The performance comparison of different RRT variants in the static straight road and curved road scenarios is conducted to verify the superiority of the improved heuristic Bi-RRT method. Meanwhile, interframe path planning and tracking are performed in dynamic scenarios to verify the effectiveness of the proposed algorithm. The simulations were performed on a PC with processor Intel I7 based on MATLAB R2019b and Carsim 2018. MATLAB R2019b is a mathematical software developed by Mathworks in Massachusetts, USA, and Carsim 2018 is a vehicle system simulation software developed by Mechanical Simulation Corporation in Michigan, USA.

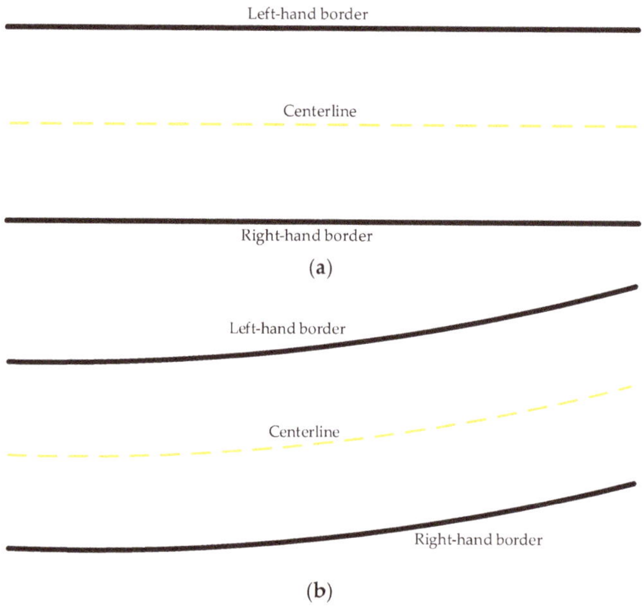

Figure 13. Road scenarios. (**a**) Straight road scenario. (**b**) Curved road scenario.

Table 1. Road parameters.

Road Type	Transverse Length (m)	Lane Width (m)	Initial Point	Target Point	Obstacle Point
Straight	130	3.75	5, −1.875	125, −1.875	65, −1.875
Curve	180	3.75	5, −1.874	175, 3.676	90, −1.095

Table 2. Simulation parameters.

Parameter	Value	Parameter	Value
Obstacle vehicle width W_o (m)	1.8	Expansion coefficient (straight) s_{f1}	$\sqrt{2}$
Obstacle vehicle length L_o (m)	4.8	Expansion coefficient (straight) s_{f2}	$\sqrt{3}$
Biased step size χ (m)	3	Expansion coefficient (curve) s_{f1}	$\sqrt{3}$
Step size λ (m)	10	Expansion coefficient (curve) s_{f2}	$\sqrt{2}$
Regulating coefficient s	1.5	Weighted coefficient ω_1	0.4
Host vehicle width W_h (m)	1.8	Weighted coefficient ω_2	0.6
Host vehicle speed v (km/h)	60	Weighted coefficient ξ_1	0.7
Friction coefficient η	0.8	Weighted coefficient ξ_2	0.3
Gravity acceleration g (m/s2)	9.8	Proportionality coefficient ρ	0.6
Constraint angle δ_f (°)	30		

4.2. Performance Measure of Path Planning

The improved heuristic Bi-RRT algorithm is compared with the basic RRT, the biased RRT, the Bi-RRT, and the RRT* in the straight and curved road scenarios, and the results are shown in Figure 14a,b, respectively. The blue dotted line represents the path planned by the basic RRT, the red dashed line represents the path planned by the biased RRT, and the black dash-dotted line represents the path planned by the RRT*. The solid green and black lines represent the paths planned by the Bi-RRT and the improved heuristic Bi-RRT, respectively. It can be seen in Figure 14a,b that the paths generated by the basic RRT, the Bi-RRT, and the biased RRT contain a large number of corners, and there are frequent large curvature changes. In contrast, the path generated by the RRT* is relatively gentle, but it is still a broken line, which does not meet the steering requirement of the vehicle. However, the path generated by the improved heuristic Bi-RRT is a continuous and smooth curve, and the path curvature is also continuous, as shown in Figure 15a,b, respectively, resulting in the convenience being well followed by the vehicle. Observing Figure 14a,b, all paths planned by the basic RRT, the biased RRT, the Bi-RRT, the RRT*, and the improved heuristic Bi-RRT can successfully avoid the obstacle vehicle. However, there are differences in obstacle avoidance modes. The paths planned by the basic RRT, the biased RRT, the Bi-RRT, and the RRT* always start emergency obstacle avoidance when approaching the obstacle vehicle. The path planned by the proposed algorithm can start obstacle avoidance in advance by a safe distance from the obstacle vehicle, ensuring safe driving, conforming to the driver's behavior habit, and not bringing excessive tension to passengers. The proposed algorithm can realize obstacle avoidance in advance because the obstacle avoidance distance is embedded when constructing the obstacle vehicle constraint.

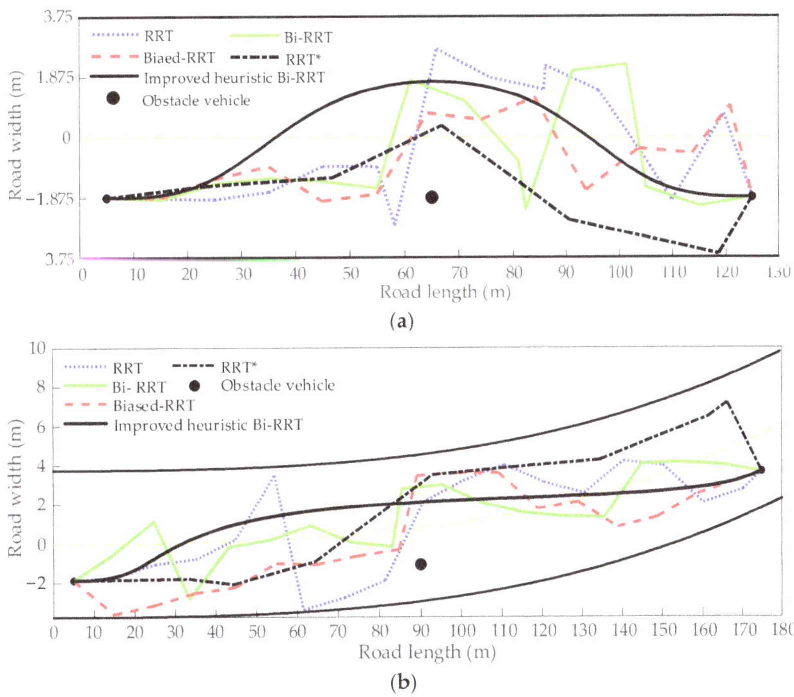

Figure 14. The simulation paths. (**a**) The planned paths in the straight road scenario. (**b**) The planned paths in the curved road scenario.

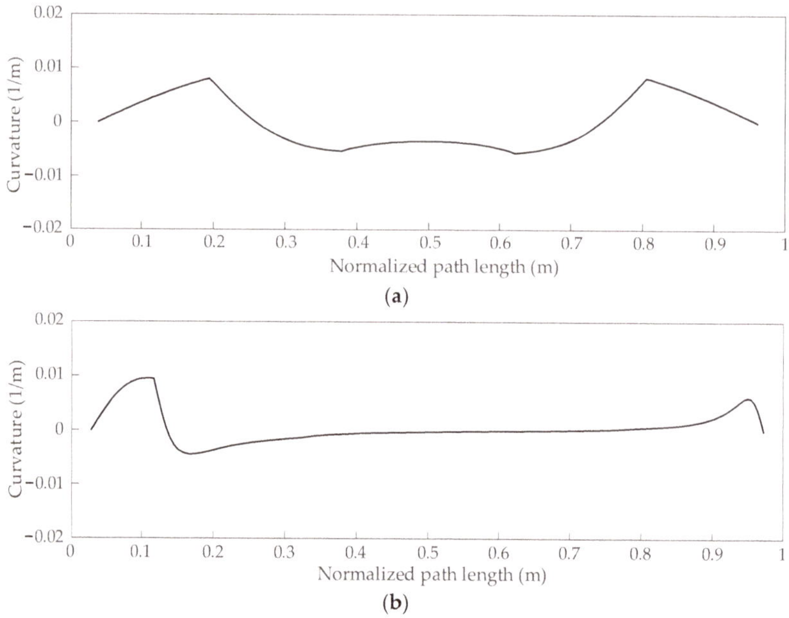

Figure 15. The simulation path curvatures. (**a**) The path curvature in straight road scenario. (**b**) The path curvature in curved road scenario.

A set of benchmarking parameters is defined to objectively compare the performance of the improved heuristic Bi-RRT and some RRT variants. The number of nodes on the mature random tree is denoted by "tree nodes." The length of the generated path is denoted by "path length." The searching time of path planning is denoted by "time." In addition, the number of segments comprising the generated path is donated by "path segments"; in particular, the path segments of the improved heuristic RRT refer to the number of segments of the path after being processed by path reconnection [30,43,44].

Thirty independent simulation experiments were implemented on each algorithm to offset the random deviation of a single experiment, and the results are shown in Tables 3 and 4. The average path length generated by the RRT* algorithm is much lower than those of the basic RRT, the biased RRT, and the Bi-RRT because the RRT * embedded with the reconnection mechanism can approach the approximate shortest path. However, the average path length generated by the improved heuristic Bi-RRT algorithm is smaller than that of the RRT* algorithm, and the average tree nodes and the average path segments of the proposed algorithm are also minimal compared with those of the other four algorithms. The decrease in path length and number of path segments is mainly due to the introduction of the path node reconnection method. The relatively small number of path segments can reduce the control difficulty of the vehicle and make the further smoothed path not contain unnecessary turns. In terms of the average planned time, the performance of the improved heuristic Bi-RRT algorithm is optimal compared with those of the other four algorithms, and especially in the straight road scenario, the planning time of the improved heuristic Bi-RRT algorithm can reduce significantly.

Table 3. Performance measures in the straight road scenario.

Algorithm	RRT	Biased RRT	Bi-RRT	RRT*	Improved Heuristic Bi-RRT
Average tree nodes	38.733	28.567	17.667	52.967	6.033
Average path segments	14.267	13.433	13.367	6.233	3.000
Average path length (m)	123.977	122.176	122.781	120.961	120.299
Average time (s)	0.026	0.022	0.020	0.043	0.010

Table 4. Performance measures in the curve scenario.

Algorithm	RRT	Biased RRT	Bi-RRT	RRT*	Improved Heuristic Bi-RRT
Average tree nodes	41.500	30.733	23.067	39.067	20.500
Average path segments	18.333	18.233	18.267	8.267	3.000
Average path length (m)	173.834	172.414	172.899	170.848	170.152
Average time (s)	3.363	2.432	0.732	12.757	0.626

As a result of the performance measure, the improved heuristic Bi-RRT algorithm shows superior performance regarding path quality and planning time compared with the basic RRT, the biased RRT, the Bi-RRT, and the RRT*.

4.3. Path Coherence Validation

The dynamic environment can usually be treated as a static environment in a dynamic path planning process. That is, the re-planning of paths needs to be conducted in each frame environment region obtained by the perception module. Thus, the running speed of the path planning algorithm should not be only considered, but also the interframe connection of the planned paths should be concerned. In order to further verify the superior performance of the improved heuristic Bi-RRT algorithm, the planning effect of the corresponding algorithm at a specific frame is described in detail.

In order to express the coherence between the front and back frame paths, the real-time re-planning is carried out in two dynamic driving scenarios, where the host vehicle and the obstacle vehicle move in opposite and the same directions, respectively. The red triangle represents the position of the obstacle vehicle in the current frame, the yellow arrow represents the driving direction of the obstacle vehicle, and the green solid point and the black solid point represent the start point and end point of the current frame, respectively.

Figure 16a–c show the dynamic path planning process of the improved heuristic Bi-RRT algorithm for three consecutive frames in the dynamic scenario with driving in opposite directions. Observing Figure 16a–c, the paths obtained in the first, second, and third frames are smooth and successfully bypass the moving obstacle vehicle. It can be seen from Figure 16d that the paths, magenta lines, obtained in the first frame, the second frame, and the third frame can be smoothly connected to form a smooth coherence path, as shown in Figure 16e. Moreover, the curvature of the coherence path is continuous and varies in a small range, which is convenient for it to be tracked by the host vehicle, as shown in Figure 16f.

Figure 17a–f show the dynamic path planning process of the improved heuristic Bi-RRT algorithm for six consecutive frames in the dynamic scenario with driving in the same direction. It can be seen from Figure 17a–f that the paths obtained in the first, second, third, fourth, fifth, and sixth frames are smooth and also successfully bypass the moving obstacle vehicle. As seen in Figure 17g, the paths, magenta lines, obtained in the first frame, the second frame, the third frame, the fourth frame, the fifth frame, and the sixth frame can be smoothly connected to generate a smooth coherence path, as shown in Figure 17h. In addition, the curvature of the obtained coherence path is continuous and within the range of 0.02 1/m, as shown in Figure 17i, resulting in the condition that the host vehicle can easily track the coherence path.

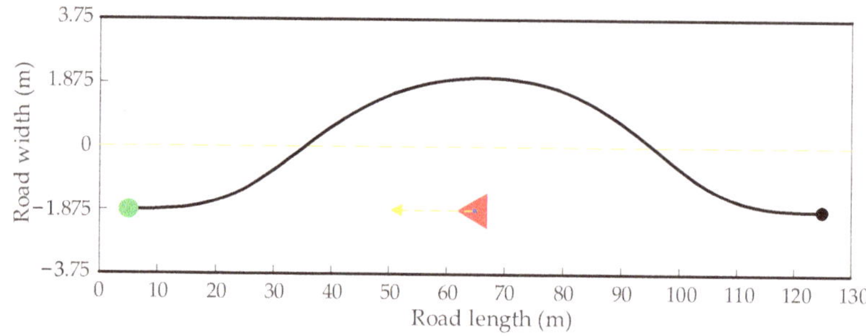

(a)

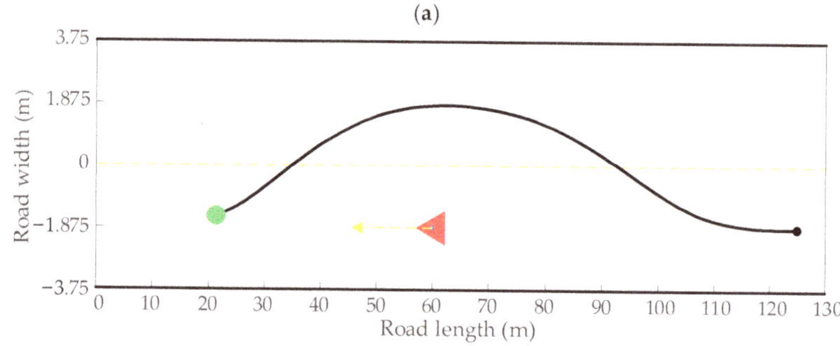

(b)

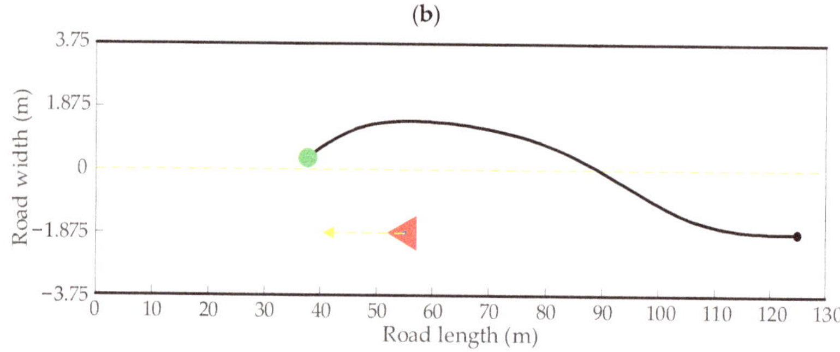

(c)

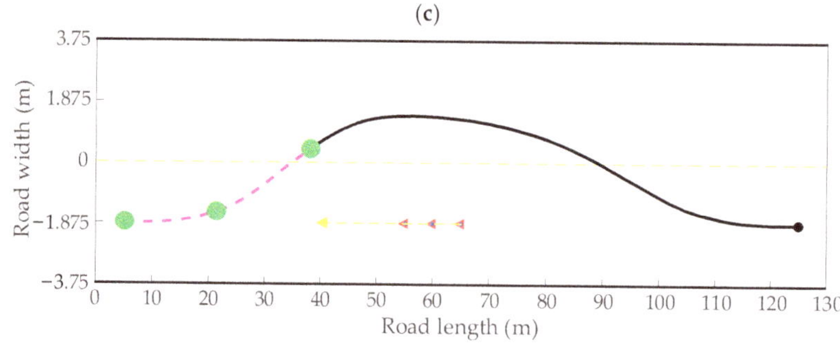

(d)

Figure 16. *Cont.*

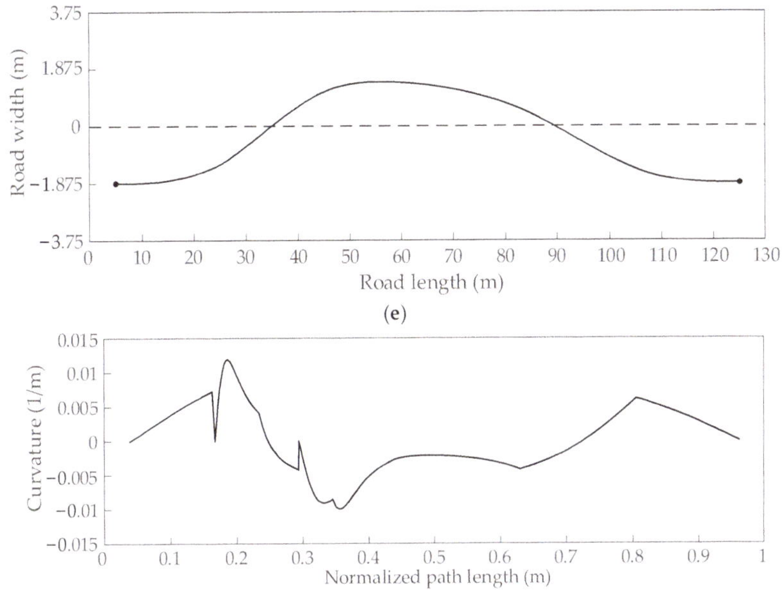

(e)

(f)

Figure 16. The path planning results in a dynamic scenario with driving in opposite directions. (**a**) The planned path in the first frame. (**b**) The planned path in the second frame. (**c**) The planned path in the third frame. (**d**) The path coherence process of three frames. (**e**) The coherence path of three frames. (**f**) The curvature of the coherence path.

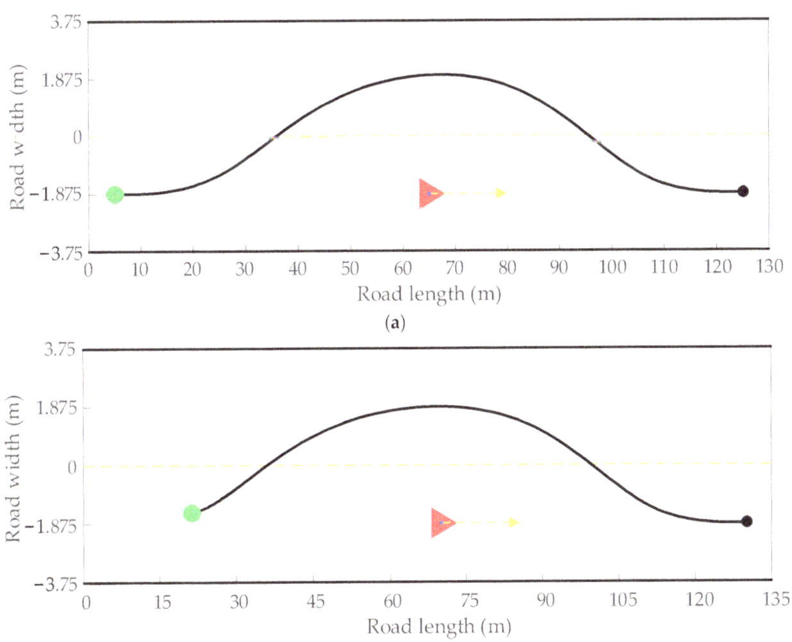

(a)

(b)

Figure 17. *Cont.*

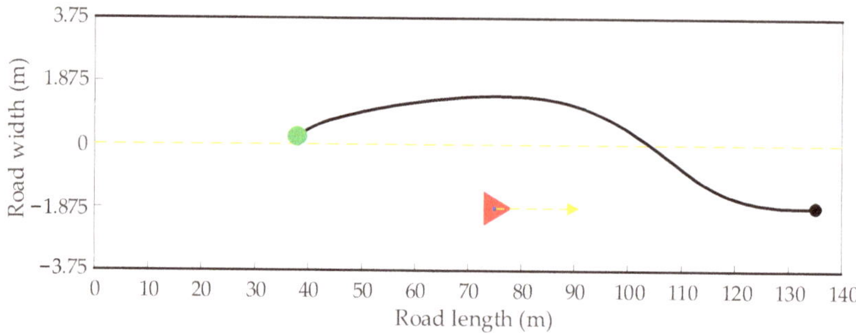

(c)

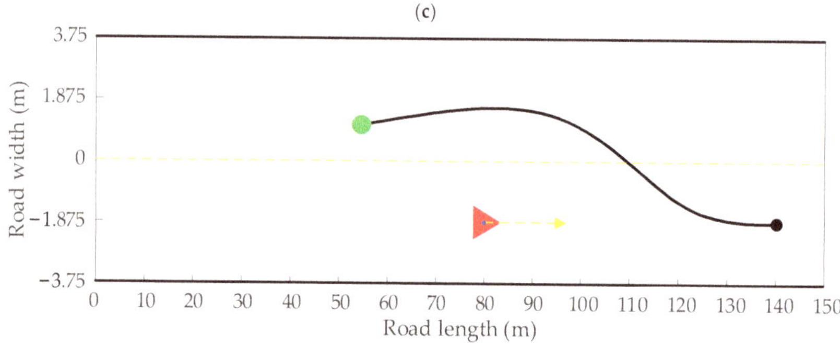

(d)

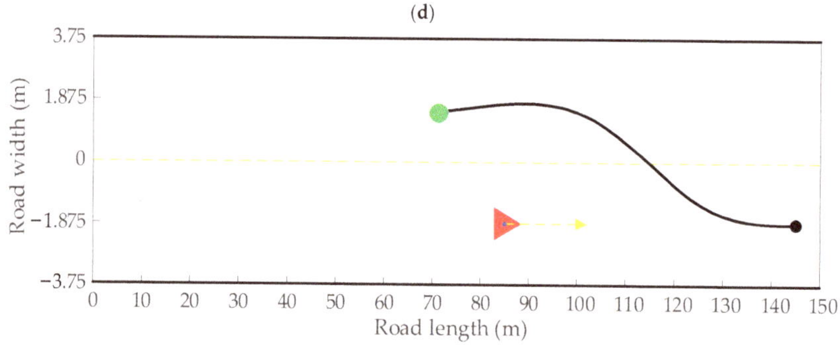

(e)

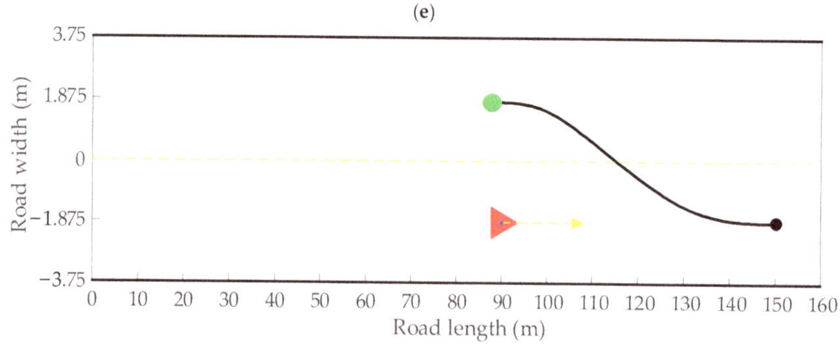

(f)

Figure 17. *Cont.*

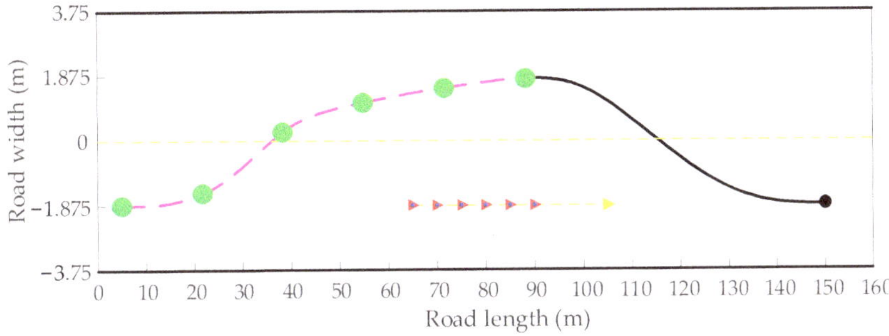

(g)

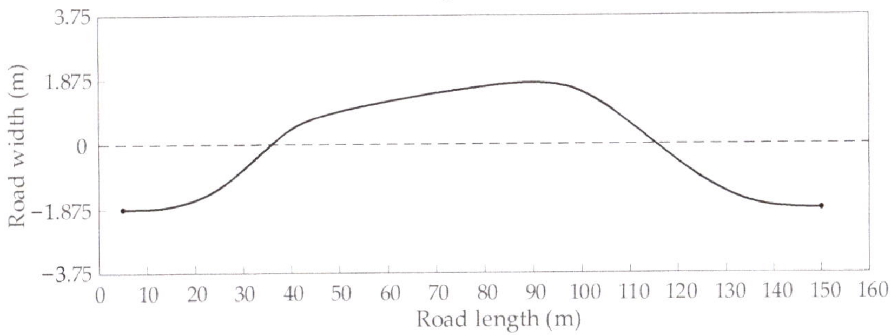

(h)

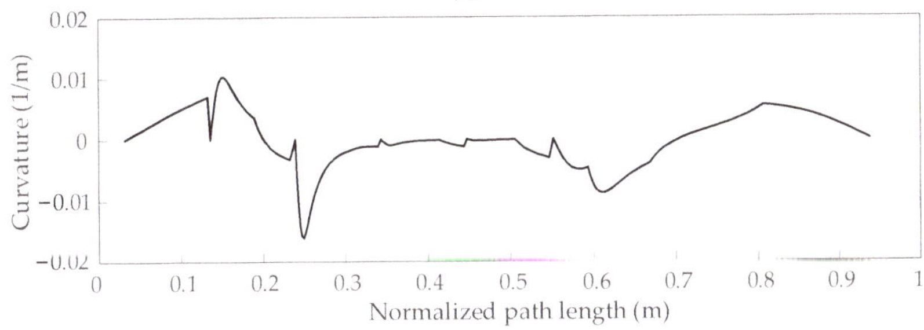

(i)

Figure 17. The path planning results in a dynamic scenario with driving in the same direction. (**a**) The planned path in the first frame. (**b**) The planned path in the second frame. (**c**) The planned path in the third frame. (**d**) The planned path in the fourth frame. (**e**) The planned path in the fifth frame. (**f**) The planned path in the sixth frame. (**g**) The path coherence process of six frames. (**h**) The coherence path of six frames. (**i**) The curvature of the coherence path.

For further verifying the tracking performance of the obtained coherence paths shown in Figures 16e and 17h, the path following experiments can be carried out in the Carsim simulation platform. The appropriate vehicle and driver models are selected to simulate the driving state of the real vehicle. The solid black and red dashed lines represent the target and followed paths, respectively.

The path following experiment in the dynamic scenario with driving in opposite directions is conducted. The result shown in Figure 18a represents the target path and the followed path. The following error between the followed path and the target path is relatively small and within the range of 0.06 m, as shown in Figure 18b, resulting in making the following effect acceptable. The yaw rate is within the range of 8 deg/s, as shown in Figure 18c, and the lateral acceleration is within the range of 0.25 g and less than the usual value of 0.8 g of the maximum lateral acceleration of steady driving, as shown in Figure 18d. These two indicators prove that the dynamic performance of the host vehicle is stable in the simulation experiment. Based on these facts, the obtained coherence path is satisfactory and effective.

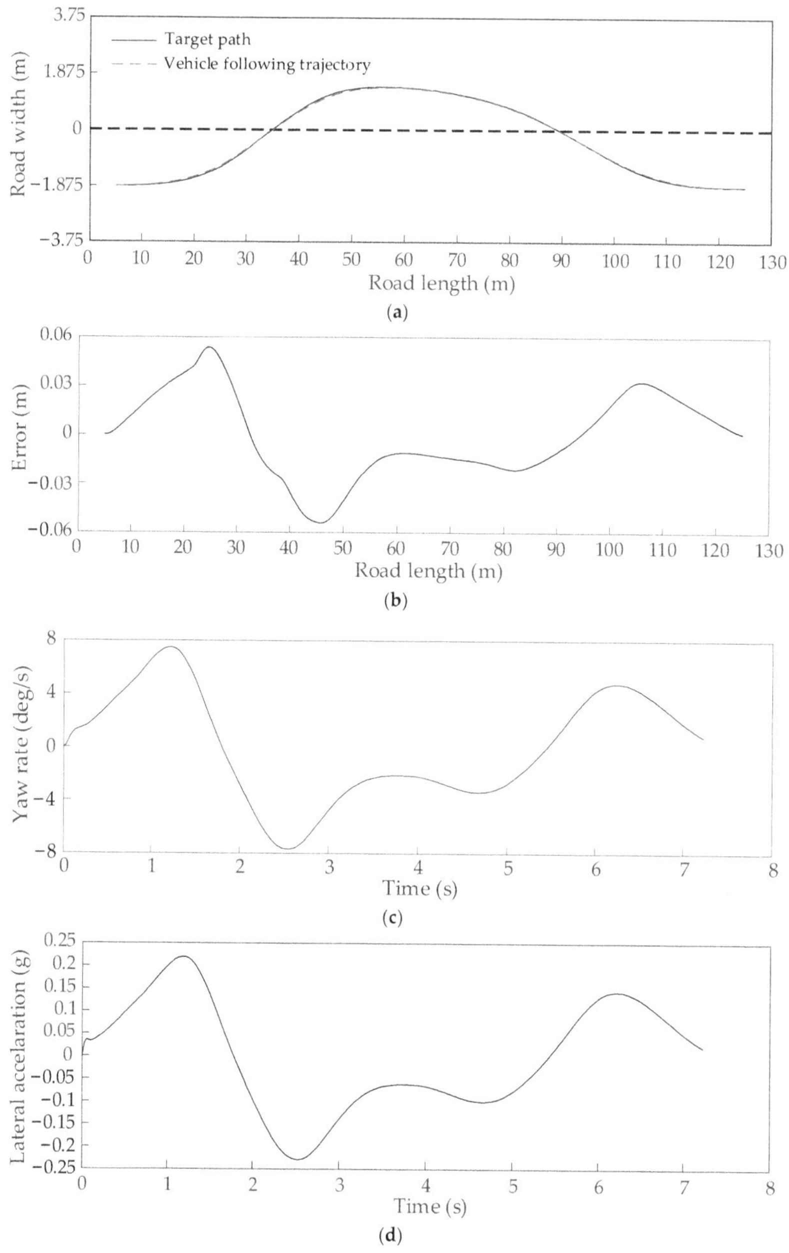

Figure 18. The path tracking results of the coherence path in a dynamic scenario with driving in opposite directions. (**a**) The path following result. (**b**) The path following error. (**c**) The yaw velocity. (**d**) The lateral acceleration.

The path following experiment in the dynamic scenario with driving in the same direction is conducted. The target path and the followed path are shown in Figure 19a. The following error between the followed path and the target path is within the range of 0.07 m, as shown in Figure 19b, and is relatively small, which shows that the coherence path has a satisfactory tracking effect. The yaw rate and the later acceleration of the host vehicle when following the trajectory are shown in Figure 19c,d, respectively. The yaw rate is within the range of 7 deg/s, and the lateral acceleration is within the range of 0.2 g and less than the usual value of 0.8 g of the maximum lateral acceleration of steady driving, which shows the excellent dynamic performance of the host vehicle in the simulation process. Thus, the obtained coherence path is feasible and acceptable.

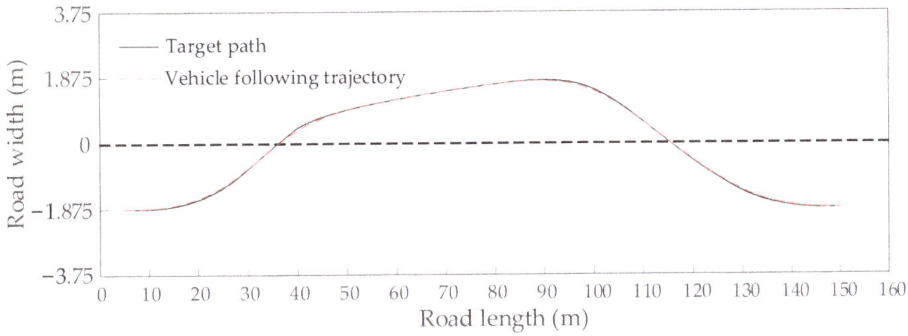

(**a**)

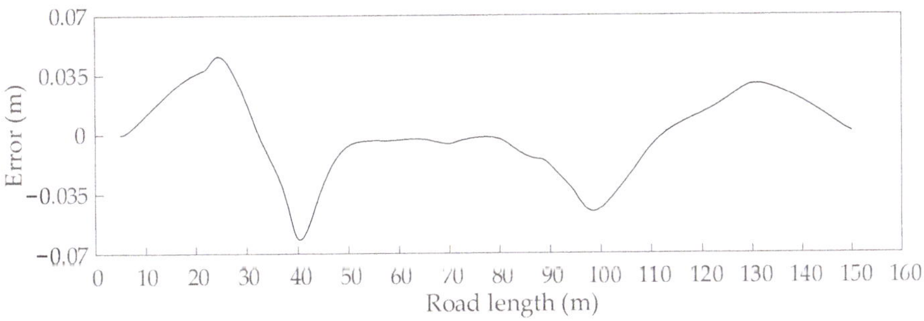

(**b**)

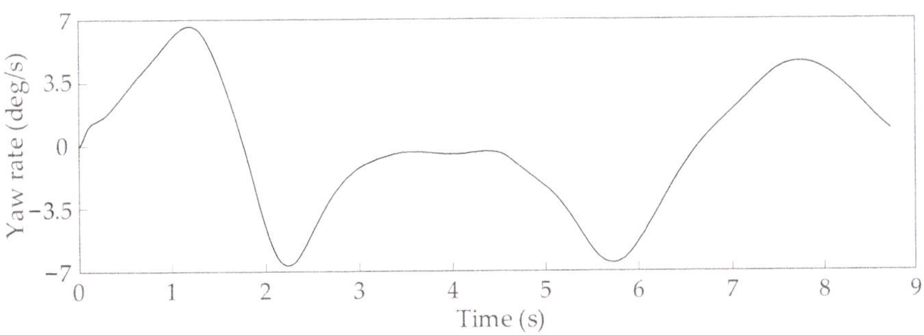

(**c**)

Figure 19. *Cont.*

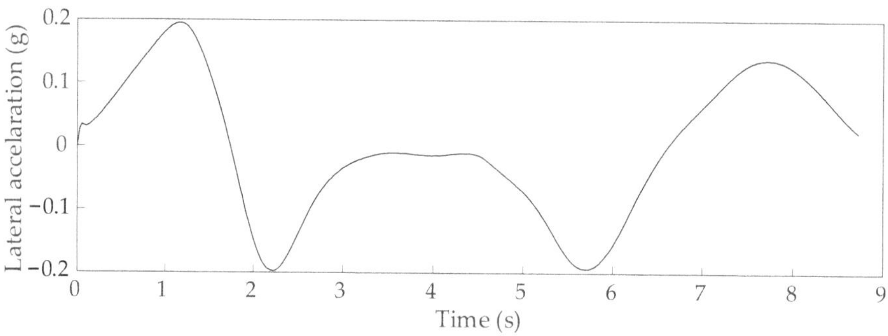

(**d**)

Figure 19. The path tracking results of the coherence path in a dynamic scenario with driving in the same direction. (**a**) The path following result. (**b**) The path following error. (**c**) The yaw velocity. (**d**) The lateral acceleration.

It can be seen from the test results that the algorithm proposed in this article can plan a path with smooth transition connections and continuous curvature. Furthermore, it is applied successfully in the dynamic driving scenarios of opposite driving and traveling in the same direction, which are the most common in vehicle driving.

5. Discussion

The path planning of the vehicle in dynamic scenarios often pays attention not only to the quality of each frame path but also to the difference in paths between frames. In addition, the Bi-RRT algorithm, a variant of the basic RRT algorithm, is often used for path planning because of its probability completeness and rapidity. However, its planned path is not differentiable and relatively poor in length. Based on those facts, some improved heuristic methods are introduced to the Bi-RRT algorithm to make it suitable for the dynamic path planning of the vehicle. The multi-sampling method biased towards the target state makes the growth of the random tree directional and reasonable. The adaptive greedy step size considering the target direction can increase the success rate of node expansion and make the newly extended node tend to the target state direction to a certain extent. The two random trees are directly interconnected when there are no obstacles between them, as a result, which further reduces the running time of the algorithm. Changing from only considering the Euclidean distance to considering the distance from the target state and the included angles between the connection lines of tree nodes, the parent node selection method improves the running speed of the algorithm to a certain extent while making the generated path tend to be gentle. The path reorganization method, including path node reconnection and path smoothing, can remove necessary turning points, significantly reduce the path length, and plan a path with continuous curvature. As for the problem of path connection between frames, the path coherence method can make paths between different frames smoothly connect to form a differentiable path. By way of the simulation experiment study, the improved heuristic Bi-RRT algorithm has a real-time performance, especially in a straight road scenario, and guarantees the shortest path while obeying the road constraints and the vehicle constraint and considering the driving habit of the driver. On the contrary, the vehicle constraint and the driver's driving habits are not considered when applying the basic Bi-RRT algorithm.

However, it must be admitted that the running time may not be fast enough when the proposed algorithm is applied in a high-speed and dynamic curved road scene, which may be due to a large number of numerical calculations based on graphic geometry in obstacle detection. In future research, obstacle detection of gray value comparison and parallel computing are introduced to further reduce the proposed algorithm's running time to meet the path planning requirement in a high-speed curved road scenario. After the preparation of the perception module, changing from the numerical calculation mode of graphics geometry to the comparison of gray values mode on binarized images, the obstacle detection can greatly reduce the calculation amount of the algorithm, thus speeding up the search speed of the algorithm. On this basis, the improved heuristic Bi-RRT algorithm can be applied to more complex driving scenarios, such as unstructured road scenes with a mixture of static and moving obstacles with grooves, to explore its adaptability. Furthermore, compared with other related algorithms, finding out the algorithm's shortcomings and the scenes where the algorithm cannot be applied are made to improve the algorithm and enable it to be applied to more driving scenes.

6. Conclusions

This paper is concerned with the path planning of autonomous vehicles in a dynamic environment. An improved heuristic Bi-RRT algorithm has been proposed and tested. The proposed algorithm can solve the path query problem of the basic Bi-RRT algorithm and the interconnection problem of paths between various frames in a dynamic scenario to obtain a smooth and asymptotically optimal path with continuous curvature with high efficiency and accuracy. The proposed path planning algorithm consists of the obstacle-free direct connection of two trees, the heuristic target bias sampling, the heuristic parent node selection, the heuristic node extension, the improved constraints, path reorganization, and path coherence. The obstacle-free direct connection mode can further accelerate the interconnection of the two random trees. The heuristic target bias sampling can reduce blind sampling, and the heuristic node extension can decrease invalid expansion, thereby accelerating the running speed and improving the searching efficiency of the algorithm. The heuristic parent node selection speeds up the algorithm's calculation and improves path quality to some extent. The improved environmental constraint and the improved vehicle constraint integrated with the advanced obstacle avoidance distance are considered together to make the vehicle avoid the obstacle in advance and accurately and make the vehicle drive safely. Path reorganization aims to post-process the initial path to obtain a reasonable and differentiable path with the approximate optimal length, which can be tracked by the vehicle smoothly and successfully. In addition, path coherence solves the problem of the smooth connection of paths between different frames, enabling the vehicle to run smoothly and steadily at the connection point. Through the simulation experiments, the improved heuristic Bi-RRT algorithm can generate the smoothest path and takes the shortest time compared with the other four algorithms. As a result, it is an effective local path planning algorithm for the autonomous vehicle and has practical value in the application of the wheeled robot.

In future works, the research focuses on increasing the solution speed, further reducing the calculation time, especially in the curved road scenario. The proposed algorithm will be applied in more complex driving scenarios, such as the parking scene and the drift scene with moving obstacles, to test its adaptiveness. Moreover, after the preparation of the test platform of the autonomous vehicle, an on-site experiment is conducted to test its effectiveness in practical applications.

Author Contributions: Conceptualization, X.Z.; methodology, X.Z.; software, X.Z.; validation, Y.H. and L.D.; resources, T.Z.; writing—original draft preparation, X.Z.; writing—review and editing, T.Z. and H.L.; funding acquisition, T.Z. and Y.H. All authors have read and agreed to the published version of the manuscript.

Funding: This research was funded by the National Key R&D Program of China (No. 2019YFE0108000) and the Shaanxi Provincial Basic Research Program of Natural Sciences Project (No. 2022JQ-540).

Institutional Review Board Statement: Not applicable.

Informed Consent Statement: Not applicable.

Data Availability Statement: Data is contained within the article as figures and tables.

Conflicts of Interest: The authors declare no conflict of interest.

References

1. Zhao, L.; Jia, Y.H. Intelligent transportation system for sustainable environment in smart cities. *Int. J. Elec. Eng. Educ.* **2021**, 0020720920983503. [CrossRef]
2. Verma, S.; Kaur, S.; Sharma, A.K.; Kathuria, A.; Piran, M.J. Dual Sink-Based Optimized Sensing for Intelligent Transportation Systems. *IEEE Sens. J.* **2021**, *21*, 15867–15874. [CrossRef]
3. Sleem, L.; Noura, H.N.; Couturier, R. Towards a secure ITS: Overview, challenges and solutions. *J. Inf. Secur. Appl.* **2020**, *55*, 102637. [CrossRef]
4. Goswami, P.; Mukherjee, A.; Hazra, R.; Yang, L.; Ghosh, U.; Qi, Y.; Wang, H. AI Based Energy Efficient Routing Protocol for Intelligent Transportation System. *IEEE Trans. Intell. Transp. Syst.* **2022**, *23*, 1670–1679. [CrossRef]
5. Dogra, R.; Rani, S.; Babbar, H.; Verma, S.; Verma, K.; Rodrigues, J.J.P.C. DCGCR: Dynamic Clustering Green Communication Routing for Intelligent Transportation Systems. *IEEE Trans. Intell. Transp. Syst.* **2022**, *23*, 3148471. [CrossRef]
6. Ming, Y.; Li, Y.Q.; Zhang, Z.H.; Yan, W.Q. A Survey of Path Planning Algorithm for Autonomous Vehicles. *SAE Int. J. Commer. Veh.* **2021**, *14*, 97–109. [CrossRef]
7. Sun, Y.; Ren, D.; Lian, S.; Fu, S.; Teng, X.; Fan, M. Robust Path Planner for Autonomous Vehicles on Roads with Large Curvature. *IEEE Robot. Autom. Lett.* **2022**, *7*, 2503–2510. [CrossRef]
8. Lv, X.X.; Li, W.H.; Wang, J.F. Safety-field-based Path Planning Algorithm of Lane Changing for Autonomous Vehicles. *Int. J. Control Autom. Syst.* **2022**, *20*, 564–576. [CrossRef]
9. Chen, W.; Sun, C.Y.; Liu, H.; Liu, J.J.; Tang, Y. Path Planning Scheme for Spray Painting Robot with Bezier Curves on Complex Curved Surfaces. In Proceedings of the 32nd Youth Academic Annual Conference of Chinese Association of Automation (YAC), Hefei, China, 19–21 May 2017; pp. 698–703.
10. Huang, J.; Yang, Y.F.; Ding, D.L.; Li, Y.H.; He, Y. Automatic parking paths planning research based on scattering points six-degree polynomial and easement curve. *Proc. Inst. Mech. Eng. Part D-J. Automob. Eng.* **2022**, 09544070221076594. [CrossRef]
11. Hong, Z.; Sun, P.; Tong, X.; Pan, H.; Zhou, R.; Zhang, Y.; Han, Y.; Wang, J.; Yang, S.; Xu, L. Improved A-Star Algorithm for Long-Distance Off-Road Path Planning Using Terrain Data Map. *ISPRS Int. J. Geo-Inf.* **2021**, *10*, 785. [CrossRef]
12. Liu, L.-S.; Lin, J.-F.; Yao, J.-X.; He, D.-W.; Zheng, J.-S.; Huang, J.; Shi, P. Path Planning for Smart Car Based on Dijkstra Algorithm and Dynamic Window Approach. *Wirel. Commun. Mob. Comput.* **2021**, *2021*, 8881684. [CrossRef]
13. Bresciani, M.; Ruscio, F.; Tani, S.; Peralta, G.; Timperi, A.; Guerrero-Font, E.; Bonin-Font, F.; Caiti, A.; Costanzi, R. Path Planning for Underwater Information Gathering Based on Genetic Algorithms and Data Stochastic Models. *J. Mar. Sci. Eng.* **2021**, *9*, 1183. [CrossRef]
14. Han, G.; Zhou, Z.; Zhang, T.; Wang, H.; Liu, L.; Peng, Y.; Guizani, M. Ant-Colony-Based Complete-Coverage Path-Planning Algorithm for Underwater Gliders in Ocean Areas with Thermoclines. *IEEE Trans. Veh. Technol.* **2020**, *69*, 8959–8971. [CrossRef]
15. Khatib, O. Real-Time Obstacle Avoidance System for Manipulators and Mobile Robots; The International Journal of Robotics Research. In Proceedings of the 1985 IEEE International Conference on Robotics and Automation, St. Louis, MO, USA, 25–28 March 1985; pp. 500–505.
16. Chen, X.; Li, Y.; Hong, X.; Wei, X.; Huang, Y. Unmanned Ship Path Planning Based on RRT. In Proceedings of the 14th International Conference on Intelligent Computing (ICIC), Wuhan, China, 15–18 August 2018; Volume 10954, pp. 102–110.
17. Chen, G.; Luo, N.; Liu, D.; Zhao, Z.; Liang, C. Path planning for manipulators based on an improved probabilistic roadmap method. *Robot. Comput.-Integr. Manuf.* **2021**, *72*, 102196. [CrossRef]
18. Wang, W.; Zuo, L.; Xu, X. A Learning-based Multi-RRT Approach for Robot Path Planning in Narrow Passages. *J. Intell. Robot. Syst.* **2018**, *90*, 81–100. [CrossRef]
19. Heo, Y.J.; Chung, W.K. RRT-based Path Planning with Kinematic Constraints of AUV in Underwater Structured Environment. In Proceedings of the 10th International Conference on Ubiquitous Robots and Ambient Intelligence (URAI), Jeju, Korea, 30 October–2 November 2013; pp. 523–525.
20. Gan, Y.; Zhang, B.; Ke, C.; Zhu, X.; He, W.; Ihara, T. Research on Robot Motion Planning Based on RRT Algorithm with Nonholonomic Constraints. *Neural Process. Lett.* **2021**, *53*, 3011–3029. [CrossRef]
21. Lan, W.; Jin, X.; Wang, T.L.; Zhou, H. Improved RRT Algorithms to Solve Path Planning of Multi-Glider in Time-Varying Ocean Currents. *IEEE Access* **2021**, *9*, 158098–158115. [CrossRef]
22. Wu, D.; Sun, Y.J.; Wang, X.; Wang, X.L. An Improved RRT Algorithm for Crane Path Planning. *Int. J. Robot. Autom.* **2016**, *31*, 84–92. [CrossRef]
23. Jaillet, L.; Cortes, J.; Simeon, T. Sampling-Based Path Planning on Configuration-Space Costmaps. *IEEE Trans. Robot.* **2010**, *26*, 635–646. [CrossRef]
24. Jin, X.J.; Yan, Z.Y.; Yang, H.; Wang, Q.K.; Yin, G.D. A Goal-Biased RRT Path Planning Approach for Autonomous Ground Vehicle. In Proceedings of the 2020 4th CAA International Conference on Vehicular Control and Intelligence (CVCI), Hangzhou, China, 18–20 December 2020; pp. 743–746.
25. Wang, J.K.; Li, B.P.; Meng, M.Q.H. Kinematic Constrained Bi-directional RRT with Efficient Branch Pruning for robot path planning. *Expert Syst. Appl.* **2021**, *170*, 114541. [CrossRef]
26. Chen, Y.Y.; Fu, Y.X.; Zhang, B.; Fu, W.; Shen, C.J. Path planning of the fruit tree pruning manipulator based on improved RRT-Connect algorithm. *Int. J. Agric. Biol. Eng.* **2022**, *15*, 177–188. [CrossRef]

27. Li, Y.C.; Shao, J. A Revised Gaussian Distribution Sampling Scheme Based on RRT* Algorithms in Robot Motion Planning. In Proceedings of the 3rd IEEE International Conference on Control, Automation and Robotics (ICCAR), Nagoya, Japan, 22–24 April 2017; pp. 22–26.
28. Park, C.; Kee, S.C. Online Local Path Planning on the Campus Environment for Autonomous Driving Considering Road Constraints and Multiple Obstacles. *Appl. Sci.* **2021**, *11*, 3909. [CrossRef]
29. Chen, J.; Zhang, R.; Han, W.; Jiang, W.; Hu, J.; Lu, X.; Liu, X.; Zhao, P. Path Planning for Autonomous Vehicle Based on a Two-Layered Planning Model in Complex Environment. *J. Adv. Transp.* **2020**, *2020*, 6649867. [CrossRef]
30. Li, A.; Li, S.; Du, H.; Huang, X.; Niu, C. Improved Bidirectional RRT * Path Planning Method for Smart Vehicle. *Math. Probl. Eng.* **2021**, *2021*, 6669728.
31. Qi, J.; Yang, H.; Sun, H.X. MOD-RRT*: A Sampling-Based Algorithm for Robot Path Planning in Dynamic Environment. *IEEE Trans. Ind. Electron.* **2021**, *68*, 7244–7251. [CrossRef]
32. Zou, Q.J.; Zhang, Y.; Liu, S.H. A path planning algorithm based on RRT and SARSA (lambda) in unknown and complex conditions. In Proceedings of the 32nd Chinese Control and Decision Conference (CCDC), Hefei, China, 22–24 August 2020; pp. 2035–2040.
33. Li, Y.J. An RRT-Based Path Planning Strategy in a Dynamic Environment. In Proceedings of the 7th International Conference on Automation, Robotics and Applications (ICARA), Prague, Czech Republic, 4–6 February 2021; pp. 1–5.
34. Peng, J.; Chen, Y.; Duan, Y.; Zhang, Y.; Ji, J.; Zhang, Y. Towards an Online RRT-based Path Planning Algorithm for Ackermann-steering Vehicles. In Proceedings of the 2021 IEEE International Conference on Robotics and Automation (ICRA), Xi'an, China, 30 May–5 June 2021; pp. 7407–7413.
35. Wen, N.F.; Zhang, R.B.; Wu, J.W.; Liu, G.Q. Online planning for relative optimal and safe paths for USVs using a dual sampling domain reduction-based RRT* method. *Int. J. Mach. Learn. Cybern.* **2020**, *11*, 2665–2687. [CrossRef]
36. Niu, C.A.H.; Li, A.J.; Huang, X.; Li, W.; Xu, C.A.Y. Research on Global Dynamic Path Planning Method Based on Improved A* Algorithm. *Math. Probl. Eng.* **2021**, *2021*, 4977041. [CrossRef]
37. Wu, M.; Chen, E.; Shi, Q.; Zhou, L.; Chen, Z.; Li, M. Path planning of mobile robot based on improved genetic algorithm. In Proceedings of the Chinese Automation Congress (CAC), Jinan, China, 20–22 October 2017; pp. 6696–6700.
38. Lu, X.; Wan, J.; Zhong, Y.; Wang, J. Dual Redundant UAV Path Planning and Mission Analysis Based on Dubins Curves. In Proceedings of the 2022 3rd International Conference on Geology, Mapping and Remote Sensing (ICGMRS), Zhoushan, China, 22–24 April 2022; pp. 387–390.
39. Yang, S.M.; Lin, Y.A. Development of an Improved Rapidly Exploring Random Trees Algorithm for Static Obstacle Avoidance in Autonomous Vehicles. *Sensors.* **2021**, *21*, 2244. [CrossRef]
40. Noreen, I. Collision Free Smooth Path for Mobile Robots in Cluttered Environment Using an Economical Clamped Cubic B-spline. *Symmetry* **2020**, *12*, 1567. [CrossRef]
41. Cao, H.; Zoldy, M. Implementing B-Spline Path Planning Method Based on Roundabout Geometry Elements. *IEEE Access* **2022**, *10*, 81434–81446. [CrossRef]
42. Qian, S.; Bao, K.L.; Zi, B.; Zhu, W.D. Dynamic Trajectory Planning for a Three Degrees-of-Freedom Cable-Driven Parallel Robot Using Quintic B-Splines. *J. Mech. Des.* **2020**, *142*, 4045723. [CrossRef]
43. Wang, H.; Li, G.Q.; Hou, J.; Chen, L.Y.; Hu, N.L. A Path Planning Method for Underground Intelligent Vehicles Based on an Improved RRT* Algorithm. *Electronics* **2022**, *11*, 294. [CrossRef]
44. Moses, E.D.B.; Anitha, G. Goal Directed Approach to Autonomous Motion Planning for Unmanned Vehicles. *Def. Sci. J.* **2017**, *67*, 45–49. [CrossRef]

sensors

MDPI

Article

Computational Efficient Motion Planning Method for Automated Vehicles Considering Dynamic Obstacle Avoidance and Traffic Interaction

Yuxiang Zhang [1], Jiachen Wang [2], Jidong Lv [3], Bingzhao Gao [4], Hongqing Chu [5,*] and Xiaoxiang Na [6]

1 The State Key Laboratory of Automotive Simulation and Control, Jilin University, Changchun 130025, China
2 The College of Computer Science and Technology, Jilin University, Changchun 130015, China
3 The Utopilot SAIC MOTOR, Shanghai 200438, China
4 The Clean Energy Automotive Engineering Center, Tongji University, Shanghai 201804, China
5 School of Automotive Studies, Tongji University, Shanghai 201804, China
6 The Department of Engineering, University of Cambridge, Cambridge CB2 1PZ, UK
* Correspondence: chuhongqing@tongji.edu.cn

Abstract: In complex driving scenarios, automated vehicles should behave reasonably and respond adaptively with high computational efficiency. In this paper, a computational efficient motion planning method is proposed, which considers traffic interaction and accelerates calculation. Firstly, the behavior is decided by connecting the points on the unequally divided road segments and lane centerlines, which simplifies the decision-making process in both space and time span. Secondly, as the dynamic vehicle model with changeable longitudinal velocity is considered in the trajectory generation module, the C/GMRES algorithm is used to accelerate the calculation of trajectory generation and realize on-line solving in nonlinear model predictive control. Meanwhile, the motion of other traffic participants is more accurately predicted based on the driver's intention and kinematics vehicle model, which enables the host vehicle to obtain a more reasonable behavior and trajectory. The simulation results verify the effectiveness of the proposed method.

Keywords: autonomous vehicles; trajectory planning; model predictive control

Citation: Zhang, Y.; Wang, J.; Lv, J.; Gao, B.; Chu, H.; Na, X. Computational Efficient Motion Planning Method for Automated Vehicles Considering Dynamic Obstacle Avoidance and Traffic Interaction. *Sensors* **2022**, *22*, 7397. https://doi.org/10.3390/s22197397

Academic Editor: Sindhuja Sankaran

Received: 22 August 2022
Accepted: 23 September 2022
Published: 28 September 2022

1. Introduction

Due to the complex driving scenarios and difficulty in accurately predicting the behavior of the surrounding vehicles, the automated vehicles need to adapt to great complexity and dynamics in real traffic [1,2]. Therefore, advanced motion planning algorithms should help the agent behave reasonably and respond adaptively in dynamic and complex driving scenarios with computational efficient and reliable control [3].

1.1. State-of-the-Art Review and Challenges

The performance of motion planning methods is closely related to the trajectory prediction of other traffic participants, behavior planning, and trajectory generation. As automated vehicles will frequently interact with other traffic participants, trajectory prediction will influence behavior planning and trajectory generation modules. The motion of other environment vehicles is predicted with constant longitudinal velocity or acceleration [4]. Such a prediction is inaccurate and will decrease the reasonability of behavior planning and trajectory generation modules [5,6]. Thus, more accurately predicted trajectories will promote the performance of motion planning.

In behavior planning aspects, the machine learning-based or model-based methods usually decide a human-like behavior, like lane-change and obstacle avoidance, which contains a wide range of trajectories. To ensure absolute safety, decision-making methods are always conservative. Thus, more complex behaviors should be generated in decision-making. POMDP can generate a more abundant behavior [7,8]. The road is divided into several segments and the points are connected to represent different behavioral decisions [9,10].

Regarding trajectory generation, current methods can be divided into graph-search-based methods [11], incremental search [12], interpolating curve methods [13] and numerical optimization [14,15]. Model Predictive Control (MPC) is widely used because it can explicitly deal with constraints to ensure safety with consideration of traffic interaction not only at the current time step but also in the predictive horizon [16,17]. As for the control model in MPC, the dynamic vehicle model is added to the kinematic model to realize stable motion control and enrich driving behaviors [18,19].

The high-performance motion planning methods should be computationally efficient and enable automated vehicles to behave reasonably and respond adaptively in dynamic and complex scenarios. First, the host vehicle will interact with the surrounding vehicles while driving on the road inevitably. The motion prediction of environment vehicles needs to be predicted in the planning horizon, which enables the host vehicle to behave reasonably and adapt to the dynamic traffic environment. Meanwhile, the driving process contains a large span in both time and space, which means precise driving behavior will cause huge calculations. To balance the calculation time and planning performance, the problem in the model formulation aspect should be simplified without losing reasonability.

To further promote the reasonability of motion planning, except for the dynamic vehicle model, variational velocity can be considered in the trajectory generation. Thus, the control model will change from the linear model to the nonlinear model [20]. Such a control model increases the time for online solving, which needs an additional fast solving algorithm to realize online calculation [21–23].

1.2. Work and Contributions

As shown in Figure 1, this paper proposes a computational efficient motion planning method for autonomous vehicles, which can behave reasonably and adaptively in dynamic and complex driving scenarios. After predicting the trajectory of environment vehicles, the behavior planning and trajectory generation will be done sequentially. The following improvements simplify the problem and decrease calculation to realize computational efficiency. In the behavior planning module, the road is divided into several segments with road points and different behaviors are represented with different connections between road points in each segment. Firstly, rather than only set road points in the center of the lane, the road points are also distributed on the lane line, which enables the behavior planner to generate more complex behaviors. Meanwhile, in predictive control, short-term behavior is much more complex and also should be paid more attention to. Besides, the prediction is not accurate while lengthening the predictive horizon. Thus, rather than equally dividing the road, unequal segments that the distance between these road segments is gradually increasing can further decrease calculation and raise reasonability. Secondly, based on the analysis of different traffic participants, the motion of static environment vehicles is much more simple, and can directly exclude corresponding behaviors. Therefore, static environment vehicles are used to narrow the feasible region of the solution and further decrease online calculation. Thirdly, variational longitudinal velocity and dynamic vehicle models are considered to raise the reasonability of trajectory generation. It can also speed up the trajectory following process with the resulting acceleration and steering wheel angle. We use the C/GMRES algorithm to realize on-line calculations in NMPC. The main contributions are summarized as follows

- The complex and reasonable behavior of the host vehicle is efficiently realized by connecting different points located on unequally divided road segments and lane centerlines.
- Trajectory prediction of surrounding vehicles is considered during trajectory planning. And the trajectory planning is based on both driver's intention and the kinematics vehicle model, which can increase the accuracy and rationality.

- C/GMRES is used to realize online calculation and raise the reasonability of trajectory generation and trajectory following.

The remainder of this paper is organized as follows. In Section 2, the coordinate systems and the trajectory prediction are introduced. Section 3 introduces the behavioral planning module. In Section 4, the trajectory generation module is introduced. In Section 5, the simulation process is shown, and the results are given and analyzed in detail. The simulation results verify the effectiveness of the proposed method. Section 6 is the conclusion of this paper.

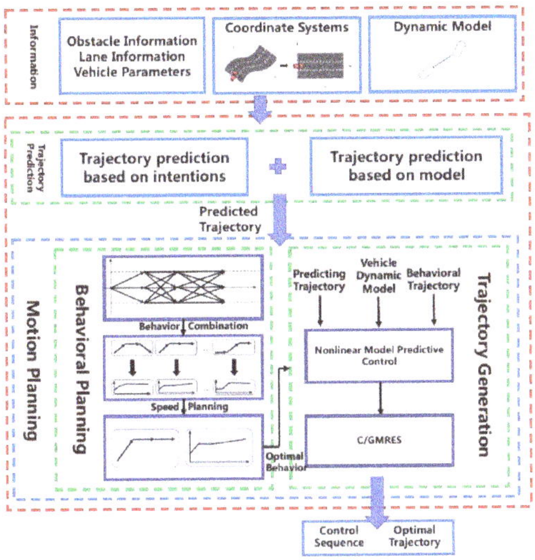

Figure 1. Diagram of the proposed motion planning frame.

2. Coordinate Systems Conversion and Trajectory Prediction

In this section, first, the conversion between the Cartesian coordinate system and the Frenet coordinate system is introduced to simplify the planning process. Then, three ways of trajectory prediction are illustrated and compared.

2.1. Coordinate Systems Conversion

To describe the relation between two coordinates, as shown in Figure 2, the Cartesian coordinate of $\vec{x}$ is $\vec{x}(x,y)$ while the Frenet coordinate of $\vec{x}$ is $\vec{x}(s,l)$, where l is the distance from $\vec{x}$ to the reference point [24]. In coordinate systems conversion, the lane centerline is extracted with a third-degree polynomial equation as the reference curve of the Frenet coordinate system, e.g., $y = ax^3 + bx^2 + cx + d$. For the conversion from Frenet coordinate $C_a : (s_a, l_a)$ to Cartesian coordinate $C_a : (x_a, y_a)$, a nearest point p_0 works as the reference point to convert p_a, whose Frenet coordinate can be written as $C_0 : (x_0, ax_0^3 + bx_0^2 + cx_0 + d)$. Since the arc length s_a is already known, x_0 can be calculated by dividing the curve and sampling from the starting point p_s. The integral value of the sampling point s'_0 can similarly be calculated. Compare s'_0 with s_a to determine whether the x'_0 is the desired coordinate point x_0, and finally find the coordinates of p_0. And the Cartesian coordinate of $p_a, C_a : (x_a, y_a)$ can be calculated with

$$x_a = x_0 - l \cdot \sin(\arctan(k)), \tag{1a}$$

$$y_a = y_0 - l \cdot \cos(\arctan(k)). \tag{1b}$$

where $k = 3ax_0^2 + 2bx_0 + c$ is the curvature of the reference curve at point p_0.

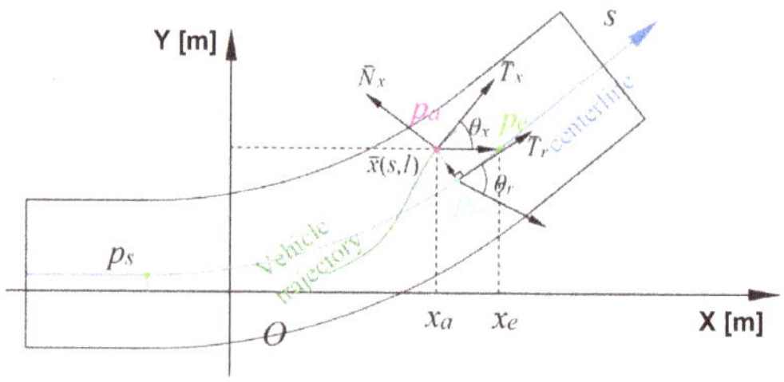

Figure 2. Diagram of coordinate systems conversion. p_s is the starting point, p_e is the ending point. p_a is the point on the trajectory of the vehicle to be converted. p_0 is the nearest point that works as the reference point to convert p_a. $\vec{T}_x$ is the reference curve tangent vector in the Frenet coordinate system, and $\vec{N}_x$ is the normal vector in the Frenet coordinate system. $\vec{x}(s,l)$ is the Frenet coordinate of $\vec{x}$.

For the conversion from Cartesian coordinate $C_a : (x_a, y_a)$ to Frenet coordinate $C_a : (s_a, l_a)$, the reference point p_0 is also needed. We use D_2 to represent the square of the distance from p_a to the reference point p_0. The horizontal coordinate value of the reference point x_0^* that minimizes D_2 satisfies $D_2'(x_0^*) = 0$, which can be solved by Newton's method [25]. By calculating D_2 and D_2', the iteration formula can be expressed as

$$x_0^{*,m+1} = x_0^{*,m} - \frac{D_2'(x_0^{*,m})}{D_2''(x_0^{*,m})}, m = 0, 1, 2, \ldots. \tag{2}$$

The iteration stops while $|x_0^{*,m+1} - x_0^{*,m}| \leqslant \delta$ and $x_0^{*,m+1}$ is the target value. The Frenet coordinate $C_a : (s_a, l_a)$ can be solved as

$$s_a = \int_{x_0}^{x_0^{*,m+1}} \sqrt{1 + (3ax^2 + 2bx + c)^2} dx, \tag{3a}$$

$$l_a = D_2(x_0^{*,m+1}) = \sqrt{(x_a - x_0^{*,m+1})^2 + (y_a - y_0^{*,m+1})^2}. \tag{3b}$$

2.2. Trajectory Prediction

The information about the surrounding vehicles and the trajectories of the surrounding vehicles in a period of time in the future is essential for motion planning. Such trajectory prediction can be done based on the driver's intention, vehicle kinematics model, or both driver's intention and vehicle kinematics model, which will be compared in this section.

2.2.1. Trajectory Prediction Based on Driver's Intention

We consider lane change and lane-keeping operations. For an operation intention, countless driving trajectories can be realized. Based on the driver of the vehicle, the actual driving trajectory may be very gentle or aggressive. In addition, the geometric environment of the road will also affect the actual trajectory. Therefore, the trajectory prediction based on the operation intention can generate a set of predicted trajectories based on the current state of the vehicle, the operation intention, and the road parameters, and then select the optimal one based on the information. Since the shape of the road has a great influence on the predicted trajectory, the predicted trajectory cluster is firstly generated in the Frenet coordinate system, then converted into the Cartesian coordinate system.

In Frenet coordinate system, $s(t)$ and $l(t)$ represent longitudinal distance and lateral distance, respectively. We use $F_0 = (s_0, \dot{s}_0, \ddot{s}_0, l_0, \dot{l}_0, \ddot{l}_0)$ and $F_1 = (s_1, \dot{s}_1, \ddot{s}_1, l_1, \dot{l}_1, \ddot{l}_1)$ to represent the initial state and the end state of the vehicle trajectory. To ensure the continuity of the trajectory and provide unique expressions for different trajectories, high-order polynomials are used for fitting to represent the trajectory of longitudinal $s(t)$ and lateral distance $l(t)$ over time t. For the initial state F_0, each state variable can be obtained by converting the current kinematic parameters in the Cartesian coordinate system, which can be expressed as

$$\begin{cases} l_0 = l_0^*,\ \dot{l}_0 = v_0 \sin(\theta_0 - \theta_{T_0}), \\ \ddot{l}_0 = \sqrt{(a_{0^2} + \gamma_0 v_{0^2})}\sin(\theta_0 - \theta_{T_0}), \\ s_0 = 0,\ \dot{s}_0 = v_0 \cos(\theta_0 - \theta_{T_0}), \\ \ddot{s}_0 = \sqrt{(a_{0^2} + \gamma_0 v_{0^2})}\cos(\theta_0 - \theta_{T_0}), \end{cases} \tag{4}$$

where l_0^* is the distance from the initial position to the centerline of the lane. θ_{T_0} is the orientation of the tangent vector $\vec{T}_0$. $\gamma_0 v_0^2$ is the current value of the normal acceleration of the vehicle and γ is the curvature of the reference point.

We assume that the vehicle is on the center line of its target lane after finishing the intended operation and it remains the same longitudinal acceleration throughout the operation. Therefore, some of the operation termination state F_1 can be calculated as

$$\begin{cases} l_1 = l_1^*,\ \dot{l}_1 = 0, \\ \ddot{l}_1 = 0,\ \ddot{s}_1 = a_0\,. \end{cases} \tag{5}$$

$|l_1^*| = d$, where d is the width of the lane while the vehicle is changing the current lane and the sign is determined by the lane change direction. For lane keeping operation, $l_1^* = 0$.

For a complete lane change operation, the time to complete the operation t_{end} is about 6 s, and the length of t_{end} can be adjusted according to the driver's driving style. For lane-keeping operations, the time t_{end} is significantly shorter. Use $t_1 \in [0, t_{end}]$ to represent the end time of the operation, that is, take a fixed step size and sample start from 0 to t_{end} with K steps. Since it is assumed that the longitudinal acceleration of the vehicle remains unchanged, the longitudinal speed at the termination state is $\dot{s}_1 = v_0 + a_0 \cdot t_1$. For the lateral distance at the termination state, a fifth-degree polynomial fit for time t can be used, which is calculated as

$$l(t) = c_5 t^5 + c_4 t^4 + c_3 t^3 + c_2 t^2 + c_1 t + c_0, \tag{6}$$

where $c_0, c_2, \ldots, c_5$ are the parameters of the curve. Then, $\dot{l}(t)$ and $\ddot{l}(t)$ can be calculated respectively. Therefore, the lateral state values at the initial state ($t = 0$) and the terminal lateral state ($t = t_1$) can be written as

$$\begin{cases} l_0 = l(0), \dot{l}_0 = \dot{l}(0), \ddot{l}_0 = \ddot{l}(0), \\ l_1 - l(t_1), \dot{l}_1 = \dot{l}(t_1), \ddot{l}_1 = \ddot{l}(t_1). \end{cases} \tag{7}$$

And the parameters c_i can be obtained by solving the following matrix

$$\begin{bmatrix} 0 & 0 & 0 & 0 & 0 & 1 \\ t_1^5 & t_1^4 & t_1^3 & t_1^2 & t_1 & 1 \\ 0 & 0 & 0 & 0 & 1 & 0 \\ 5t_1^4 & 4t_1^3 & 3t_1^2 & 2t_1 & 1 & 0 \\ 0 & 0 & 0 & 2 & 0 & 0 \\ 20t_1^3 & 12t_1^2 & 6t_1 & 2 & 0 & 0 \end{bmatrix} \cdot \begin{bmatrix} c_5 \\ c_4 \\ c_3 \\ c_2 \\ c_1 \\ c_0 \end{bmatrix} = \begin{bmatrix} l_0 \\ l_1 \\ \dot{l}_0 \\ \dot{l}_1 \\ \ddot{l}_0 \\ \ddot{l}_1 \end{bmatrix} = \begin{bmatrix} l_0^* \\ l_1 \\ v_0 \sin(\theta_0 - \theta_{T_0}), \\ \sqrt{(a_{0^2} + \gamma_0 v_{0^2})}\sin(\theta_0 - \theta_{T_0}), \\ 0 \\ 0 \end{bmatrix}. \tag{8}$$

Since the longitudinal displacement changes according to t_1, a fourth-degree polynomial with respect to time t is used to fit the longitudinal kinematic, which can be expressed as

$$s(t) = c'_4 t^4 + c'_3 t^3 + c'_2 t^2 + c'_1 t + c'_0, \tag{9}$$

where $c'_0, c'_2, \ldots, c'_5$ are the parameters. The initial states and the terminal states can be represented by replacing t in the above formula with 0 and t_1, respectively.

$$\begin{cases} s_0 = s(0), \dot{s}_0 = \dot{s}(0), \ddot{s}_0 = \ddot{s}(0), \\ \dot{s}_1 = \dot{s}(t_1), \ddot{s}_1 = \ddot{s}(t_1). \end{cases} \tag{10}$$

Therefore, the parameters c'_i can be obtained by solving the following matrix

$$\begin{aligned} \begin{bmatrix} 0 & 0 & 0 & 0 & 1 \\ 0 & 0 & 0 & 1 & 0 \\ 4t_1^3 & 3t_1^2 & 2t_1 & 1 & 0 \\ 0 & 0 & 2 & 0 & 0 \\ 12t_1^2 & 6t_1 & 2 & 0 & 0 \end{bmatrix} \cdot \begin{bmatrix} c'_4 \\ c'_3 \\ c'_2 \\ c'_1 \\ c'_0 \end{bmatrix} &= \begin{bmatrix} s_0 \\ \dot{s}_0 \\ \dot{s}_1 \\ \ddot{s}_0 \\ \ddot{s}_1 \end{bmatrix}, \\ &= \begin{bmatrix} 0 \\ v_0 \cos(\theta_0 - \theta_{T_0}), \\ \sqrt{(a_{02} + \gamma_0 v_{02})} \cos(\theta_0 - \theta_{T_0}), \\ v_0 + a_0 \cdot t_1 \\ 0 \end{bmatrix}. \end{aligned} \tag{11}$$

Different $t_1 \in [0, t_{end}]$ corresponds to different fitting parameters c and different driving trajectories. When selecting the optimal trajectory, first convert the trajectory to the Cartesian coordinate system. The principles to be followed in the selection process include: t_1 is as short as possible, the driving process is comfortable, and the lateral displacement is reduced as much as possible during the lane change operation. Therefore, for the selection of the optimal predicted trajectory based on the driver's intention, the maximum normal acceleration value in the driving trajectory and the time to complete the operation t_1 are mainly considered

$$C(traj_i) = \mathrm{w}_1 \max(\bar{a}(t)) + \mathrm{w}_2 t_i, \tag{12}$$

where $traj_i$ represents the i_{th} trajectory and $C(traj_i)$ represents the cost of the i_{th} trajectory. $\bar{a}(t)$ is the normal acceleration at time t. t_i is the total time of the i_{th} trajectory. w_1 and w_2 are two coefficients. The resulting trajectory with the smallest cost value is used as the optimal prediction trajectory based on the operation intention prediction

$$T_{\mathrm{man}} = \arg\min(C(traj_i))_{i=1,\ldots,K}. \tag{13}$$

2.2.2. Trajectory Prediction Based on Vehicle Kinematics Model

The trajectory predicted based on the driver's intention is more accurate at a longer time horizon, but the accuracy is lower on a shorter time horizon. The trajectory predicted using the current kinematic parameters of the vehicle is more accurate in a shorter time. So, it is necessary to put both the intention and kinematics model into consideration. We assume that the vehicle acceleration and yaw rate remain unchanged. Therefore, in the Cartesian coordinate system, the vehicle speeds along the x and y axes at time t can be represented as

$$v_x(t) = v(t) \cdot \cos(w_0 t + \theta_0), \tag{14a}$$

$$v_y(t) = v(t) \cdot \sin(w_0 t + \theta_0), \tag{14b}$$

where the velocity at time t is $v(t) = a_0 t + v_0$. The predicted trajectory based on the kinematic model can be obtained by integrating the vehicle velocities,

$$traj_{mdl}: \begin{cases} x(t) = \frac{a_0}{w_0^2}\cos(\theta(t)) + \frac{v(t)}{w_0}\sin(\theta(t)) + c_x, \\ y(t) = \frac{a_0}{w_0^2}\sin(\theta(t)) - \frac{v(t)}{w_0}\cos(\theta(t)) + c_y, \end{cases} \tag{15}$$

where c_x and c_y are two parameters determined by the initial state and can be expressed as

$$c_x = x_0 - \frac{v_0}{w_0}\cos(\theta_0) - \frac{a_0}{w_0^2}\sin(\theta_0), \tag{16a}$$

$$c_y = y_0 + \frac{v_0}{w_0}\sin(\theta_0) - \frac{a_0}{w_0^2}\cos(\theta_0). \tag{16b}$$

In particular, when the initial yaw rate $w_0 = 0$, the predicted trajectory changes to

$$traj_{mdl}: \begin{cases} x(t) = (\frac{1}{2}\cdot a_0 \cdot t^2 + v_0)\cos(\theta_0) + x_0, \\ y(t) = (\frac{1}{2}\cdot a_0 \cdot t^2 + v_0)\sin(\theta_0) + y_0. \end{cases} \tag{17}$$

2.2.3. Trajectory Prediction Based on Both Driver's Intention and Vehicle Kinematics Model

Since the predicted trajectory based on the kinematics model is more accurate only in a shorter prediction time, and the trajectory based on the driver's intention has higher accuracy in a longer period of time, the predicted trajectory obtained by combining the two will be more accurate. Let the coefficient of the predicted trajectory based on the kinematics model be $\mathrm{w}(t)$, and the predicted trajectory model can be changed to

$$traj(t) = \mathrm{w}(t)traj_{mdl}(t) + (1 - \mathrm{w}(t))traj_{man}(t), \tag{18}$$

where $\mathrm{w}(t) \in [0,1]$ is a time-varying variable that is designed to predict trajectory with high accuracy. Here, $\mathrm{w}(t)$ is designed and tuned with multiple simulations as

$$\mathrm{w}(t) = \begin{cases} \frac{2}{27}t^3 - \frac{1}{3}t^2 + 1 & \text{for } 0 \le t < 3, \\ 0 & \text{for } t \ge 3. \end{cases} \tag{19}$$

Combined with the driver's intention and the vehicle kinematics model, a more accurate predicted trajectory can be obtained. The results of trajectory prediction are shown in Figure 3. The trajectory that is predicted based on both operation intention and the vehicle kinematics model is more accurate than the other two methods.

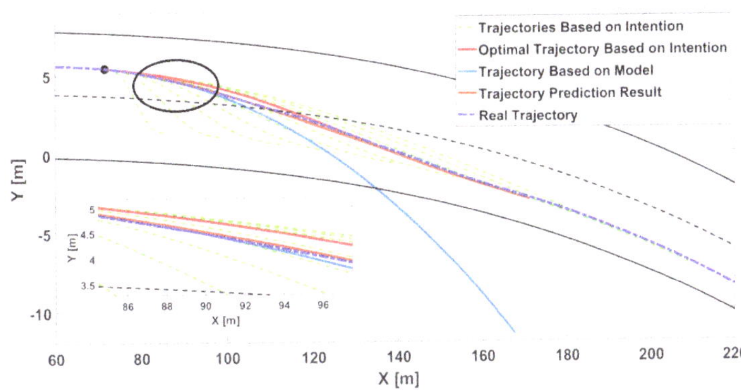

Figure 3. The results of trajectory prediction based on operation intention, vehicle kinematics model, and both operation intention and vehicle kinematics model.

3. Behavioral Planning

In the behavioral planning, by comprehensively considering the host vehicle information, road information, and surrounding environment information, an optimal behavioral trajectory is selected, which will be used for trajectory generation.

3.1. Generation of Candidate Paths

To reduce the size of the search space for trajectory planning and speed up the calculation, a series of candidate paths need to be generated first. The optimal path that does not collide is selected from the candidate paths. The candidate path is a candidate set that can represent the behavior that the vehicle can take in the planning process, and the road space of the entire prediction time is reasonably divided into multiple road segments. Comprehensively considering the prediction length and calculation time, the road space in the prediction time is divided into three road segments. The candidate points at the same s coordinates are called a layer of candidate point sets. The diagram of this division is shown in Figure 4. The road space occupied by each road segment can be represented as

$$s_i = \begin{cases} N_i \cdot \Delta s & \text{for } v > v_{\min}, \\ N_i \cdot \Delta s_{min} & \text{for } v \le v_{\min}, \end{cases} \tag{20}$$

where Δs refers to the distance traveled by the vehicle in 1 s, s_i refers to the length of the i_{th} road space. To prevent the predicted length from being too short, a minimum interval Δs_{min} needs to be set so that the vehicle can plan the trajectory even if the current vehicle speed is too slow. The value of N_i is set based on the comprehensive consideration of the calculation time and the predicted length. The step length near the current position is short and the one farther away from the current position is large.

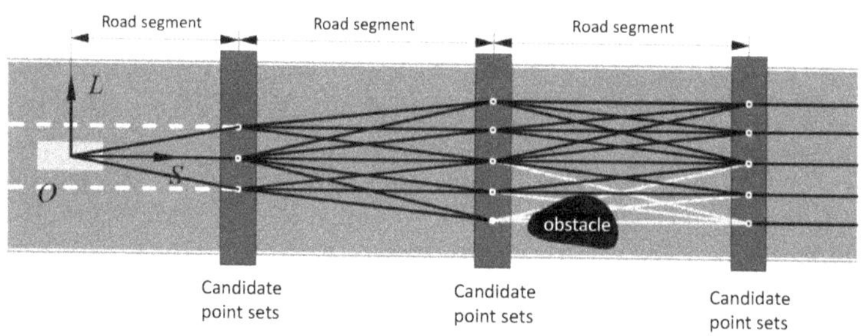

Figure 4. The diagram of Generation of Candidate Paths.

All candidate roads together constitute a candidate set of trajectories in the future. Each layer of candidate point sets contains the center point of both the current lane and the adjacent lane and the lane change point of the two lanes. By connecting the points in the candidate point set, a series of path candidate sets can be generated. To further narrow the search range, considering that the vehicle has a high risk of completing the lane change operation in a short period of time, the set of candidate points of the first layer contains only the road center point of the current lane and the lane change point with the adjacent lane.

The path candidate set is a connection of a series of points in the search space, but not every road can be driven in the path candidate set. There will inevitably be static and dynamic obstacles on the road. Regardless of the dynamic obstacles, the paths that pass through the static obstacles are firstly removed. When driving along these paths, the vehicle will inevitably collide with a static obstacle at any time. After removing, all paths in the remaining set of paths candidates will become candidate paths in the behavior decision layer. Speed planning is needed for these candidate paths to select the optimal one.

3.2. Speed Profile

The selection of vehicle speed needs comprehensive consideration of traffic and road information and restrictions, vehicle restrictions, dynamic obstacles, and other information. Traffic road information mainly includes traffic lights, traffic signs, stop lines, maximum and minimum speed limits, etc, which is simply shown in Figure 5. When selecting the optimal speed sequence, traffic road information must be extracted as the first constraint. Since the influence of road traffic information on vehicle speed mainly acts on the road driving direction, that is, the S direction, the limit curve based on the traffic road information on the vehicle speed can be represented in a v-s profile.

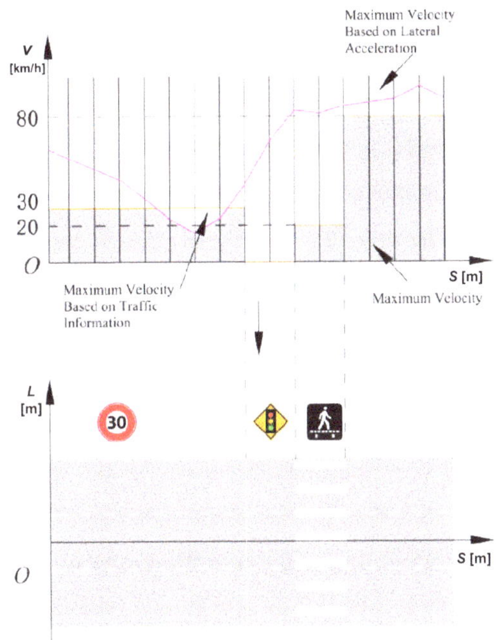

Figure 5. Maximum velocity considering traffic information and lateral acceleration.

The maximum lateral acceleration of the vehicle while driving needs to consider the physical limitations of the vehicle and the impact on comfort. Based on the maximum lateral acceleration and the road curvature, the speed limit based on the lateral acceleration can be calculated as

$$v_{a_{y,max}} = \sqrt{\frac{a_{y,max}}{C_{Lane}}}. \tag{21}$$

Since the road curvatures C_{Lane} of adjacent lanes are the same, the limitation of the maximum lateral acceleration on the vehicle speed still only exists in the S direction. By comparing the value of the above two speeds, the limit curve of the vehicle speed in the S direction can be obtained as shown in Figure 5.

For dynamic obstacles, if time and space are considered at the same time, the problem of speed selection is a problem of optimal selection in S-L-T three-dimensional space, which is extremely high in calculation complexity. To reduce the dimension, each of the above candidate paths can be subjected to speed planning once. The coordinates L can be ignored to reduce the difficulty of calculation. An S-T profile is used to plan speed.

In an *S*-*T* profile, the *T* axis represents the time axis for predicting the future along the candidate road, and the *S* axis represents the space axis extending from the origin along the candidate road. We assume that the dynamic obstacles have constant acceleration within the prediction. The information on dynamic obstacles can be displayed in blue blocks and a safe distance is reserved too. The vehicle needs to travel in the space-time area corresponding to the blank grid. The origin of the *S*-*T* map is the current position of the host vehicle, and how to get to the target *S* position is the goal of speed planning.

Some restrictions and objective functions are necessary. First, the speed of the vehicle should be as fast as possible, which is calculated as

$$f_v = |v_{max}(s_i) - v_i|. \tag{22}$$

In addition, large acceleration will reduce driving comfort, so acceleration needs to be limited as

$$f_a = |a_i| = \frac{|v_i - v_{i-1}|}{\Delta t}. \tag{23}$$

Finally, it is necessary to avoid frequent acceleration and deceleration of the vehicle during driving, which is

$$f_{jerk} = \frac{|a_i - a_{i-1}|}{\Delta t} = \frac{|v_i - 2v_{i-1} + v_{i-2}|}{\Delta t^2}. \tag{24}$$

In summary, the cost function is

$$f_{speed} = w_v f_v + w_a f_a + w_j f_{jerk}, \tag{25}$$

where w_v, w_a and w_j are coefficients corresponding to f_v, f_a and f_{jerk}. The optimal speed sequence for a certain behavior can be obtained through the cost function as shown in Figure 6.

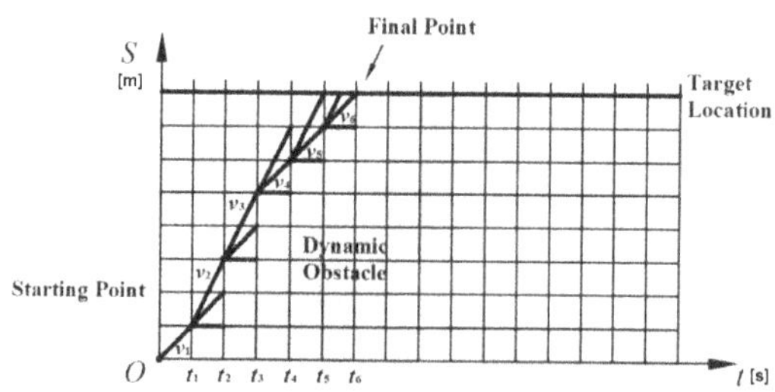

Figure 6. Speed profile.

3.3. Optimal Behavioral Trajectory Selection

Each candidate trajectory in the behavior planning candidate set represents a behavioral operation, and it is especially important to select the optimal one. It is necessary to consider efficiency, comfort, energy consumption, and other aspects.

First, the number of lane changes needs to be considered. C_{LC} is a cost function factor affected by the number of lane changes. Since lane changing greatly increases energy consumption and greatly reduces comfort, it is necessary to avoid unnecessary and frequent lane changing operations. Thus,

$$C_{LC} = \sum_{k=1}^{N_{layer}} |Lane(n_k) - Lane(n_{k-1})|, \tag{26}$$

where N_{layer} represents the layer number, n_k represents the k_{th} candidate point. $Lane(n_{(.)})$ represents the lane at $n_{(.)}$ candidate point.

In addition, since the S coordinate of the endpoint of each candidate route is the same, the shorter the travel time, the higher the efficiency. $C_T = T$ is only determined by the time. Frequent changes in behavior can reduce comfort and increase control difficulty. To reduce unnecessary changes in behavior planning, a consistency coefficient is introduced to indicate the difference between the candidate behaviors at the current time and the previously executed behavior. Thus, C_{con} can be formulated as

$$C_{con} = \sum_{j=1}^{N_b - k} [(S_t^j - S_{t-\Delta t}^{j+k})^2 + (L_t^j - L_{t-\Delta t}^{j+k})^2]. \tag{27}$$

When generating the speed profile in the previous step, the candidate trajectory has been discretized. The step size after discretization is Δt_{ST}, and the total number of discrete points is $N_b = T/\Delta t_{ST} + 1$. Δt_{re} is the discrete step size of the model used for resampling performed. Then, $k = \Delta t_{re}/\Delta t_{ST}$. Therefore, S_t^j and L_t^j represent the coordinate values of the j_{th} point on the candidate trajectory in Frenet coordinate system at time t. $S_{t-\Delta t}^{j+k}$ and $L_{t-\Delta t}^{j+k}$ represent the coordinate values of the j_{th} point on the candidate trajectory in Frenet coordinate system at time t, which are the same corresponding point with S_t^j and L_t^j. In this way, by constraining the difference between the same corresponding points in the two behavior planning paths taken at adjacent moments, the consistency and continuity of the behavior planning path can be constrained.

Last but not least, to make the vehicle location in the center of the road as much as possible while lane-keeping, $C_{boun} = \sum n_{boun}$ is used to represent the number of nodes where the vehicle is on the road boundary in the planned behavior trajectory, where n_{boun} represents nodes where the vehicle is at the road boundary in the behavior trajectory.

In summary, a behavior path that is optimal in terms of efficiency, comfort, and energy consumption is selected according to the following cost function, and this behavior path is used as a reference for motion planning

$$C(t) = w_{LC}C_{LC}(t) + w_T C_T(t) + w_{con}C_{con}(t). \tag{28}$$

3.4. Resampled Behavioral Trajectory

The step length of the behavior trajectory is longer. After selecting an optimal behavior trajectory, the trajectory needs to be converted from the Frenet coordinate system to the Cartesian coordinate system. The diagram of this process is shown in Figure 7. The step length of the candidate trajectory is Δt_{ST}, and the time sequence is $\Delta t_{ST}^0, \Delta t_{ST}^1, \ldots$. The step size used in resampling is Δt_{re}, and the time sequence is $\Delta t_{re}^0, \Delta t_{re}^1, \ldots$. Since Δt_{ST} is larger than Δt_{re}, the optimal behavior trajectory needs to be interpolated to shorten the step size. In the process of interpolating, it is assumed that the vehicle speed remains unchanged within Δt_{ST}, and the number of Δt_{re} contained in each Δt_{ST} is $N = \Delta t_{ST}/\Delta t_{re}$. The time sequence of resampled trajectory can be written as $\Delta t_{ST}^0, \Delta t_{re}^{01}, \ldots, \Delta t_{re}^{0(N-1)}, \Delta t_{ST}^1, \Delta t_{re}^{11}, \ldots$, where Δt_{ST}^i represents the time at the i_{th} step, and Δt_{re}^{ij} represents the j_{th} resampled point in the i_{th} step.

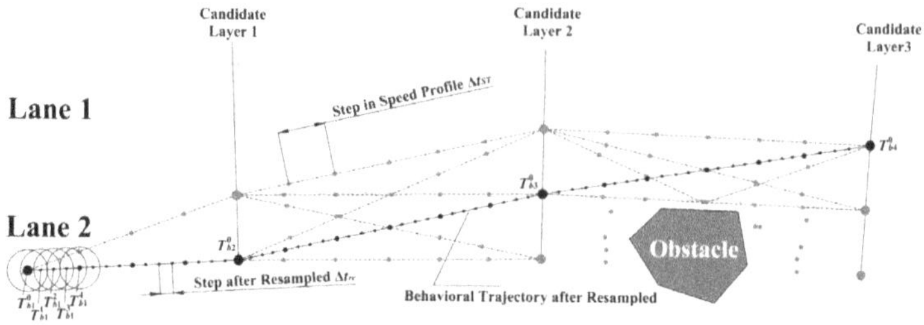

Figure 7. Resampled behavioral trajectory.

4. Trajectory Generation

With selected behavior from the behavioral planning module, nonlinear model predictive control is used in the trajectory generation module, which considers variational longitudinal velocity and dynamic vehicle model, and uses the C/GMRES algorithm to speed up the calculation process.

4.1. Vehicle Dynamic Model

To describe the vehicle dynamics characteristics more accurately, this paper uses a vehicle dynamics model. For the vehicle system, the coordinate system is the right-handed coordinate system, and the origin is the position of the center of mass of the vehicle. According to Newton's second law, the dynamic characteristics can be represented as

$$2(F_{yf} + F_{yr}) = m \cdot a_y\,, \tag{29a}$$

$$\dot{w} = 2\frac{F_{yf} \cdot l_f - F_{yr} \cdot l_r}{I_z}\,, \tag{29b}$$

where m is the mass of the vehicle and I_z is the inertia of the vehicle. F_{yf} and F_{yr} are the lateral forces of the front wheel and rear wheel, respectively. a_y is the lateral acceleration. l_f and l_r are the lengths from the center of mass to the front axle and the rear axle, respectively. w is the yaw rate. β is used to represent the ratio of lateral speed to the longitudinal speed,

$$\beta = \frac{v_y}{v_x}\,. \tag{30}$$

From the geometric relationship, the slip angles of the front and rear wheels can be represented as

$$\alpha_f = \arctan(\frac{v_y + l_f w}{v_x}) \approx -\delta + \frac{w l_f}{v_x} + \beta\,, \tag{31a}$$

$$\alpha_r = \arctan(\frac{v_y - l_r w}{v_x}) \approx \beta - \frac{w l_r}{v_x} + \beta\,. \tag{31b}$$

where δ is the steering angle. When the lip angle is small, the tire characteristics can be regarded as linear, which is calculated as

$$F_{yf} = k_f \alpha_{yf}\,, \tag{32a}$$

$$F_{yr} = k_r \alpha_{yr}\,. \tag{32b}$$

The derivative of longitudinal acceleration is the acceleration of the vehicle.

Using $X = [x, y, \phi, v_x, v_y, w]^T$ as the state vector of the vehicle dynamics system and $U = [a, \delta]^T$ as the input vector of the vehicle dynamics system, the state equation of the vehicle dynamics system is

$$\dot{X} = f(X,U) = \begin{bmatrix} v_x \cos\varphi - v_y \sin\varphi \\ v_x \sin\varphi + v_y \cos\varphi \\ w \\ a \\ \frac{2k_f(v_f + wl_f - \delta v_x) + 2k_r(v_y - wl_r)}{mv_x} - v_x w \\ \frac{2l_r k_f(v_y + wl_f - \delta v_x) - 2l_r k_r(v_y - wl_r)}{I_z v_x} \end{bmatrix}. \tag{33}$$

Transform the state equation of the continuous form vehicle dynamics system into a discrete form

$$X_{k+1} = X_k + f(X_k, U_k)\Delta t. \tag{34}$$

4.2. Controller Design

To complete the task of trajectory planning, it is necessary to rationally design the objective functions and constraints of model predictive control. The input value of the vehicle system cannot exceed the limit of the physical structure, so the upper and lower limits of the input variable need to be limited as

$$|a_k| \leq a_{max}, \tag{35a}$$
$$|\delta_k| \leq \delta_{max}. \tag{35b}$$

To ensure that the planned trajectory does not collide with obstacles, it is necessary to maintain a certain safety distance between the trajectory at each moment and the obstacle at the corresponding moment. Since the longitudinal speed of the vehicle is often much higher than the lateral speed, a larger safe distance is needed in the longitudinal direction than in the lateral direction. The safety range around the host autonomous vehicle is designed as an ellipse, where the long axis of the ellipse is the longitudinal direction so that the safety constraint can be expressed as

$$(\frac{x_{host} - x_{obs}}{r_x})^2 + (\frac{y_{host} - y_{obs}}{r_y})^2 \geq 1, \tag{36}$$

where r_x and r_y represent safety distance in the longitudinal direction and the lateral direction, respectively. The larger the safety distance is, the further the host vehicle acts.

To make the final trajectory planning result close to the behavioral planning path, the following objective function is needed

$$J_{coor} = w_1(x_k - x_{ref,k})^2 + w_2(y_k - y_{ref,k})^2, \tag{37}$$

where x_k and y_k are the coordinates of the trajectory at time-step k. $x_{ref,k}$ and $y_{ref,k}$ are the coordinates of the resampling behavioral planning path at time-step k.

As mentioned before, excessive acceleration and a small turning radius will greatly reduce driving comfort, and so it's necessary to control their value of them.

$$J_{com} = w_3 a_k^2 + w_4 \delta_k^2. \tag{38}$$

A terminal constraint is added to ensure the final trajectory matches the planned behavioral path better.

$$J_{end} = \frac{1}{2}(X_N - X_{ref,N})^T S_f (X_N - X_{ref,N}). \tag{39}$$

When there are two environment vehicles, the expression of the controller for motion planning is

$$\min_{\lambda_{M\times N}} J_{U_{i-N}} = \frac{1}{2}(X_N - X_{ref,N})^T S_f (X_N - X_{ref,N}) + \sum_{k=1}^{N}\Big(w_1(x_k - x_{ref,k})^2 + w_2(y_k - y_{ref,k})^2 + w_3 a_k^2 + w_4 \delta_k^2\Big)\Delta t \tag{40a}$$

$$s.t.\ \dot{X}_{k+1} = X_k + f(X_k, U_k)\Delta t, \tag{40b}$$

$$|a| \le a_{max}, |\delta| \le \delta_{max}, \tag{40c}$$

$$(\frac{x_k - x^j_{obs,k}}{r_x})^2 + (\frac{y_i - y^j_{obs,k}}{r_y})^2 \ge 1\,, j = 1,2. \tag{40d}$$

Since the vehicle dynamics model used is a non-linear model, a suitable solving algorithm is necessary. In this paper, a Continuation/GMRES algorithm is used to solve the nonlinear model predictive control problem.

4.3. C-GMRES

In the nonlinear model predictive control problem, the small sampling period of mechanical systems will bring a great burden to the computing platform. Based on Generalized Minimum Residual Method (GMRES), Ohtsuka introduced the concept of the Continuation Generalized Minimum Residual Method (C/GMRES), which solves the linear equations involved in the differential equations at each sampling instant, thereby solving the control input sequence [26]. The detailed C/GMRES algorithm that is used in this paper is depicted in Algorithm 1.

Algorithm 1 C/GMRES Algorithm

//Initialize $t = 0$, $l = 0$, initial state $x_0 = x(0)$ and find U_0 analytically or numerically such that $||F(U_0, x_0, 0)|| \le \delta$ for some positive δ, maximum iteration number k_{max}.
1. For $t' \in [t, t+\Delta t]$, compute the real control input by $u(t') = P_0(U_l)$.
2. At next sampling instant $t + \Delta t$, measure the state $x_{l+1} = x(t+\Delta t)$, set $\Delta x_l = x_{l+1} - x_l$.
3. $\dot{U}_l = FDGMRES(U_l, x_l, \Delta x_l/\Delta t, t, \hat{U}_l, h, k_{max})$, where $\hat{U}_l = \hat{U}_{l-1}$ with $\hat{U}_{-1} = 0$.
4. $U_{l+1} = U_k + \hat{U}_k + \Delta t$.
5. Update $t = t + \Delta t$, $k = k + 1$.

To solve the trajectory planning problem, first, the dummy inputs are used to convert inequality constraints into equality constraints.

$$(\delta_k^2 + u_{d1,k}^2 - \delta_{max}^2)/2 = 0\,, \tag{41a}$$

$$(a_k^2 + u_{d2,k}^2 - a_{max}^2)/2 = 0\,, \tag{41b}$$

$$(\frac{x_k - x^1_{obs,k}}{r_x})^2 + (\frac{y_k - y^1_{obs,k}}{r_y})^2 - 1 - u_{d3,i}^2 = 0\,, \tag{41c}$$

$$(\frac{x_k - x^2_{obs,k}}{r_x})^2 + (\frac{y_k - y^2_{obs,k}}{r_y})^2 - 1 - u_{d4,i}^2 = 0\,. \tag{41d}$$

Meanwhile, to prevent multiple solutions of dummy inputs, adding a small dummy penalty to the objective function

$$\min_{\lambda_{M\times N}} J_{U_{i-N}} = \frac{1}{2}(X_N - X_{ref,N})^T S_f (X_N - X_{ref,N}) + \sum_{k=1}^{N}\Big(w_1(x_k - x_{ref,k})^2 + w_2(y_k - y_{ref,k})^2 + w_3 a_k^2 + w_4 \delta_k^2 - w_{U_d} U_{d,k}\Big)\Delta t$$

where w_{U_d} is a small positive constant.

Then, define the Hamiltonian by

$$H(x, \lambda, u, \mu, p) = L(x, u, p) + \lambda^T f(x, u, p) + \mu^T C(x, u, p), \tag{42}$$

where $\lambda \in \mathbb{R}^n$ represents costate and $\mu \in \mathbb{R}^{m_c}$ represents language multiplier. m_c represents the dimension of constraints.

For an optimal control $\{u_k^*\}_{k=0}^{N-1}$, it exists $\{\lambda_k^*\}_{k=0}^{N}$ and $\{\mu_k^*\}_{k=0}^{N-1}$, the following conditions should be satified

$$x_{k+1}^* = x_k^* + f(x_k^*, u_k^*, p_k)\Delta t, \tag{43a}$$
$$\lambda_k^* = \lambda_{k+1}^* + H_x^T(x_k^*, \lambda_{k+1}^*, u_k^*, \mu_k^*, p_k)\Delta t, \tag{43b}$$
$$\lambda_N^* = \varphi_x^T(x_N^*, p_N), \tag{43c}$$
$$x_k^* = x(t_0), \tag{43d}$$
$$H_u(x_k^*, \lambda_{k+1}^*, u_k^*, \mu_k^*, p_k) = 0, \tag{43e}$$
$$C(x_k^*, u_k^*, p_k) = 0. \tag{43f}$$

Here, Equaton (43) can be summarized as

$$F(U(t), x(t), t) = \begin{bmatrix} H_u^T(x_0^*, \lambda_1^*, u_0^*, \mu_0^*, p_0) \\ C(x_0^*, u_0^*, p_0) \\ \cdots \\ H_u^T(x_{N-1}^*, \lambda_N^*, u_{N-1}^*, \mu_{N-1}^*, p_{N-1}) \\ C(x_{N-1}^*, u_{N-1}^*, p_{N-1}) \end{bmatrix}, \tag{44}$$
$$= 0,$$

Then, $\dot{F}(U, x, t)$ can be expressed as

$$\dot{F}(U, x, t) = A_s F(U, x, t). \tag{45}$$

Here, A_s is an introduced stable matrix that stabilizes $F(U, x, t)$ at the origin. Then, $\dot{U}$ can be computed with

$$\dot{U} = F_U^{-1}(A_s F - F_x \dot{x} - F_t). \tag{46}$$

The solution curve $U(t)$ is approximated by forward difference if an initial solution $U(0)$ satisfying $F(U(0), x(0), 0) = 0$ can be found. Here, generalized minimal residual (GMRES) method is applied to solve the linear equation $F_U \dot{U} = A_s F - F_x \dot{x} - F_t$. The combination of forward difference approximation and GMRES is called FDGMRES.

5. Simulation

In this section, the proposed computational efficient motion planning method is verified in three different environments. According to the information of the obstacle and the predicted trajectory, the behavior is selected, and the trajectory is optimized using NMPC, which is fast solved by the C/GMRES algorithm.

5.1. Obstacle Avoidance on Straight Lane

As shown in Figure 8, the host vehicle travels on a straight lane with an initial speed of 10 m/s. The maximum speed limit of the road is 15 m/s and each lane width is 4 m. An obstacle vehicle is 30 m ahead of the host vehicle at speed of 5 m/s. The trajectory of the host vehicle and the obstacle vehicle is shown in Figure 8a. The speed profile of two vehicles and the steering wheel angle of the host vehicle is shown in Figure 8b,c. In this driving scenario, the host vehicle executes two consecutive lane-changing operations smoothly and quickly to avoid dynamic obstacles and keep the speed under the maximum speed.

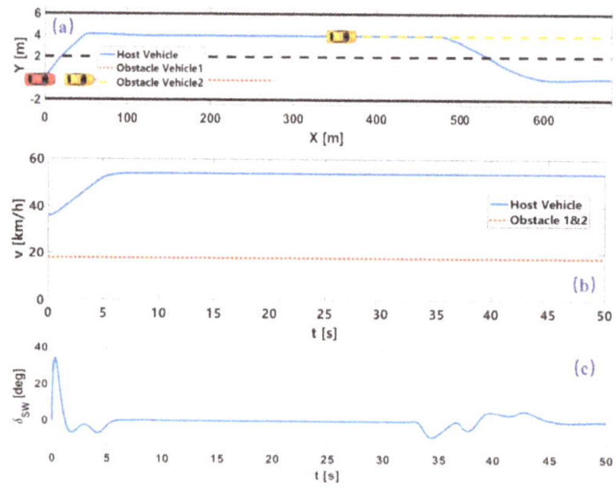

Figure 8. Simulation results of obstacle avoidance on a straight lane. Here, (**a**) is vehicle trajectory. (**b**) is vehicle velocity. (**c**) is steering wheel angle.

The processor of the computer is Intel (R) Core (TM) i7-6700hq CPU@ 2.60GHZ. The time step size Δt for simulation is 0.05s and the nonlinear model predictive control problem using the C/GMRES algorithm, which is also compared with fmincon function in MATLAB. As shown in Figure 9, the calculating time of solving is much less than the time step size Δt, and also much less than the calculating time of the fmincon function in MATLAB which is more than 10 min. It shows that the local path planning module solved by C/GMRES can better meet the requirements of solving speed.

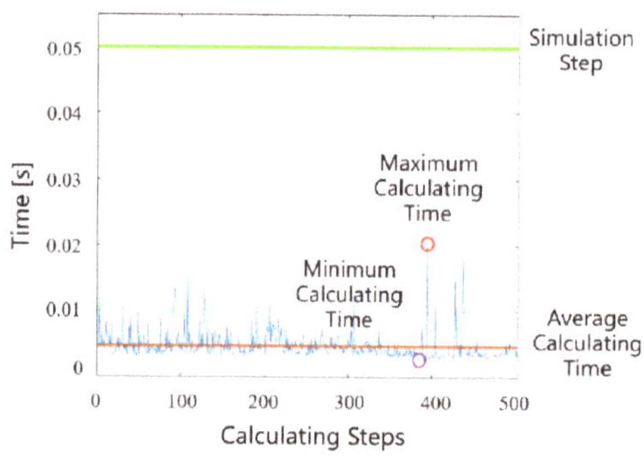

Figure 9. Calculating time.

5.2. Obstacle Avoidance on Winding Lane

In the scenario of a winding road, the center line of the initial lane is $y = 10^{-6}x^3 + 10^{-5}x^2$ The obstacle vehicle travels at a speed of 5 m/s at 50 m in front of the host vehicle. The trajectory and speed profile of the two vehicles and the steering wheel angle of the host vehicle is shown in Figure 10. The host vehicle chooses to accelerate to overtake the preceding vehicle and avoid the obstacle.

5.3. Lane-Changing Obstacle Avoidance

A much more complex scenario is verified in the third simulation, in which the intention of the obstacle changed. The trajectory of the obstacle vehicle is selected from the open data set, which executes a lane change to the left lane. The trajectory and speed profile of the two vehicles and the steering wheel angle of the host vehicle is shown in Figure 11.

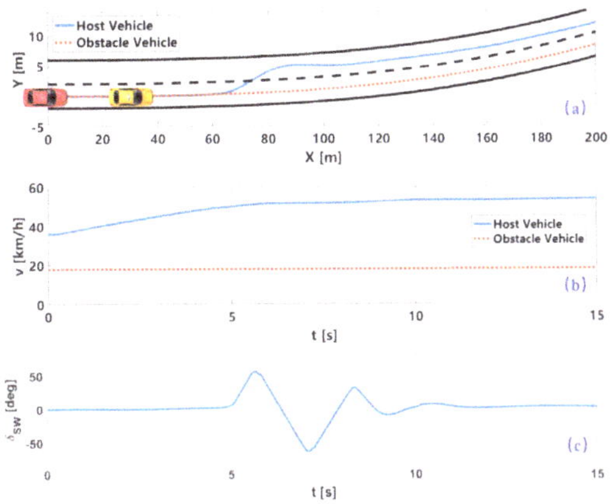

Figure 10. Simulation results of obstacle avoidance on winding lane. Here, (**a**) is vehicle trajectory. (**b**) is vehicle velocity. (**c**) is steering wheel angle.

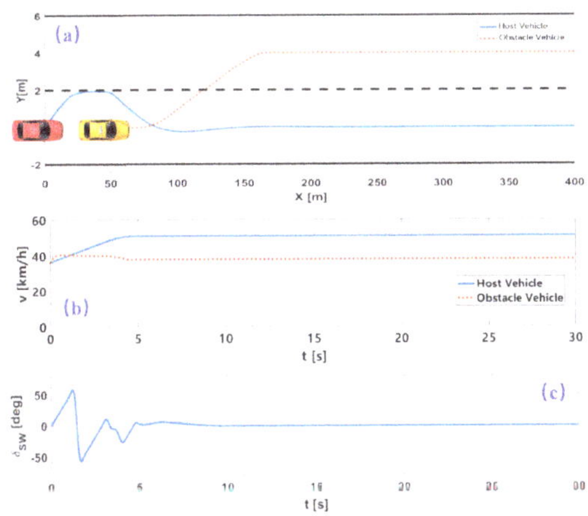

Figure 11. Simulation results of lane-changing obstacle avoidance. Here, (**a**) is vehicle trajectory. (**b**) is vehicle velocity. (**c**) is steering wheel angle.

As shown in Figure 11a, at first, the obstacle vehicle chooses lane-keeping before changing lanes to the left lane. In this process, the host vehicle tries to change lanes. After the obstacle vehicle decides to change to the target lane of the host vehicle, the host vehicle decides to change back to its original lane. From the above simulation results, the proposed motion planning method considers safety, computational efficiency, and comfort simultaneously and obtains good system performance.

6. Conclusions

This paper proposes a computationally efficient motion planning method for autonomous vehicles, which considers dynamic obstacle avoidance and traffic interaction. The decision process for complex behavior is reasonably simplified in both time and space span. Different points located on unequally divided road segments and lane centerlines are connected to represent behavior. And C/GMRES algorithm is used to accelerate the calculation of the NMPC problem in the trajectory generation module. The trajectories of other traffic participants are more accurately predicted with known intention and vehicle models, which enables the movement to be more reasonably planned. Finally, three groups of simulation experiments are carried out to verify the rationality and superiority of the algorithm.

In future works, the interactive intention prediction will be considered in the intention predictive layer, which can extend the motion planning method from reacting adaptively to predicting adaptively. By considering the interactions between the ego vehicle and surrounding drivers socially via implicit and/or explicit communications, the behavior of the autonomous vehicle can be more human-like and facilitate safety performance under complex and dynamic environments [27].

Author Contributions: Conceptualization, B.G. and H.C.; methodology, J.L. and Y.Z.; validation, J.L. and J.W.; investigation, X.N.; writing—original draft preparation, J.L. and Y.Z.; writing—review and editing, Y.Z. and J.W.; supervision, H.C.; funding acquisition, B.G. All authors have read and agreed to the published version of the manuscript.

Funding: This research was supported in part by the International Technology Cooperation Program of Science and Technology Commission of Shanghai Municipality (21160710600), in part by the Shanghai Pujiang Program (21PJD075), in part by the China Automobile Industry Innovation and Development Joint Fund (U1864206).

Institutional Review Board Statement: Not applicable.

Informed Consent Statement: Not applicable

Data Availability Statement: The data presented in this study are available on request from the corresponding author. The data are not publicly available due to privacy reason.

Conflicts of Interest: The authors declare no conflict of interest.

References

1. Paden, B.; Čáp, M.; Yong, S.Z.; Yershov, D.; Frazzoli, E. A survey of motion planning and control techniques for self-driving urban vehicles. *IEEE Trans. Intell. Veh.* **2016**, *1*, 33–55. [CrossRef]
2. González, D.; Pérez, J.; Milanés, V.; Nashashibi, F. A review of motion planning techniques for automated vehicles. *IEEE Trans. Intell. Transp. Syst.* **2015**, *17*, 1135–1145. [CrossRef]
3. Schwarting, W.; Alonso-Mora, J.; Rus, D. Planning and decision-making for autonomous vehicles. *Annu. Rev. Control. Robot. Auton. Syst.* **2018**, *1*, 187–210. [CrossRef]
4. Nilsson, J.; Sjöberg, J. Strategic decision making for automated driving on two-lane, one way roads using model predictive control. In Proceedings of the 2013 IEEE Intelligent Vehicles Symposium (IV), Gold Coast, Australia, 23–26 June 2013; pp. 1253–1258.
5. Xie, G.; Gao, H.; Qian, L.; Huang, B.; Li, K.; Wang, J. Vehicle trajectory prediction by integrating physics-and maneuver-based approaches using interactive multiple models. *IEEE Trans. Ind. Electron.* **2017**, *65*, 5999–6008. [CrossRef]
6. Houenou, A.; Bonnifait, P.; Cherfaoui, V.; Yao, W. Vehicle trajectory prediction based on motion model and maneuver recognition. In Proceedings of the 2013 IEEE/RSJ International Conference on Intelligent Robots and Systems, Tokyo, Japan, 3–7 November 2013; pp. 4363–4369.
7. Liu, W.; Kim, S.W.; Pendleton, S.; Ang, M.H. Situation-aware decision making for autonomous driving on urban road using online POMDP. In Proceedings of the 2015 IEEE Intelligent Vehicles Symposium (IV), Seoul, Korea, 28 June–1 July 2015; pp. 1126–1133.
8. Song, W.; Xiong, G.; Chen, H. Intention-aware autonomous driving decision-making in an uncontrolled intersection. *Math. Probl. Eng.* **2016**, *2016*, 1–15. [CrossRef]
9. Lim, W.; Lee, S.; Sunwoo, M.; Jo, K. Hierarchical trajectory planning of an autonomous car based on the integration of a sampling and an optimization method. *IEEE Trans. Intell. Transp. Syst.* **2018**, *19*, 613–626. [CrossRef]
10. Wei, J.; Snider, J.M.; Gu, T.; Dolan, J.M.; Litkouhi, B. A behavioral planning framework for autonomous driving. In Proceedings of the 2014 IEEE Intelligent Vehicles Symposium Proceedings, Dearborn, MI, USA, 8–11 June 2014; pp. 458–464.
11. Schmerling, E.; Janson, L.; Pavone, M. Optimal sampling-based motion planning under differential constraints: The driftless case. In Proceedings of the 2015 IEEE International Conference on Robotics and Automation (ICRA), Seattle, WA, USA, 26–30 May 2015; pp. 2368–2375.
12. Li, Y.; Littlefield, Z.; Bekris, K.E. Sparse methods for efficient asymptotically optimal kinodynamic planning. In *Algorithmic Foundations of Robotics XI*; Springer: Berlin/Heidelberg, Germany, 2015; pp. 263–282.
13. Broggi, A.; Medici, P.; Zani, P.; Coati, A.; Panciroli, M. Autonomous vehicles control in the VisLab intercontinental autonomous challenge. *Annu. Rev. Control.* **2012**, *36*, 161–171. [CrossRef]
14. Dolgov, D.; Thrun, S.; Montemerlo, M.; Diebel, J. Path planning for autonomous vehicles in unknown semi-structured environments. *Int. J. Robot. Res.* **2010**, *29*, 485–501. [CrossRef]
15. Ziegler, J.; Bender, P.; Schreiber, M.; Lategahn, H.; Strauss, T.; Stiller, C.; Dang, T.; Franke, U.; Appenrodt, N.; Keller, C.G.; et al. Making bertha drive?aan autonomous journey on a historic route. *IEEE Intell. Transp. Syst. Mag.* **2014**, *6*, 8–20. [CrossRef]

16. Rasekhipour, Y.; Fadakar, I.; Khajepour, A. Autonomous driving motion planning with obstacles prioritization using lexicographic optimization. *Control. Eng. Pract.* **2018**, *77*, 235–246. [CrossRef]
17. Gil, A.F.A.; Ruíz, A.M.; Espinosa, J.J. Nonlinear Model Predictive Control of a Passenger Vehicle for Automated Lane Changes. *Revista Ingeniería Electrónica Automática y Comunicaciones* **2019**, *38*, 56. ISSN: 1815-5928.
18. Nolte, M.; Rose, M.; Stolte, T.; Maurer, M. Model predictive control based trajectory generation for autonomous vehicles? An architectural approach. In Proceedings of the 2017 IEEE Intelligent Vehicles Symposium (IV), Los Angeles, CA, USA, 11–14 June 2017; pp. 798–805.
19. Arikere, A.; Yang, D.; Klomp, M.; Lidberg, M. Integrated evasive manoeuvre assist for collision mitigation with oncoming vehicles. *Veh. Syst. Dyn.* **2018**, *56*, 1577–1603. [CrossRef]
20. Zhang, Y.; Gao, B.; Guo, L.; Guo, H.; Cui, M. A Novel Trajectory Planning Method for Automated Vehicles Under Parameter Decision Framework. *IEEE Access* **2019**, *7*, 88264–88274. [CrossRef]
21. Tajeddin, S.; Vajedi, M.; Azad, N.L. A Newton/GMRES approach to predictive ecological adaptive cruise control of a plug-in hybrid electric vehicle in car-following scenarios. *IFAC-PapersOnLine* **2016**, *49*, 59–65. [CrossRef]
22. Shen, C.; Buckham, B.; Shi, Y. Modified C/GMRES algorithm for fast nonlinear model predictive tracking control of AUVs. *IEEE Trans. Control. Syst. Technol.* **2016**, *25*, 1896–1904. [CrossRef]
23. Deng, H.; Ohtsuka, T. A parallel Newton-type method for nonlinear model predictive control. *Automatica* **2019**, *109*, 108560. [CrossRef]
24. Kim, J.; Jo, K.; Lim, W.; Lee, M.; Sunwoo, M. Curvilinear-coordinate-based object and situation assessment for highly automated vehicles. *IEEE Trans. Intell. Transp. Syst.* **2015**, *16*, 1559–1575. [CrossRef]
25. Wang, H.; Kearney, J.; Atkinson, K. Robust and efficient computation of the closest point on a spline curve. In Proceedings of the 5th International Conference on Curves and Surfaces, Saint-Malo, France, 27 June–3 July 2002; pp. 397–406.
26. Ohtsuka, T. A continuation/GMRES method for fast computation of nonlinear receding horizon control. *Automatica* **2004**, *40*, 563–574. [CrossRef]
27. Wenshuo, W.; Letian, W.; Chengyuan, Z.; Changliu, L.; Lijun, S. Social Interactions for Autonomous Driving: A Review and Perspective. *arXiv* **2022**, arXiv:2208.07541.

sensors

MDPI

Article

Driver Assisted Lane Keeping with Conflict Management Using Robust Sliding Mode Controller

Gabriele Perozzi [1], Mohamed Radjeb Oudainia [1], Chouki Sentouh [1], Jean-Christophe Popieul [1] and Jagat Jyoti Rath [2,*]

1 LAMIH Laboratory UMR CNRS 8201, Université Polytechnique Hauts-de-France, 59300 Valenciennes, France
2 Department of Mechanical and Aero-Space Engineering, Institute of Infrastructure Technology Research and Management (IITRAM), Ahmedabad 380026, India
* Correspondence: jagat.rath@iitram.ac.in

Abstract: Lane-keeping assistance design for road vehicles is a multi-objective design problem that needs to simultaneously maintain lane tracking, ensure driver comfort, provide vehicle stability, and minimize conflict between the driver and the autonomous controller. In this work, a cooperative control strategy is proposed for lane-keeping keeping by integrating driving monitoring, variable level of assistance allocation, and human-in-the-loop control. In the first stage, a time-varying physical driver loading pattern is identified based on a relationship between lateral acceleration, road curvature, and the measured maximum driver torque. Together with the monitored driver state that indicates driver mental loading, an adaptive driver activity function is then formulated that replicates the levels of assistance required for the driver in the next stage. To smoothly transition authority between various modes (from manual to autonomous and vice versa) based on the generated levels of assistance, a novel higher-order sliding mode controller is proposed and closed-loop stability is established. Further, a novel sharing parameter (which is proportional to the torques coming from the driver and from the autonomous controller) is used to minimize the conflict. Experimental results on the SHERPA high-fidelity vehicle simulator show the real-time implementation feasibility. Extensive experimental results provided on the Satory test track show improvement in cooperative driving quality by 9.4%, reduction in steering workload by 86.13%, and reduced conflict by 65.38% when compared with the existing design (no sharing parameter). These results on the cooperative performance highlight the significance of the proposed controller for various road transportation challenges.

Keywords: human-machine shared control; lane keeping assistance; higher order sliding mode; conflict minimization; ADAS; driver assist system

Citation: Perozzi, G.; Oudainia, M.R.; Sentouh, C.; Popieul, J.-C.; Rath, J.J. Driver Assisted Lane Keeping with Conflict Management Using Robust Sliding Mode Controller. *Sensors* **2023**, *23*, 4. https://doi.org/10.3390/s23010004

Academic Editors: Yafei Wang, Chao Huang, Peng Hang, Zhiqiang Zuo and Bo Leng

Received: 17 November 2022
Revised: 13 December 2022
Accepted: 16 December 2022
Published: 20 December 2022

1. Introduction

Advanced driver assist systems (ADASs; acronyms of this manuscript are defined in the Acronyms section) such as lane keeping assist (LKA), adaptive cruise control (ACC), and collision avoidance (CA) systems have been widely employed in commercial vehicles. These systems greatly reduce the workload of human drivers and reduce the risk of accidents, and crashes by warning or supporting the driver for particular maneuvers [1]. The ADASs developed for semi-autonomous driving scenarios can be categorized into human-guided, human-supervised, and human-assisted architectures [2]. In recent works, it has been established that driver-in-the-loop (DiL) human-assisted ADAS architectures can be employed to address various human–machine interaction (HMI) challenges inclusive of authority allocation [3], the transition of authority [4], conflict management [5], and human driver workload reduction and skill enhancement [6]. Such cooperative driving architectures have been explored for adaptive cruise control, collision avoidance systems, and lane departure/keeping systems among others [7,8]. To design cooperative control architectures for ADAS, DiL architectures are typically formulated by integrating driver attributes such as workload, experience, and skill in the control design. For effective action which reflects such attributes, various driver models based on neuromuscular dynamics [9], data-driven [10], hand impedance [11], and vision/preview have been developed [5]. In this work, the avenue of cooperative control for lane-keeping assistance (LKA) systems considering the steering input (angle or torque) as a control signal is explored with a focus on HMI management and vehicle positioning error minimization.

1.1. State of the Art

Many works can be found in the literature dealing with the design of controllers for trajectory following [12]. Among of all the robust controllers, the sliding mode law is world-wide recognized as one of the most effective to reject external matched perturbations [13], so they can be used to reject perturbations that affect road vehicles. The system disturbances and parameter uncertainties introduced by human–machine cooperation driving are also inevitable. Ref. [14] proposed a control method to solve the above problems. Optimization algorithms have also been used to reduce the computational cost of implementing the control law in real-time applications [15]. Active fault-tolerant controllers have been largely used to increase plant availability and reduce the risk of safety hazards, preventing simple faults from developing into serious failure [16,17]. The last decade witnessed a great development of automated driving vehicles and vehicle intelligence. The significant increment of machine intelligence poses a new challenge to the community, which is the collaboration between human drivers and vehicle autonomy. In [18], a literature review was conducted and perspectives on the human behaviors and cognition (HBC) for ADVs toward human-autonomy (H-A) collaboration were proposed.

Various cooperative control architectures have been proposed in [5,7,8,19,20] based on DiL designs. In [21], a driver model using a weighting process of visual guidance from the road ahead and haptic guidance from a steering system for a lane-following task were proposed. In [3,22–24], haptic feedback from the steering wheel was used to ensure both driver and the autonomous controller participated in the driving action. In [25], an extended shared steering control system with an authority adaptive allocation model was proposed to improve the reliability of the shared steering control system, and weaken the influence of uncertain driver behavior on driving safety. Ref. [26] presented a shared control framework based on handling inverse dynamics and driving intention for lane changing, in particular, the influence of the driver's lane-changing start point and end point is considered in the design of the shared controller. In [6], a cooperative control approach for lane keeping based on H_2 preview control was proposed by incorporating a neuro-muscular driver model. Similarly, in [20], a haptic shared control between driver and e-copilot considered the use of driver torque as haptic feedback to design T-S fuzzy controllers for lane keeping. In [19], for varying driver steering characteristics such as delays, and preview time, a DiL gain-scheduling H_∞ robust shared controller was proposed. These approaches typically validated the cooperative performance of the DiL design for lane-keeping tasks in presence of driver parameter uncertainty and environmental disturbances such as crosswinds, and road curvature. Although efficient lane-keeping performance under various driving conditions was validated, issues of conflict between human driver and autonomous controller, driver workload management and performance enhancement were not explicitly addressed.

Driver workload typically characterizes the driving action required by the human driver to perform a typical task. Based on monitored cognitive states (mental workload) and physical driving effort (physical workload) applied by the driver, the workload can be categorized into under-load, normal and over-load regions [5,27]. The mental workload of the driver reflects the state of involvement of the driver in the driving task. Typically, driver state of drowsiness [5,28], the intention of driving action [28], and meticulous steering action [29] are employed as indicators of the mental workload. The physical workload of the driver can be determined by monitoring the driver torque/steer input applied, and the steer reversal rate. The objective of a cooperative LKA strategy is then adapting the driver activity in terms of workload into the controller design for effective management of HMI and keeping vehicles on the lane. In [30], an optimal modulation policy was designed with a cost function, then a nonlinear stochastic model predictive approach was used to solve the cost function subjected to probabilistic uncertainties in human's biomechanics. In [27], the relationship between driver workload and level of assistance required was explored for the design of an LKA controller to improve driver performance. Takagi-Sugeno (T-S) models [31,32] used driver activity functions considering driver state, torque, and intention, which replicate the level of assistance required during a typical task [27].

The conflict between the human driver and the autonomous controller typically occurs when both agents have different actions for the same driving task. Such scenarios arise during the transition of authority between the agents, sudden maneuvers executed by driver/automation which is not predicted by the other agent, and during extreme maneuvers i.e., sharp curve negotiation. In [4,6], based on cooperative status detection, a conflict-free smooth transition of authority between human driver and autonomous controller was proposed. Similarly, in [23], conflict mitigation by adapting the parameters of the controller with respect to individual drivers was proposed. Extending the work of [31], a cooperative control approach employing T-S models was proposed in [5] to perform lane keeping and conflict minimization simultaneously. In [33], a haptic control architecture was developed for the smooth transition of control authority with adaptation to driver cognitive workload. In the works of [6,19,20,31,32], the controllers designed were based on the linear bicycle model which did not account for varying tire friction forces. The works in [6,19] assumed constant longitudinal speed in the design of lane-keeping controllers. Further, conflicts between the driver and the automated driving system were not explicitly addressed in [19,32]. In [5,31,33], by the design of shared control dependent on driver attributes, the issue of conflict between the driver and automated system was addressed for variable longitudinal speeds and fixed longitudinal speeds. However, these works were analyzed for the linear bicycle model that did not consider the aspect of saturated tire friction forces during extreme maneuvers.

1.2. Proposed Methodology

To account for tire-force non-linearities and environmental disturbances, management of HMI between human drivers and autonomous controller with respect to driver workload, and conflict management, a robust cooperative control approach is proposed in this work. Based on the non-linear representation of tire friction dynamics [34] integrated with a human driver model developed using visual cues [5], a DiL design is formulated. The HMI between the human driver and the driver torque is then developed based on adaptation to driver workload and subsequent driver performance. For adaptation to driver performance, a non-linear representation of driver activity based on physical and cognitive workload is formulated. For quantifying adaptive physical workload, a rule-based logic is used to explore the relationship between lateral acceleration, predicted road curvature, and maximum driver torque. Based on the developed DiL model dynamics, a novel robust nonlinear feedback controller based on adaptive higher order sliding mode (HOSM) [35,36] is developed for the system. The conflict is managed by the introduction of a sharing parameter, which is a function for driver and assistance torques in the input-dependent sliding surface. The developed feedback control is then modulated using the non-linear function developed on the relationship of driver performance-level of assistance required, for effective HMI management. The closed-loop stability of the time-varying system dynamics involving the non-linear modulating function, DiL dynamics, and environmental disturbances is then established.

1.3. Contribution

The main contributions of this work are:

- The introduction of a shared control parameter into the control design to minimize conflict between the human driver and automated driving system.

- The design of a novel higher-order sliding mode control algorithm with linear and nonlinear terms.
- The addressing of multiple objectives of lane position error reduction, enhancement of driver satisfaction, and conflict management.

The manuscript is organized as follows. Section 2 introduces the driver–vehicle–lane model. Section 3 focuses on the design of the proposed controller. Extensive discussions about the performance of the proposed approach with regard to lane position error reduction, driver satisfaction, and the influence of the conflict parameter are provided in Section 4.

2. Problem Formulation: Driver Adapted Lane Keeping

The time-varying dynamics governing a DiL vehicle model in the presence of environmental disturbances for lateral control and the problem of designing a closed-loop controller to manage the HMI between a driver and an autonomous controller are discussed in this section. The symbols of this manuscript are defined in the Nomenclature section.

2.1. DiL Modeling: Vehicle-Road-Driver Dynamics

The DiL model development is carried out by integrating the vehicle's lateral and yaw motion dynamics with the steering column dynamics, the lane tracking dynamics, and a linear model of the human driver's torque. The governing dynamics for the lateral motion of the vehicle under assumptions of negligible influence of the longitudinal dynamics can be efficiently represented using the non-linear bicycle dynamic model [1,37] as in Equation (2).

$$Mv_x\dot{\beta} = F_{yr} + F_{yf}cos(\delta_f) - Mv_x\dot{\psi}_v + F_w \tag{1}$$

$$I_z\ddot{\psi}_v = l_fF_{yf}cos(\delta_f) - l_rF_{yr} + M_w \tag{2}$$

where β is the side slip angle, δ is the steering angle, ψ is the heading angle, F_{yf}, F_{yr} are the front and rear friction forces, F_w is the crosswinds force, and v_x is the longitudinal velocity. To represent the tire–road friction conditions, several linear, adaptive, uncertain, and nonlinear models like the Brush-Tire (BT) friction model, LuGre friction model among others are employed [38]. Although the nonlinear models represent the dynamic characteristics of tire–road friction, these models are not easily applicable in control approaches due to their highly complex behavior and dynamics. The linear uncertain friction model [39] has been employed in this work for controller development. The lateral tire friction forces and the self-align torque of the steering wheel are then given as in Equations (3) and (4).

$$F_{yi} = 2C_{pi}\alpha_i + \Delta F_i \tag{3}$$

$$T_s = \frac{K_pt_pF_{yf}}{R_s} \tag{4}$$

with α_f, α_r denoting the front and rear slip angles, T_s denoting the self-aligning torque, and ΔF_i denoting the lumped uncertain part of the tire friction forces indicative of the effects of changing road conditions, tire pressure variations, saturation, etc., which can be modeled using any of the above-discussed dynamic friction models. The variable $K_p \in (0,1]$ is a ratio denoting the level of assistance from the active steering system. In the absence of any active steering support, the value of $K_p = 1$. The front and rear slip angles under small angle assumptions are given as in Equation (6).

$$\alpha_f = \delta_f - \frac{\beta v_x + l_f\dot{\psi}_v}{v_x} \tag{5}$$

$$\alpha_r = \frac{\beta v_x - l_r\dot{\psi}_v}{v_x} \tag{6}$$

Under the small angle assumptions, the above non-linear bicycle model dynamics appropriately represent the vehicle motion for low curvature roads and have been widely employed for shared lateral control [5,7].

The vehicle's lane tracking performance can be modeled using two error variables, y_l and Ψ_l, which indicate the lateral deviation error and the orientation error of the vehicle with respect to the lane center-line at a specified look-ahead distance as shown in Figure 1.

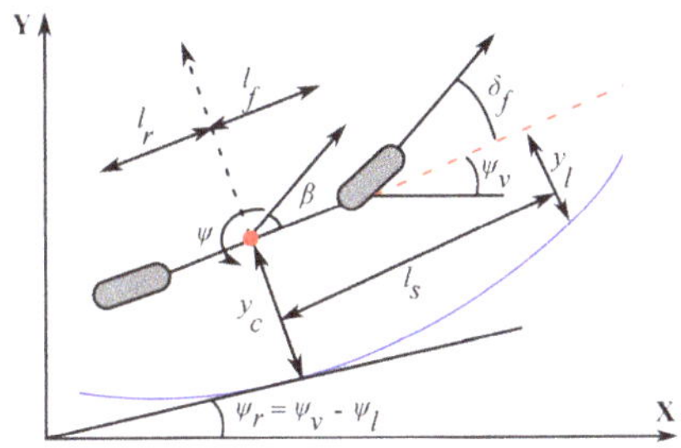

Figure 1. The nonlinear bicycle model of vehicle.

These lane errors are readily obtained using vision-based sensors from the vehicle perception unit. The dynamics of these error variables are given, as [5], in Equation (8).

$$\dot{y}_l = \beta v_x + l_s \dot{\psi}_v + \Psi_l v_x; \tag{7}$$

$$\dot{\Psi}_l = \dot{\psi}_v - \rho_r v_x \tag{8}$$

with y_l, Ψ_l representing the lateral offset error and the heading error respectively. With the road-vehicle dynamics considered, the interaction between the human driver and the vehicle is then modeled by considering the steering-column dynamics with only basic assist provided [5,7] as in Equation (9).

$$I_s \ddot{\delta}_d = T_d + T_a - T_s - B_u \dot{\delta}_d \tag{9}$$

where T_d, T_a represent the driver and the automation torques, respectively. Integrating the dynamics (8) and (9), an autonomous controller can be designed to generate the assistance torque T_a which can maintain the vehicle on the lane. Further, the consideration of the steering column dynamics also helps in informing the human driver of the external road conditions directly.

2.2. HMI Management: Driver Workload-Level of Assistance

Driver-adaptive LKA systems intend to provide assistance to human drivers for difficult and adverse scenarios [5,40–42]. Specifically, adaptation techniques are designed such the physical and mental workload of drivers during driving can be easily managed. Using measured vehicle responses such as steering torque, steering wheel reversal rate, and jerk among others, the physical workload of a driver is quantified [5,27,42]. Similarly, based on measured driver responses such as gaze monitoring, drowsiness, and intention to perform a maneuver, the mental workload of a driver can be quantified [5,7,9]. Integrating both these indicators via a nonlinear mapping and relating them to driver performance, various adaptive functions have been proposed by our research group for shared lane-keeping tasks [5,28,42]. On similar lines, we consider the use of normalized driver torque and driver distraction levels as indicators of the driver's physical and cognitive workloads, respectively. The entire procedure is carried out in three steps as shown below:

- *Identification of driver workload*: The measured driver torque at the steering wheel is typically dependent on many factors such as road curvature, lateral acceleration, the preview time, and the far point distance, and dynamics of the human arm among others. In this work, the adaptive driver torque T_{dm} for various drivers/driving scenarios is computed using a simple rule-based logic with the inputs being lateral acceleration and predicted road curvature [43]. With the increase in lateral acceleration and road curvature, the value of the T_{dm} increases, to show more physical workload of the driver. Mathematically, the normalized maximum driver torque is represented in Equation (10).

$$T_{dn} = |T_d / T_{dm}| \tag{10}$$

Similarly, the mental workload is accounted for by the driver state ($DS \in [0,1]$) which categorizes the driver's involvement into different levels such as attentive, sleepy, drowsy, and distracted. With the increase in DS the driver is more involved in the driving task and vice-versa. In the case when $DS = 0$, the driver is completely distracted, and when $DS = 1$, the driver is actively involved in the driving task. For practical purposes, the DS is obtained from the driver monitoring unit (DMU) installed in vehicles comprising of a vision system to monitor driver activity [44]. It is of note that, although different states of driver are monitored, generally the output of the DMU is binary indicating an active driver or a distracted driver [28].

- *Mapping driver workload to activity*: In the context of driver workload, effective driver performance decreases with an increase in workload levels. Similarly, for low activity (corresponding low workload) level, also the performance of the driver is low, as the driver is not significantly involved in the driving task. Analytically, this relationship is expressed as in Equation (11).

$$\gamma = 1 - e^{(\sigma_1 T_{dN})^{\sigma_2} DS^{\sigma_3}} \tag{11}$$

where $\gamma \in [0,1]$ indicates driver activity, $\sigma_1 = 2$, $\sigma_2 = 3$, and $\sigma_3 = 3$ selected appropriately to consider the degree of influence of the physical and cognitive components on the driver activity. This relationship is presented graphically in form of a U-shaped function in [27].

- *Activity-based level of assistance generation*: The level of assistance (LOA) required to complete a driving task can be determined similarly to [27], by using the inverse-U relationship between driver performance and LOA. Considering the objective of providing high assistance to the driver during under-load and over-load (i.e., low activity) regions, an analytical mapping for driver performance-LOA is defined as in Equation (12).

$$\mu(\gamma) = \frac{1}{1 + |\frac{\gamma - p_3}{p_1}|^{2p_2}} + \mu_{min} \tag{12}$$

The time-varying parameter $\mu(\gamma) \in [\mu_{min}, 1]$ represents a modulation factor that relates the driver workload-based performance with the LOA for task completion. The parameters $p_1 = 0.355$, $p_2 = -2$, $p_3 = 0.5$ are chosen to replicate the U-shaped relationship as discussed in [27] and shown in Figure 2. A minimum assistance level of $\mu_{min} = 0.2$ is used to consider the influence of sensor noise, drift, etc.

The computed level of assistance function can be then used to modulate the assistance torque T_a and thus adapt the autonomous control action to the driver as in Equation (13).

$$T_a = \mu(\gamma) T_{fb} \tag{13}$$

where T_{fb} is a robust feedback control torque to be designed. Employing the modulated assistance torque, the HMI between the driver and the autonomous controller can be effectively managed for completing a specific driving task.

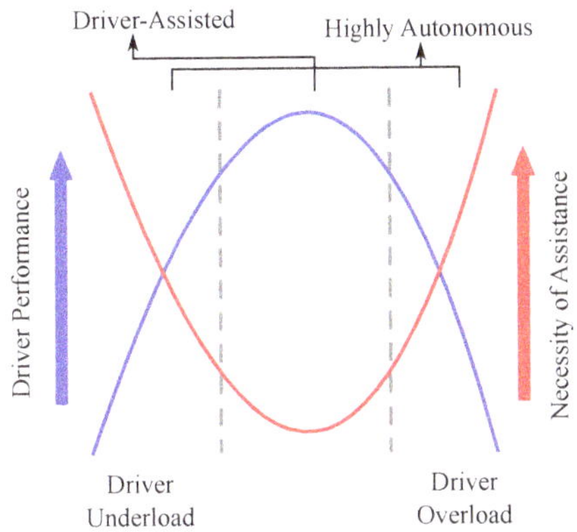

Figure 2. The driver workload and corresponding level of assistance required.

3. Robust DiL Lane Keeping Control: A HOSM approach

The shared control between the human driver and an LKA controller typically focuses on tracking the desired reference while improving the driver comfort [7,20,42].

3.1. Control Oriented DiL Modeling

For DiL tasks, we incorporate the influence of driver effort by using a two-point visual driver torque model [5] for developing the control specific model as in Equation (14).

$$T_d = K_c\theta_{near} + K_a\theta_{far} \tag{14}$$

with $\theta_{near}, \theta_{far}$ representing the near and far visual points of the driver along a road curvature. Based on information of these angles, the driver generates anticipatory action and compensatory action corresponding to the near and far angles respectively. Subsequently, he/she predicts the future road and generates the anticipated steering action before entering the curve based on the far visual angle. The compensatory behavior of the driver is emphasized for lane-keeping aspects. This driver behavior is represented using the anticipatory and compensatory gains K_a and K_c respectively as shown in (14). For further details, please refer to [5].

Integrating the above dynamics in Equations (2) and (14), a DiL lane-keeping model of the following form can be formulated in Equation (15).

$$\dot{x}(t) = A(t)x(t) + B(t)T_a(t) + E(t)\omega(t) \tag{15}$$

with the states as $x = \begin{bmatrix} x_1 & x_2 & x_3 & x_4 & x_5 & x_6 \end{bmatrix}^T = \begin{bmatrix} \beta & \dot{\psi}_v & y_l & \Psi_l & \delta_d & \dot{\delta}_d \end{bmatrix}^T$. The system matrices are given as in matrices (17).

$$A(t) = \begin{bmatrix} a_{11} & a_{12} & 0 & 0 & a_{15} & 0 \\ a_{21} & a_{22} & 0 & 0 & a_{25} & 0 \\ 0 & 1 & 0 & 0 & 0 & 0 \\ v_x & l_s & v_x & 0 & 0 & 0 \\ 0 & 0 & 0 & 0 & 0 & 1 \\ a_{61} & a_{62} & a_{63} & a_{64} & a_{65} & a_{66} \end{bmatrix}, B = \begin{bmatrix} 0 \\ 0 \\ 0 \\ 0 \\ 0 \\ b_1 \end{bmatrix} \tag{16}$$

$$E(t) = \begin{bmatrix} e_1 & e_2 & 0 & 0 & 0 & 0 \\ 0 & 0 & 0 & -v_x & 0 & 0 \\ e_1 & e_2 & 0 & 0 & 0 & e_3 \\ e_4 & e_5 & 0 & 0 & 0 & 0 \end{bmatrix}^T, w(t) = \begin{bmatrix} F_w \\ \rho_r \\ \Delta F_{yf} \\ \Delta F_{yr} \end{bmatrix}^T \tag{17}$$

with $a_{11} = -2(C_f + C_r)/Mv_x$, $a_{12} = 2(l_r C_r - l_f C_f)/Mv_x^2$, $a_{15} = 2C_f/Mv_x R_s$, $a_{21} = 2(l_r C_r - l_f C_f)/I_z$, $a_{22} = -2(l_r^2 C_r + l_f^2 C_f)/I_z v_x$, $a_{25} = 2C_f l_f/I_z R_s$, $a_{61} = (2C_f \eta_t/(I_s R_s)) + K_c \tau_a^2 a_{21}$, $a_{62} = (2C_f l_f \eta_t/(I_s R_s v_x)) + K_c(\tau_a + \tau_a^2 a_{22})$, $a_{63} = K_a/I_s$, $a_{64} = K_a/(I_s v_x \tau_p)$, $a_{65} = (-2C_f \eta_t/(I_s R_s^2)) + K_c \tau_a^2 a_{25}$, $a_{66} = -B_s/I_s$, $b_1 = 1/I_s$, $e_1 = 1/Mv_x$, $e_2 = l_w/I_z$, $e_3 = l_f/I_z$, $e_4 = -l_r/I_z$, and $e_5 = -K_p \eta_t/I_s R_s$.

The autonomous assistance torque T_a for completing the driving task in the presence of disturbances ω and the uncertainties Δ can be now designed. Integrating the assistance modulation factor developed earlier, the DiL model used for controller design can be expressed as in Equation (18).

$$\dot{x} = A(t)x(t) + B_1(t)T_{fb}(t) + E(t)\omega(t) \tag{18}$$

with $B_1 = B\mu(\gamma)$ and T_{fb} as the control torque to be designed for stabilizing the DiL system.

3.2. Control Objectives for LKA

The control objectives for the above DiL lane-keeping task are formulated as:

- Minimization of lane tracking errors: The lane tracking errors as given in Equation (8) comprise the errors lateral deviation and the heading angle. To quantify the lane error at a look-ahead distance, the parameter e_l is defined as in Equation (19).

$$e_l = y_l + l_s \Psi_l \tag{19}$$

The control objective is then to ensure that the front wheels of the vehicle are simultaneously located in strip ($\pm d = 1.5$ m) along the lane center line. In other words, the following condition in Equation (20).

$$|e_l| \leq \frac{2d - w_r}{2} \tag{20}$$

where w_r denotes the width of the vehicle.

- Improvement of driver comfort: The comfort of the driver while navigating the road can be understood as a measure of the vibrations or oscillations at the steering wheel. As such, the steering rate $\dot{\delta}_d$ or the lateral acceleration can be used as a measure to quantify the driver comfort [45].
- Conflict Minimization: The mismatch of control actions between the human driver and the autonomous controller categorized as conflict, must be minimized for having a good shared control performance [5]. This can be achieved by passing over the authority to the human driver. Accordingly, the following fictional state is introduced to achieve the above action in Equation (21).

$$\dot{x}_{cf} = T_d^s - \lambda_c T_a \tag{21}$$

where λ_c is any positive parameter reflecting the level of sharing, and T_d^s represents the driver torque measured at the steering wheel. In case of conflict, the value of $\dot{x}_{cf} \longrightarrow 0$. In such a case, it can be deduced that $\lambda_c T_a \approx T_d^s$. Hence, by the appropriate design of the parameter λ_c, the influence of the assistance torque can be reduced.

For the above control objectives, we now propose a robust HOSM controller for the DiL dynamics in Equation (18) to design the torque T_{fb} and T_a subsequently.

3.3. Robust HOSM Controller

Integrating the above control objectives, a linear error surface to be regulated can be defined as in Equation (22).

$$\sigma_c = k_1 e_l + k_2 \dot{e}_l + k_3 \dot{\delta}_d + k_4 x_{cf} \tag{22}$$

for the gains $k_i > 0$, $i = 1 \ldots 4$ designed to ensure convergence of the error surface. To stabilize the DiL system and ensure that the tracking error σ_c converges to a stable equilibrium, the following finite time controller is proposed.

Theorem 1. *For the DiL system in Equation (18), the feedback control T_{fb} which ensures that the tracking error σ_c in Equation (22) converges to a practically stable equilibrium can be designed as in Equation (23).*

$$T_{fb} = \frac{1}{\Omega_u \mu(\gamma)} \Big(-\Omega_c + \nu(\sigma_c) \Big) \tag{23}$$

where Ω_c, Ω_u are defined later and a novel robust HOSM control $\nu(\sigma_c)$ to reject the effect of disturbances is defined as in Equation (24).

$$\nu(\sigma_c) = -\alpha_1 \nu_1(\sigma_c) - \alpha_2 \int_0^t \nu_2(\sigma_c)\, dt \tag{24}$$

with, $\nu_1(\sigma_c) = |\sigma_c|^{\eta_1} sign(\sigma_c)$ and $\nu_2(\sigma_c) = |\sigma_c|^{\eta_2} sign(\sigma_c)$, $1 - 2\eta_1 + \eta_2 = 0$, $1 > \eta_1 \geq 0.5$, and $\alpha_1, \alpha_2, \alpha_3 > 0$ are positive constants.

Proof. The dynamics of the tracking error σ_c can be expressed as in Equation (25).

$$\begin{aligned} \dot{\sigma}_c &= k_1 \dot{e}_l + k_2 \ddot{e}_l + k_3 \ddot{\delta}_d + k_4 \lambda_c \dot{x}_{cf} \\ &= \beta f_1 + \dot{\psi}_v f_2 + \Psi_l f_3 + \delta_d f_4 + f_5 + \Delta_t + (k_3 b_1 - k_4 \lambda_c) \mu(\gamma) T_{fb} \\ &= \Omega_c + \Omega_u \mu(\gamma) T_{fb} + \Delta_t \end{aligned} \tag{25}$$

where $\Omega_c = \beta f_1 + \dot{\psi}_v f_2 + \Psi_l f_3 + \delta_d f_4 + f_5$, $\Omega_u = k_3 b_1 - k_4 \lambda_c$, $f_1 = k_1 v_x + k_2 v_x a_{11} + 2k_2 l_s a_{21} + k_3 a_{61}$, $f_2 = 2k_1 l_s + k_2 v_x a_{12} + 2k_2 l_s a_{22} + k_2 v_x + k_3 a_{62}$, $f_3 = k_1 v_x + k_3 a_{64}$, $f_4 = k_2 v_x a_{15} + 2k_2 l_s a_{25} + k_3 a_{65}$, $f_5 = k_3 a_{63} y_l + k_3 a_{66} \dot{\delta}_d + k_4 T_d^s$, $\Delta_t = f_{dt} + f_{dt1} - k_1 l_s v_x \rho_r + e_3 \Delta F_{yf}$, $f_{dt} = k_2 [\beta \frac{\partial v_x}{\partial dt} + \Psi_l \frac{\partial v_x}{\partial dt} - l_s \frac{\partial v_x}{\partial dt} \rho_r - l_s v_x \frac{\partial \rho_r}{\partial dt}]$, and $f_{dt1} = k_2 v_x (e_1 F_w + e_1 \Delta F_{yf} + e_4 \Delta F_{yr}) + 2k_2 l_s (e_2 F_w + e_2 \Delta F_{yf} + e_5 \Delta F_{yr}) - k_2 \rho_r v_x^2$.

Substituting for the feedback control designed in Equation (23), the error dynamics can be now expressed as in Equation (26).

$$\dot{\sigma}_c = \nu(\sigma_c) + \Delta \tag{26}$$

The lumped disturbance Δ consists of the effects of road curvature, crosswinds, and uncertain tire friction forces. For all practical operating conditions, these disturbances and their time derivatives can be assumed to be bounded. It can be further shown that the lumped disturbance can be divided as $\Delta = \Delta_1(\sigma_c) + \Delta_2$ with simplifications of the expression in Equation (25). The disturbance terms can be shown to be bounded as in Equation (28).

$$\|\Delta_1\| \leq \chi_1 \|\sigma_c\| \tag{27}$$

$$|\dot{\Delta}_2| \leq \chi_2 \tag{28}$$

where $\chi_1, \chi_2 > 0$ are any positive parameters.

Now consider the following Lyapunov function in Equation (29).

$$V_c = \Sigma^T Q_c \Sigma \tag{29}$$

with $\Sigma = \begin{bmatrix} \nu_1 & \sigma_c & \int_0^t \nu_2(\sigma_c)\, dt \end{bmatrix}^T$ and the matrix $Q_c = Q_c^T > 0$ denoting a positive definite matrix defined, as [35], in Equation (30).

$$Q_c = \frac{1}{2}\begin{bmatrix} (4\alpha_2 + \alpha_1^2) & \alpha_1\alpha_3 & -\alpha_1 \\ \alpha_1\alpha_3 & (1+\alpha_3^2) & -\alpha_3 \\ -\alpha_1 & -\alpha_3 & 2 \end{bmatrix} \quad (30)$$

The above Lyapunov function satisfies the condition in Equation (31).

$$\lambda_{min}||\Sigma||^2 \leq V_c \leq \lambda_{max}||\Sigma||^2 \quad (31)$$

with $\lambda_{min}, \lambda_{max}$ representing the minimum singular value and the maximum eigenvalue respectively. The rate of evolution of this Lyapunov function can be computed, similarly to [35], as in Equation (32).

$$\dot{V}_c = -\frac{1}{||\sigma_c||^{(1-n_1)}}\Sigma^T Q_{c1}\Sigma - \Sigma^T Q_{c2}\Sigma \quad (32)$$

where Q_{c1} and Q_{c2} are two positive definite matrices. By the choice of the gains as $\alpha_1 > [(2\alpha_2\chi_1 + \alpha_3\chi_2 - \alpha_2\alpha_3)/(2\alpha_3 - 0.5\chi_1)]^{0.5}$, $\alpha_2 > (2\chi_2 - \alpha_1^2)/2$, and $\alpha_3 > \chi_1(\alpha_1^2/2 + 2\alpha_2)/(\alpha_2 + 2\alpha_1^2 - \chi_2)$, it can be shown, similar to [35], that Equation (33) is valid.

$$\dot{V}_c = -\frac{1}{||\sigma_c||^{(1-n_1)}}\lambda_{min}Q_{c1}||\Sigma||^2 - \lambda_{min}Q_{c2}||\Sigma||^2 \quad (33)$$

Thus, with the proper selection of the gains $\alpha_i > 0$, the Lyapunov function $\dot{V}_c$ is negative definite and the sliding surface converges to attain practical bounded stability. □

In the designed closed loop shared control in Theorem 1, the sharing parameter $\mu(\gamma)$ is directly accounted for in the design of the feedback input T_{fb} as shown in Equation (23). Thus, the stability of the DiL closed-loop system in Equation (18) in the presence of road disturbances and tire-friction uncertainties for any authority transfer or shared driving between the driver and the automation system can be ensured.

Remark 1. *In the designed feedback control T_{fb}, singularity condition can arise when $\Omega_u\mu(\gamma) \rightarrow 0$, i.e., if $(k_3b_1 - k_4\lambda_c) \rightarrow 0$ or if $\mu(\gamma) \rightarrow 0$. However, the modulation factor is a positive bounded entity i.e., $\mu(\gamma) \in [\mu_{min}, 1]$ as presented earlier, and will not result in a singularity condition for the controller. Further, by the selection of the gains κ, λ_c such that $k_3b_1 \neq k_4\lambda_c$, the design of the control input would always be feasible.*

A flowchart for the methodology of implementation of the proposed control scheme is presented in Figure 3.

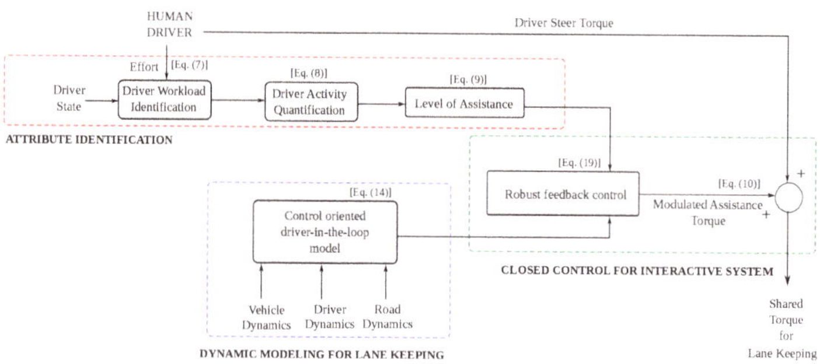

Figure 3. Flow chart of the methodology.

4. Validation and Results

The proposed driver activity adapted cooperative LKA controller was validated on a MATLAB-SIMULINK platform and the SHERPA vehicle simulator for real-time testing.

4.1. Simulation Studies

The performance of the proposed approach was evaluated to satisfy the control objectives under the following constraints for safe vehicle operation in Equation (34).

$$|\dot{\psi}| \leq \dot{\psi}_{max}, |\Psi_l| \leq \Psi_{l_{max}}, |y_l| \leq y_{l_{max}}, |\delta_f| \leq \delta_{f_{max}}, |\dot{\delta}_f| \leq \dot{\delta}_{f_{max}}, |T_a| \leq T_{a_{max}} \tag{34}$$

where $\dot{\psi}_{max}$ = 0.55 rad/s, $\Psi_{l_{max}}$ = 0.1 rad, $y_{l_{max}}$ = 1.5 m, $\delta_{f_{max}}$ = 0.2 rad, $\dot{\delta}_{f_{max}}$ = 0.15 rad/s and $T_{a_{max}}$ = 20 Nm.

For performance evaluations the following controllers were compared:

- **Auto-HOSM**: Autonomous controller (i.e., $T_d = 0$) with proposed HOSM control law.
- **CLKA-HOSM**: Shared controller with proposed HOSM control law.

The sliding surface gains, defined in Equation (22), without the sharing parameter term were obtained using particle swarm optimization (PSO) [46] for optimal results. Accordingly, each particle was defined as $X = \begin{bmatrix} k_1 & k_2 & k_3 \end{bmatrix}$. Consequently, the particles were able to obtain the optimal solutions for the gains based on an objective function which was formulated to minimize the lane tracking errors and satisfy the system constraints in Equation (34) discussed earlier. We considered particle size as 20 and a total of 100 iterations for the PSO algorithm. Using the PSO approach, the sliding surface gains were computed as $k_1 = 3.6085$, $k_2 = 10.5804$, and $k_3 = 0.9706$. Subsequently, the gains of the novel STA controller were selected as α_1 = 33.9379, $\alpha_2 = 150$, $\alpha_3 = 11.2697$ and $\beta = 0.6383$ for normal road conditions with unity road friction. The conflict parameter gains were chosen as $k_4 = 0.001$, $\lambda_c = 1.5$, respectively.

To replicate the human driver torque for the simulation study, a dynamic model based on neuromuscular attributes, time-lags, etc., as discussed in [9,43] was employed. Employing this driver model with varying parameters, the virtual driver torque for simulations was replicated. For all validation purposes, the driver gains were considered as $K_c = 8.57$ and $K_a = 15.75$ respectively. Accounting for the mental workload, two driver states i.e., watchful and distracted to compute the driver state variable DS were considered. During the distracted mode, the external driver torque input was scaled by a factor of 0.2 to represent a distracted driver.

The simulations were performed on the Satory test track [5] as shown in Figure 4a under variable longitudinal velocity conditions i.e., $v_x \in [5, 25]$ m/s shown in Figure 4b. The lateral acceleration of the vehicle is limited to $|a_y|_{max} \leq 2\,\text{m/s}^2$, indicating normal driving conditions as shown in Figure 4c. To evaluate the shared control performance, we considered the human driver to be distracted between $t \in [40, 80]$ s while during the rest of the driving cycle, the driver was watchful. Accordingly, the input driver torque reflecting such conditions is shown in Figure 4d.

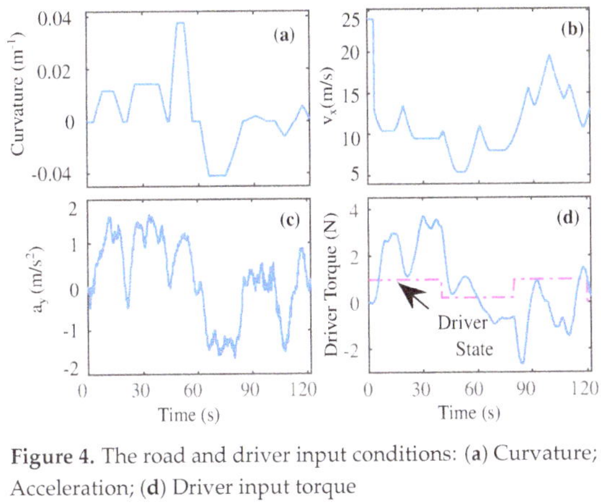

Figure 4. The road and driver input conditions: (**a**) Curvature; (**b**) Longitudinal Velocity; (**c**) Lateral Acceleration; (**d**) Driver input torque

The performance of the Auto-HOSM and CLKA-HOSM controllers are presented in Figure 5a–d along with that of the Auto-HOSM. Both controllers ensured that the lane tracking errors and the steering rate were within the prescribed limits discussed earlier. As the shared controller incorporates human action in the control process, the above performance indicators of the shared controller have a higher magnitude that their autonomous counterpart. The root mean square (rms) and maximum values of the above indicators for the Auto-HOSM controller were computed as $y_{l_{rms}} = 0.57$, $\Psi_{l_{rms}} = 0.0131$, $\dot{\delta}_{d_{rms}} = 2.5917$ and $|y_l|_{max} = 1.19$, $|\Psi_l|_{max} = 0.0469$, $|\dot{\delta}_d|_{max} = 7.6202$, respectively. Similarly, the performance metrics of the CLKA-HOSM controller were $y_{l_{rms}} = 0.5267$, $\Psi_{l_{rms}} = 0.0162$, $\dot{\delta}_{d_{rms}} = 2.0609$, and $|y_l|_{max} = 1.2750$, $|\Psi_l|_{max} = 0.0446$, $|\dot{\delta}_d|_{max} = 5.9638$. Such performance metrics indicate good lane-keeping performance for both controllers. Further, the steering rate performance shows improvement under the proposed CLKA-HOSM controller.

Along with such lane-keeping performance, the conflict between the human driver and the autonomous controller for the CLKA-HOSM controller is also presented in Figure 5d. Using the proposed controller, the conflict is kept within limits such that, $T_d T_a > -5\ N^2$.

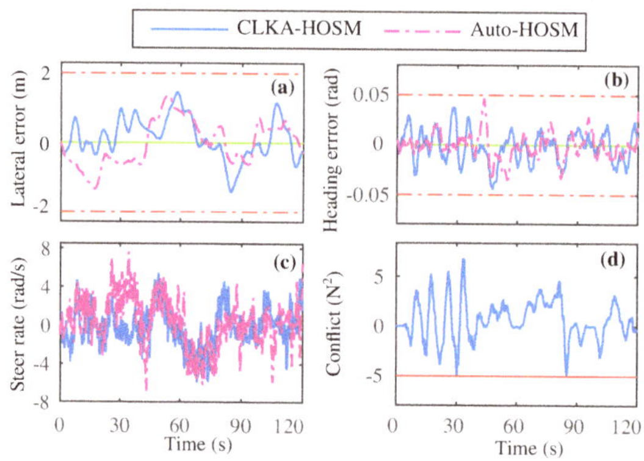

Figure 5. The lane tracking and driver comfort performance for the proposed controller (CLKA-HOSM) and the autonomous controller (Autonomous-HOSM). (**a**) Lateral deviation error; (**b**) Heading error; (**c**) steering rate; (**d**) Conflict product of driver and automation torques.

For further illustration of the shared control performance, the torques generated by the human driver and autonomous agent along with the driver activity–performance indicators are presented in Figure 6. Based on the driver's activity, the level of assistance torque generated varies for completing the driving task.

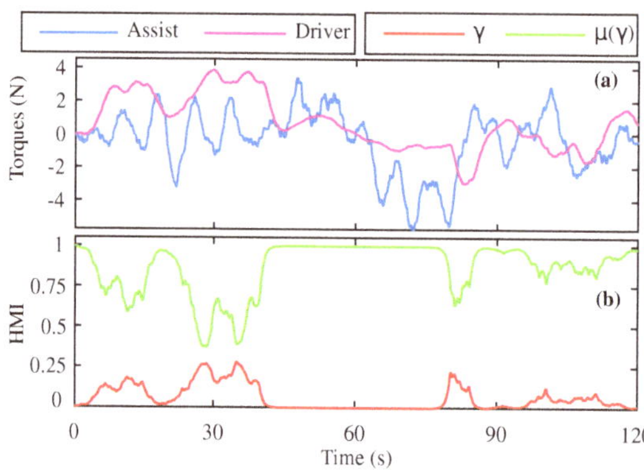

Figure 6. The HMI under the CLKA-HOSM controller. (**a**) Driver and Assistance Torque; (**b**) Driver activity and the level of assistance provided.

To assess the performance of shared control activity, the following metrics [42] were also a time interval η:

$$\text{AFac} = \frac{T_{d_{pow}}}{T_{a_{pow}}}, \ \text{SW} = \frac{1}{\eta}\int_0^{\eta} T_a(t)T_d(t)\dot{\delta}_d(t)dt, \tag{35}$$

- AFac: Denotes the ratio between efforts generated by the automation and human driver for completing the driving task i.e., in Equation (36).

$$\text{AFac} = \frac{T_{d_{pow}}}{T_{a_{pow}}} \tag{36}$$

  If the values of AFac> 1, the assistance provided by the automation is less than that of the driver, and inversely for AFac< 1.
- SW: This indicates the steering workload and is representative of the effort generated by both agents simultaneously for completing the driving task i.e., in Equation (37).

$$\text{SW} = \frac{1}{\eta}\int_0^{\eta} T_a(t)T_d(t)\dot{\delta}_d(t)dt \tag{37}$$

  A larger magnitude of negative steering workload indicates that the assistance provided by the automation to the human driver is not good for shared control.

For efficient shared control, the AFac should be less than 1 and the negative steering workload should be low. Using the proposed CLKA-HOSM controller, these metrics are computed as AFac = 0.8192 and Negative SW = −206.6476, indicating a good quality of shared control. To assess the shared control performance further, performance analysis was performed for a shared controller based on the proposed HOSM control law, but with no conflict parameter i.e., $k_4 = 0$ and $\lambda_c = 0$ (**SC-NoK4**) and is presented in Table 1. Please note that the values of Neg SW (i.e., negative steer workload) and $T_d T_{a_{min}}$ (i.e., maximum value of conflict) is less than zero.

Table 1. Influence of k_4 and λ_c on HMI.

Case	λ_c	$\lvert y_l \rvert_{max}$ (m)	AFac	Neg. SW (N^2m^2rad/s)	$T_d T_{a_{min}}$ (N^2m^2)
$k_4 = 0$	0	1.219	0.735	219.3	9.132
$k_4 = 0.001$	0.5	1.624	0.843	202.1	5.242
	1.5	1.275	0.819	206.6	5.22
	2	1.389	0.818	209.6	5.92
$k_4 = 0.01$	0.5	1.241	0.824	216.7	8.35
	1.5	1.889	0.763	223.2	6.577
	2	1.228	0.723	219.6	8.308

With the increase in the magnitude of k_4, the negative steer workload and the maximum values of conflict increase showing deteriorating shared control performance. Further, the lateral error also increases, from a minimum of 1.219 m to a maximum of 1.889 m, as more control is passed on to the human driver, from $K_4 = 0$ to $K_4 = 0.01$. Similar performance is seen with the increase in values of λ_c as well, from $\lambda_c = 0$ to $\lambda_c = 2$. From the presented results, the best performance in terms of lane errors, $|y_l|_{max} = 1.624$, and conflict reduction, $T_d T_a = 5.242$, is obtained for $k_4 = 0.001$ and $\lambda_c = 0.5$. Further, the presence of the gains k_4 and λ_c improves the performance of the controller, in terms of conflict minimization and negative SW, in comparison to the case when $k_4 = 0$ and $\lambda_c = 0$ across all aspects.

To ascertain the robustness of the proposed CLKA-HOSM controller, random parametric uncertainties in the vehicle and driver parameters were considered. Specifically, uncertain values of M, I_z, and I_s which are susceptible to the payload, wear, tear, etc. are employed. Similarly, the uncertainty in driver model parameters K_a and K_c to account for various driver behaviors are also considered. The lane-keeping and conflict reduction performance of the CLKA-HOSM (i.e., C1) and SC-NoK4 (i.e., C2) controllers under influence of such uncertainties are presented in Table 2.

Table 2. Influence of uncertainties on controller performance.

Unct.	Cont.	$\lvert y_l \rvert_{rms}$	$\lvert \Psi_l \rvert_{rms}$	$\lvert \dot{\delta}_d \rvert_{rms}$	$T_d T_{a_{min}}$	Neg. SW	AF
5%	C1	0.508	0.019	2.054	6.514	226.4	0.797
	C2	0.659	0.022	2.375	11.624	235.2	0.811
15%	C1	0.554	0.016	2.378	8.352	211.5	0.722
	C2	0.585	0.016	2.252	7.599	206.8	1.01
20%	C1	0.5221	0.0172	2.165	6.705	204.5	0.822
	C2	0.472	0.0182	2.037	7.169	208.2	0.599

Under the influence of vehicle and driver uncertainties up to 20%, the proposed CLKA-HOSM controller performs well in ensuring lane keeping ($\lvert y_l \rvert_{rms}$ = 0.52, $\lvert \Psi_l \rvert_{rms}$ = 0.017 for CLKA-HOSM, against $\lvert y_l \rvert_{rms}$ = 0.47, $\lvert \Psi_l \rvert_{rms}$ = 0.018 for SC-NoK4) and also minimizing the conflict between driver and autonomous system ($T_d T_{a_{min}}$ = 6.705 for CLKA-HOSM, against $T_d T_{a_{min}}$ = 7.169 for SC-NoK4). The CLKA-HOSM controller outperforms the SC-NoK4 controller in handling uncertainties, and thus establishes the significance of the gains k_4 and λ_c in performance enhancement.

4.2. Experimental Results: SHERPA Vehicle Simulator

The shared DiL-LKA approach was validated in real-time on the SHERPA vehicle simulator shown in Figure 7.

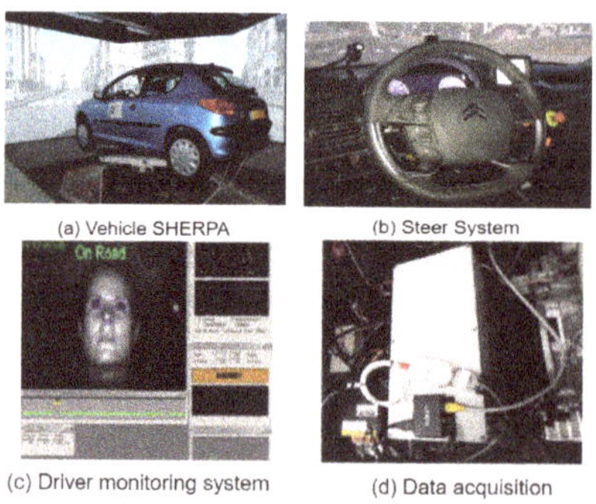

Figure 7. Experimental setup for the SHEPRA vehicle simulator.

The SHERPA simulator is built using a modified Peugot 206 vehicle on a Stewart platform and is composed of multiple modules for handling driving-related tasks such as perception, path planning, driver monitoring, and human–machine interface management. For more details on the SHERPA simulator, refer to [5]. Using the driving monitoring unit, the driver state is directly available as a binary input while the torque is measured via a sensor on the steering wheel. With haptic feedback via the steering wheel provided, this simulator setup has been used for validation of direct shared control works [5,43] similar to that proposed in this work.

Using the SHERPA setup (with a discretization time of 0.01 s), we now present illustrative results to highlight the lane-keeping and conflict-reduction performance of the proposed shared DiL controller in this work to further support our earlier presented simulation-based analysis. All performance evaluations on the SHERPA simulator are made on a test track represented in Figure 8 that comes from the CoCoVeA project (Cooperation Conductor-Véhicule Automatisé).

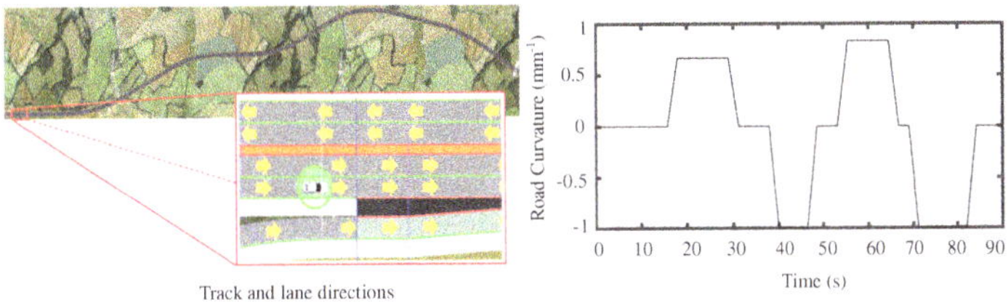

Figure 8. CoCoVeA track and lanes directions along with road curvature of the sections.

The results for the Auto-HOSM controller robustness against longitudinal speed and the friction variations are first presented to highlight the robustness of the proposed novel control law. For multiple driving tests performed, the aggregated results are presented in Table 3. It can be seen that the variations in the longitudinal speed of the vehicle and the road friction do not affect the performance of the proposed controller. The controller ensures good trajectory tracking by maintaining the lateral deviation below $|y_l|_{max} = 0.5824 < 1.5$ m, the maximum heading error below $|\Psi_l|_{max} = 0.0074 < 0.1$ rad, without saturating the motor control of the steering system $|T_a|_{max} = 1.2104 < 20$ N·m.

Table 3. Influence of speed variation and friction on the performance of the lane-keeping controller.

v_x (m/s)	Friction	$\lvert y_l\rvert_{max}$ (m)	$\lvert\Psi_l\rvert_{max}$ (rad)	$\lvert T_a\rvert_{max}$ (N·m)
14	1	0.1116	0.0024	0.2523
	0.6	0.1188	0.0063	0.2679
	0.4	0.1289	0.0024	0.2557
20	1	0.3213	0.0045	0.6807
	0.6	0.3211	0.0044	0.6783
	0.4	0.3111	0.0044	0.6663
25	1	0.5727	0.0074	1.2104
	0.6	0.5824	0.0072	1.1257
	0.4	0.5792	0.0073	1.1432

In the second case, the proposed CLKA-HOSM controller for an obstacle avoidance scenario is tested with sharing parameter values as $k_4 = 15$ and $\lambda_c = 0.5$. Accordingly, as shown in Figure 9, three obstacles were placed on the road, and the driver was asked to avoid them by changing the lane. For comparisons, the same test was also repeated with the Auto-HOSM controller weighted by the LOA function presented in Equation (12). The performance results for both controllers are presented in Figure 9.

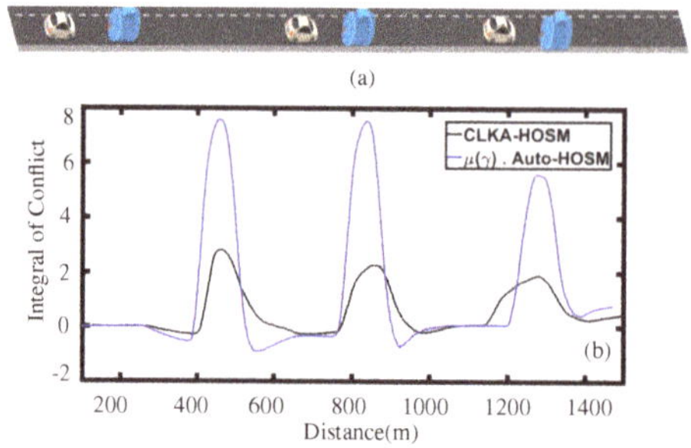

Figure 9. (**a**) The obstacle avoidance scenario; (**b**) comparison between the LKA controller weighted with the LOA function and the CLKA-HOSM controller for the minimization of the conflict for an obstacle avoidance scenario using a metric Integral of Conflict.

In Figure 9b, the metric *Integral of Conflict* is defined as $IOC = -\frac{1}{\tau}\int_0^T T_a(t)T_d(t)dt$ for a time period τ. It was observed that the CLKA-HOSM controller is more efficient in terms of conflict minimization i.e., the maximum value of the integral of the conflict is 2.7. On the other hand, the maximum value of the integral of the conflict for the Auto-HOSM controller weighted with the LOA function was 7.8. In the case of the proposed CLKA-HOSM controller, AFac = 0.8818, Negative SW = 14.9646, and $(T_dT_a)_{min} = -4.8727$ was obtained. In contrast, for the Auto-HOSM controller weighted with the LOA function, AFac = 0.974, Negative SW = 108.218, and $(T_dT_a)_{min} = -24.539$ were obtained. Such results show that the proposed CLKA-HOSM outperforms the other design in terms of shared control performance.

For further analysis of the shared control performance, the parameters k_4 and λ_c were varied and tests were performed. Performance results for the CLKA-HOSM controller under such variations are shown in Table 4.

Table 4. Influence of k_4 and λ_c on HMI.

Case	λ_c	AFac	Neg. SW (N^2m^2rad/s)	$T_dT_{a_{min}}$ (N^2m^2)
$k_4 = -5$	0.5	0.5923	348.7971	−61.0736
	0.8	0.4794	118.1967	−70.3217
$k_4 = -1$	0.5	0.9754	91.0878	−35.5465
	0.8	0.9507	138.8792	−36.7558
$k_4 = 0$	0	1.0192	106.8268	−27.2865
$k_4 = 5$	0.5	1.0240	108.8057	−18.0648
	0.8	1.0604	108.0537	−15.0333
$k_4 = 10$	0.5	0.9909	40.4668	−7.5843
	0.8	0.9515	36.0443	−8.4723
$k_4 = 15$	0.5	0.8818	14.9646	−4.8727
	0.8	0.9499	18.4843	−7.2032

It can be seen in Table 4 that the shared parameters have a significant impact on the AFac metric, from 0.4794 to 1.0604, and SW metric, from −14.9646 to −348.7971. The chosen best combination values of these metrics using the proposed CLKA-HOSM controller are AFac = 0.8818 and SW = −14.9646, indicating a good quality of shared control. From the presented results, the best performance in terms of conflict reduction is obtained for the combination $k_4 = 15$ and $\lambda_c = 0.5$.

5. Conclusions

In this work, a novel robust shared controller for a DiL-lane-keeping assistance system was proposed and evaluated. The HMI was managed via an adaptive mapping which reflected driver performance corresponding to the identified physical and mental workload of the driver. Along with lane tracking errors and driver comfort enhancement, the issue of conflict between the driver and autonomous controller was also addressed by the introduction of a novel sharing parameter. Addressing such objectives, a novel higher-order sliding mode control algorithm was proposed and its stability for the closed-loop DiL system affected by disturbances was established.

The performance of the proposed controller was evaluated via simulations and experiments on the SHERPA vehicle simulator for different longitudinal velocity, different road friction conditions, time-varying road curvatures of the Satory test track, parametric uncertainties, and for obstacle avoidance scenarios. Comparison between the fully autonomous controller, the proposed sharing control law without the introduction of the novel parameter for conflict reduction, and the proposed sharing control law with the introduction of this minimization parameter was extensively discussed. From the experimental results, it can be seen that the fully autonomous controller achieved the best lane tracking and heading error performances (30% better than the sharing control law), but the sharing control law achieved the best conflict minimization (65.38% better than the sharing control law without the introduction of this novel term). Further, the cooperative driving quality improved by 9.4%, and the negative steering workload was reduced by 86.13% in comparison to the Auto-HOSM controller showing the efficiency of the proposed controller.

The proposed controller was constructed in order to deal with the goals of lane maintenance, driver comfort improvement, and conflict reduction, which fill a particular need in improving the driving experience for road vehicle transportation. In the future, the driver activity function will be enhanced by including the driving style, skill, and other attributes reflecting a wider variety of driver behaviors. An expansion of the proposed cooperative architecture to the cruise and integrated longitudinal–lateral control will be carried out.

Author Contributions: Conceptualization: J.J.R.; Methodology: J.J.R.; Software production: G.P. and M.R.O.; Experiments: G.P., J.J.R. and M.R.O.; Writing—original draft preparation: J.J.R., G.P. and C.S.; Writing—review and editing: G.P., J.J.R., C.S. and J.-C.P. All authors have read and agreed to the published version of the manuscript.

Funding: This work has been done in the framework of the CoCoVeIA research program (ANR-19-CE22-0009-01), funded by the French National Research Agency. This work was also sponsored by the French Regional Delegation for Research and Technology, the French Ministry of Higher Education and Research, and the French National Center for Scientific Research.

Institutional Review Board Statement: Not applicable.

Informed Consent Statement: Not applicable.

Data Availability Statement: Not applicable.

Conflicts of Interest: The authors declare no conflict of interest.

Nomenclature

Symbol	Description	Value
M	total mass of the vehicle	2025 [kg]
l_f	distance from CoG to front axle	1.3 [m]
l_r	distance from CoG to rear axle	1.6 [m]
t_p	tire length contact	0.052 [m]
l_s	look-ahead distance	5 [m]
I_z	vehicle yaw moment of inertia	2800 [kgm^2]
I_s	steering moment of inertia	0.05 [kgm^2]

R_s	steering gear ratio	16.3 [-]
B_u	steering system damping	2.5 [N/rad]
C_{pf}	front cornering stiffness	42,500 [N/rad]
C_{pr}	rear cornering stiffness	57,000 [N/rad]
R_s	rate driver's—vehicle's wheel angles	
w_r	width of the vehicle	
β	side slip angle	
δ	steering angle	
δ_d	steering angle	
$\dot{\delta}_d$	steering rate	
ψ	heading angle	
F_{yf}	front friction force	
F_{yr}	rear friction force	
v_x	longitudinal velocity	
α_f	front slip angle	
α_r	rear slip angle	
T_s	self-aligning torque	
ΔF_i	uncertainty of tire friction force	
K_p	level of assistance	
y_l	lateral deviation error	
Ψ_l	orientation error w.r.t the lane center-line	
T_d	driver torque	
T_a	automation assistance torque	
$\mu(\gamma)$	rate driver workload-based performance—LOA	
T_{fb}	feedback control torque	
K_a	anticipatory gain	
K_c	compensatory gain	
θ_{near}	near visual points of the driver	
θ_{far}	far visual points of the driver	
λ_c	the level of sharing	
T_d^s	driver torque measured at the steering wheel	
$\dot{x}_{cf}$	conflict dynamics	
σ_c	linear error surface of the SMC	
$AFac$	ratio between automation and human	

Acronyms

Symbol	Description
ADAS	Advanced driver assist system
LKA	Lane keeping assistance
ACC	Adaptive cruise control
CA	Collision avoidance
HMI	Human machine interaction
DiL	Driver-in-the-loop
HOSM	High order sliding mode
SMC	Sliding mode control
BT	Brush-Tire
DS	Driver state
DMU	Driver monitoring unit
LOA	Level of assistance
Auto-HOSM	Autonomous controller with proposed HOSM control
CLKA-HOSM	Shared controller with proposed HOSM control
SC-NoK4	Shared controller with proposed HOSM control with $K_4 = 0$
PSO	Particle swarm optimization
SW	Steering workload
IOC	Integral of conflict
SHERPA	Simulateur Hybride d'Etude et de Recherche Pour l'Automobile

References

1. Rajamani, R. *Vehicle Dynamics and Control*; Springer: Boston, MA, USA, 2012.
2. Flemisch, F.; Kelsch, J.; Loper, C.; Schieben, A.; Schindler, J.; Heesen, M. Cooperative control and active interfaces for vehicle assistance and automation. In Proceedings of the FISITA World Automotive Congress, Munich, Germany, 14–19 September 2008.
3. Abbink, D.A.; Mulder, M.; Boer, E.R. Haptic shared control: Smoothly shifting control authority? *Cogn. Technol. Work* **2011**, *14*, 19–28. [CrossRef]
4. Saito, T.; Wada, T.; Sonoda, K. Control Authority Transfer Method for Automated-to-Manual Driving Via a Shared Authority Mode. *IEEE Trans. Intell. Veh.* **2018**, *3*, 198–207. [CrossRef]
5. Nguyen, A.T.; Sentouh, C.; Popieul, J.C. Driver-automation cooperative approach for shared steering control under multiple system constraints: Design and experiments. *IEEE Trans. Ind. Electron.* **2017**, *64*, 3819–3830. [CrossRef]
6. Wada, T.; Sonoda, K.; Tada, S. Simultaneous Achievement of Supporting Human Drivers and Improving Driving Skills by Shared and Cooperative Control. *IFAC-PapersOnLine* **2016**, *49*, 90–95. [CrossRef]
7. Saleh, L.; Chevrel, P.; Claveau, F.; Lafay, J.F.; Mars, F. Shared steering control between a driver and an automation: Stability in the presence of driver behavior uncertainty. *IEEE Trans. Intell. Transp. Syst.* **2013**, *14*, 974–983. [CrossRef]
8. Schnelle, S.; Wang, J.; Su, H.; Jagacinski, R. A Driver Steering Model with Personalized Desired Path Generation. *IEEE Trans. Syst. Man, Cybern. Syst.* **2017**, *47*, 111–120. [CrossRef]
9. Sentouh, C.; Chevrel, P.; Mars, F.; Claveau, F. A sensorimotor driver model for steering control. In Proceedings of the IEEE International Conference on Systems, Man and Cybernetics, San Antonio, TX, USA, 11–14 October 2009; pp. 2462–2467.
10. Li, L.; Liu, Y.; Wang, J.; Deng, W.; Oh, H. Human dynamics based driver model for autonomous car. *IET Intell. Transp. Syst.* **2016**, *10*, 545–554. [CrossRef]
11. Tanaka, Y.; Kashiba, Y.; Yamada, N.; Suetomi, T.; Nishikawa, K.; Nouzawa, T.; Tsuji, T. Active-steering control system based on human hand impedance properties. In Proceedings of the 2010 IEEE International Conference on Systems, Man and Cybernetics, Istanbul, Turkey, 10–13 October 2010. [CrossRef]
12. Sharma, O.; Sahoo, N.C.; Puhan, N.B. Recent advances in motion and behavior planning techniques for software architecture of autonomous vehicles: A state-of-the-art survey. *Eng. Appl. Artif. Intell.* **2021**, *101*, 104211. [CrossRef]
13. Gambhire, S.J.; Kishore, D.R.; Londhe, P.S.; Pawar, S.N. Review of sliding mode based control techniques for control system applications. *Int. J. Dyn. Control* **2021**, *9*, 363–378. [CrossRef]
14. Wu, J.; Zhang, J.; Tian, Y.; Li, L. A Novel Adaptive Steering Torque Control Approach for Human–Machine Cooperation Autonomous Vehicles. *IEEE Trans. Transp. Electrif.* **2021**, *7*, 2516–2529. [CrossRef]
15. Kumar, V.; Naresh, R.; Sharma, V.; Kumar, V. State-of-the-Art Optimization and Metaheuristic Algorithms. In *Handbook of Intelligent Computing and Optimization for Sustainable Development*; John Wiley & Sons, Ltd.: Hoboken, NJ, USA, 2022; pp. 509–536. [CrossRef]
16. Brizuela-Mendoza, J.A.; Sorcia-Vázquez, F.D.J.; Rumbo-Morales, J.Y.; Lozoya-Ponce, R.E.; Rodríguez-Cerda, J.C. Active fault tolerant control based on eigenstructure assignment applied to a 3-DOF helicopter. *Asian J. Control* **2021**, *23*, 673–684. [CrossRef]
17. Rumbo Morales, J.Y.; Brizuela Mendoza, J.A.; Ortiz Torres, G.; Sorcia Vázquez, F.d.J.; Rojas, A.C.; Pérez Vidal, A.F. Fault-Tolerant Control implemented to Hammerstein–Wiener model: Application to Bio-ethanol dehydration. *Fuel* **2022**, *308*, 121836. [CrossRef]
18. Xing, Y.; Lv, C.; Cao, D.; Hang, P. Toward human-vehicle collaboration: Review and perspectives on human-centered collaborative automated driving. *Transp. Res. Part C Emerg. Technol.* **2021**, *128*, 103199. [CrossRef]
19. Wang, J.; Zhang, G.; Wang, R.; Schnelle, S.C.; Wang, J. A Gain-Scheduling Driver Assistance Trajectory-Following Algorithm Considering Different Driver Steering Characteristics. *IEEE Trans. Intell. Transp. Syst.* **2017**, *18*, 1097–1108. [CrossRef]
20. Soualmi, B.; Sentouh, C.; Popieul, J.; Debernard, S. Automation-driver cooperative driving in presence of undetected obstacles. *Control Eng. Pract.* **2014**, *24*, 106–119. [CrossRef]
21. Wang, Z.; Zheng, R.; Nacpil, E.J.C.; Nakano, K. Modeling and analysis of driver behaviour under shared control through weighted visual and haptic guidance. *IET Intell. Transp. Syst.* **2022**, *16*, 648–660. [CrossRef]
22. Mars, F.; Deroo, M.; Hoc, J.M. Analysis of Human-Machine Cooperation When Driving with Different Degrees of Haptic Shared Control. *IEEE Trans. Haptics* **2014**, *7*, 324–333. [CrossRef]
23. Boink, R.; van Paassen, M.M.; Mulder, M.; Abbink, D.A. Understanding and reducing conflicts between driver and haptic shared control. In Proceedings of the 2014 IEEE International Conference on Systems, Man, and Cybernetics (SMC), San Diego, CA, USA, 5–8 October 2014. [CrossRef]
24. Huang, C.; Lv, C.; Hang, P.; Hu, Z.; Xing, Y. Human–Machine Adaptive Shared Control for Safe Driving Under Automation Degradation. *IEEE Intell. Transp. Syst. Mag.* **2022**, *14*, 53–66. [CrossRef]
25. Li, X.; Wang, Y.; Su, C.; Gong, X.; Huang, J.; Yang, D. Adaptive Authority Allocation Approach for Shared Steering Control System. *IEEE Trans. Intell. Transp. Syst.* **2022**, *23*, 19428–19439. [CrossRef]
26. Deng, H.; Zhao, Y.; Feng, S.; Wang, Q.; Lin, F. Shared Control for Intelligent Vehicle Based on Handling Inverse Dynamics and Driving Intention. *IEEE Trans. Veh. Technol.* **2022**, *71*, 2706–2720. . [CrossRef]
27. Flemisch, F.; Nashashibi, F.; Rauch, N.; Schieben, A.; Glaser, S.; Temme, G.; Resende, P.; Vanholme, B.; Löper, C.; Thomaidis, G.; et al. Towards highly automated driving: Intermediate report on the HAVEit-joint system. In Proceedings of the 3rd European Road Transport Research Arena, Brussels, Belgium, 7–10 June 2010.

28. Sentouh, C.; Nguyen, A.T.; Benloucif, M.A.; Popieul, J.C. Driver-Automation Cooperation Oriented Approach for Shared Control of Lane Keeping Assist Systems. *IEEE Trans. Control Syst. Technol.* **2019**, *27*, 1962–1978. [CrossRef]
29. Shimizu, Y.; Kawai, T.; Yuzuriha, J. *Improvement in Driver-Vehicle System Performance by Varying Steering Gain with Vehicle Speed and Steering Angle: VGS (Variable Gear-Ratio Steering System)*; SAE Technical Paper Series; SAE International: Warrendale, PA, USA, 1999. [CrossRef]
30. Izadi, V.; Ghasemi, A.H. Modulation of control authority in adaptive haptic shared control paradigms. *Mechatronics* **2021**, *78*, 102598. [CrossRef]
31. Nguyen, A.T.; Sentouh, C.; Popieul, J.C.; Soualmi, B. Shared Lateral Control with Online Adaptation of the Automation Degree for Driver Steering Assist System: A Weighting Design Approach. In Proceedings of the IEEE 54th Annual Conference on Decision and Control (CDC), Osaka, Japan, 15–18 December 2015; pp. 857–862.
32. Oufroukh, N.A.; Mammar, S. Integrated driver co-pilote approach for vehicle lateral control. In Proceedings of the 2014 IEEE Intelligent Vehicles Symposium Proceedings, Dearborn, MI, USA, 8–11 June 2014. [CrossRef]
33. Lv, C.; Wang, H.; Cao, D.; Zhao, Y.; Sullman, M.; Auger, D.J.; Brighton, J.; Matthias, R.; Skrypchuk, L.; Mouzakitis, A. A Novel Control Framework of Haptic Take-Over System for Automated Vehicles. In Proceedings of the 2018 IEEE Intelligent Vehicles Symposium (IV), Changshu, China, 26–30 June 2018. [CrossRef]
34. Ahn, C.; Peng, H.; Tseng, H.E. Robust estimation of road friction coefficient. In Proceedings of the 2011 American Control Conference, San Francisco, CA, USA, 29 June–1 July 2011. [CrossRef]
35. Rath, J.J.; Veluvolu, K.C.; Defoort, M.; Soh, Y.C. Higher-order sliding mode observer for estimation of tyre friction in ground vehicles. *IET Control Theory Appl.* **2014**, *8*, 399–408. [CrossRef]
36. Moreno, J.A.; Osorio, M. Strict Lyapunov Functions for the Super-Twisting Algorithm. *IEEE Trans. Autom. Control* **2012**, *57*, 1035–1040. [CrossRef]
37. Ahn, C.; Peng, H.; Tseng, H.E. Robust Estimation of Road Frictional Coefficient. *IEEE Trans. Control Syst. Technol.* **2013**, *21*, 2170838. [CrossRef]
38. Kiencke, U.; Nielsen, L. *Automotive Control Systems*; Springer: Berlin/Heidelberg, Germany, 2005. [CrossRef]
39. Baffet, G.; Charara, A.; Lechner, D.; Thomas, D. Experimental evaluation of observers for tire–road forces, sideslip angle and wheel cornering stiffness. *Veh. Syst. Dyn.* **2008**, *46*, 501–520. [CrossRef]
40. Nguyen, A.T.; Sentouh, C.; Popieul, J.C. Online Adaptation of the Authority Level for Shared Lateral Control of Driver Steering Assist System Using Dynamic Output Feedback Controller. In Proceedings of the 41st Annual Conference of the IEEE Industrial Electronics Society, Yokohama, Japan, 9–12 November 2015; pp. 3767–3772.
41. Nguyen, A.T.; Rath, J.J.; Lv, C.; Guerra, T.M.; Lauber, J. Human-Machine Shared Driving Control for Semi-Autonomous Vehicles Using Level of Cooperativeness. *Sensors* **2021**, *21*, 4647. [CrossRef]
42. Rath, J.J.; Senouth, C.; Popieul, J.C. Personalised lane keeping assist strategy: Adaptation to driving style. *IET Control Theory Appl.* **2019**, *13*, 106–115. [CrossRef]
43. Benloucif, M.; Nguyen, A.T.; Sentouh, C.; Popieul, J. Cooperative Trajectory Planning for Haptic Shared Control between Driver and Automation in Highway Driving. *IEEE Trans. Indus. Electron.* **2019**, *66*, 9846–9857. [CrossRef]
44. Dong, J.; Yang, G.H. Control synthesis of T-S fuzzy systems based on a new control scheme. *IEEE Trans. Fuzzy Syst.* **2011**, *19*, 323–338. [CrossRef]
45. Reymond, G.; Kemeny, A.; Droulez, J.; Berthoz, A. Role of Lateral Acceleration in Curve Driving: Driver Model and Experiments on a Real Vehicle and a Driving Simulator. *Hum. Factors J. Hum. Factors Ergon. Soc.* **2001**, *43*, 483–495. [CrossRef] [PubMed]
46. Poli, R.; Kennedy, J.; Blackwell, T. Particle swarm optimization. *Swarm Intell.* **2007**, *1*, 33–57. [CrossRef]

MDPI
St. Alban-Anlage 66
4052 Basel
Switzerland
Tel. +41 61 683 77 34
Fax +41 61 302 89 18
www.mdpi.com

Sensors Editorial Office
E-mail: sensors@mdpi.com
www.mdpi.com/journal/sensors

www.ingramcontent.com/pod-product-compliance
Lightning Source LLC
Chambersburg PA
CBHW040809250926
51850CB00007BA/1120
9783036573304